SCIENCE IN TRANSLATION

Science in Translation

MOVEMENTS OF KNOWLEDGE THROUGH CULTURES AND TIME

Scott L. Montgomery

THE UNIVERSITY OF CHICAGO PRESS
CHICAGO AND LONDON

The University of Chicago Press, Chicago 60637
The University of Chicago Press, Ltd., London

Printed in the United States of America

24 23 22 21 20 19 18 17 16 15 2 3 4 5

ISBN-10: 0-226-53480-4 (cloth)
ISBN-13: 978-0-226-53481-7 (paper)

Library of Congress Cataloging-in-Publication Data

Montgomery, Scott L.
Science in translation : movements of knowledge through cultures and time / Scott L. Montgomery.
p. cm.
Includes bibliographical references and index.
ISBN 0-226-53480-4 (cloth : alk. paper)
1. Science—Translating—History. 2. Science—Language. I. Title.
Q124 .M66 2000
500—dc21 99-053389

⊗ The paper printed in this publication meets the minimum requirements of the American National Standard for Information Sciences—Permanence of Paper for Printed Library Materials, ANSI Z39.48-1992.

TO DOM,

all times best remembered

CONTENTS

Preface ix

Introduction: Transfers of Learning, Questions of Influence 1

PART I THE HEAVENS THROUGH TIME AND SPACE
A History of Translating Astronomy in the West

1 The Era of Roman Translation
FROM GREEK SCIENCE TO MEDIEVAL MANUSCRIPT
17

2 Astronomy in the East
THE SYRIAC AND PERSIAN-INDIAN CONVERSIONS
60

3 The Formation of Arabic Science, Eighth through Tenth Centuries
TRANSLATION AND THE CREATION OF INTELLECTUAL TRADITIONS
89

4 Era of Transfer into Latin
TRANSFORMATIONS OF THE MEDIEVAL COSMOS
138

PART II SCIENCE IN THE NON-WESTERN WORLD
Levels of Adaptation

5 Record of Recent Matters
TRANSLATION AND THE ORIGINS OF MODERN JAPANESE SCIENCE
189

6 Japanese Science in the Making
OF TEXTS AND TRANSLATORS
227

PART III THE CONTEMPORARY CONTEXT
Realities of Change and Difference

7 Issues and Examples for the Study of Scientific Translation Today
253

8 Conclusions
GAINED IN TRANSLATION
271

References 295

Index 317

PREFACE

It has been said that books are written to satisfy two states of mind: discomfort and pleasure. In my own case, this involved, on the one hand, the nagging desire for a type of book that I found did not yet exist, and, on the other hand, the opportunity of ambition to produce such a book. I take it that this is not a rare thing; scholars who are also writers (or writers who occasionally dip into the wax of scholarship) must have often felt similarly. Nonetheless, the chance to produce this work constitutes an opportunity for which I am grateful and to which I dedicate my appreciation for the help received from two individuals in particular: Steve Fuller, who encouraged the project from the beginning, and Susan Abrams, whose intelligence, enthusiasm, and patience are models of the editorial art.

This book is a combination of several intentions and efforts. It is, first of all and most fundamentally, a proposal: simply put, that translation be allowed its crucial role in the history of scientific knowledge—ancient, medieval, and modern—as a subject worthy of diverse inquiry. Secondly, I have tried to demonstrate the validity of this claim through a range of case study examples revealing of the powers translation has commanded in the building of Western science. Many topics are touched upon or nourished along the way; it is my hope to at least suggest some of the types of study that might expand this subject area in the future.

By way of confession, I admit that for more than ten years I myself labored in the salt mines of scientific translation, if only on a part-time basis. A technical background made me prey to nearly every conceivable type of task, rendering anything from journal articles in physical chemistry and geology to patents on copying equipment, manuals of medical technology, astronomical memoirs, parts of theses, legal depositions, even advertisements. Such experience was enough to reveal that translation is involved at every level of knowledge production and distribution in the sciences. Yet, for the interested analyst or reader, it remained a ghost in the machine. There was no evidence of sustained inquiry in this area, no field or subfield that examines what translation means and has meant in the history of science. If we are to accept "the translator's invisibility" in literary studies—and there is every reason that we should—there should be a willingness to admit a double erasure in the case of scientific *translatio*. The topic represents a notable gap; this book is an attempt to begin closing it.

No work on a subject so vast and relatively uncharted can avoid errors and lacunae of its own. I have tried to gather evidence, both direct and circumstantial, from a wide range of sources and disciplines, and in so doing I have inevitably missed or skipped over subjects and works of no small importance. A notable example is the lack in chapter 5 of any significant discussion on the translations of scientific works performed by Jewish scholars (into Hebrew and Latin) during the late medieval period. I have also dealt only glancingly with Persian versions of Greek technical works as well as the rendering of Latin scientific literature into the European vernaculars. These are all serious omissions, for which I can offer only the apology of limited space, focus, and ability.

Translation is a restive subject. It involves a larger array of working parts than is commonly brought to bear upon the study of other semantic activities. It calls directly on—indeed, cannot honestly be dealt with in historical terms without—such concepts as authorship, cultural displacement, originality, textual transmission, literacy and orality, and so on. Some of these ideas have proved gelatinous in the face of detailed study; others open up into realms so great I have only been able to view them through the rhumb lines and portolan charts of summary. It is thus part of my proposal here to suggest that, as a subject, "translation" presents a wealth of opportunities to combine the insights of literary, historical, and cultural studies of science. Louis Kelley, in his widely praised book *The True Interpreter,* declares in no uncertain terms, and to the frequent applause of translation scholars everywhere, that Europe owes the greater part of its civilization to translation. Surely science must be allowed its major place in this same civilization.

The writing of this book has been largely a personal affair. Some portions of chapters 4, 6, and 7 appeared, in different form, in the journal *Science as Culture.* My thanks go to Les Levidow for his useful comments on the relevant articles. Chapters 8 and 9, meanwhile, represent reworked and expanded material from my earlier book *The Scientific Voice* (Guilford). Gratitude is also here expressed to several anonymous reviewers of the original manuscript, whose comments, criticisms, and corrections have resulted in a much superior work and have spared me any undue revelations of incompetence. In writing the final draft, I have benefited considerably from the generous support of the National Endowment for the Humanities' Fellowship Program for Independent Scholars (1998).

Finally, my deepest and most necessary appreciation extends to my family, Marilyn, Kyle, and Cameron, whose understanding and forbearance of human frailty have come to know new limits.

Introduction

Transfers of Learning, Questions of Influence

Clarifying Terms

As a textual reality, the history of science in any single tongue involves many matters of coinage, the creation of vocabularies. Each discipline, even each term, can be said to stand at the apex of such an evolution. Single terms are nearly always the result of some conscious choice, and this choice must often, of necessity, bear the marks of larger influence, above all the era-bound proclivities of the men and women who discovered the need for such choice. A nomenclature is built from thousands of such selections; it leaks history at every pore.

Consider the periodic table of the elements, for instance. How might this fundament of chemical speech appear in different languages? How diverse, if looked at closely? Here is a hint:

> The names of the elements in Japan are not a little complex in their construction. Besides those such as hydrogen (*suiso*), oxygen (*sanso*), and carbon (*tanso*) that form their names by attaching the character *so* [lit. simple, principal], there are others [based on Chinese example] that use only a single ideogram, for example gold (*kin*), silver (*gin*), copper (*doh*), and iron (*tetsu*). In this light, we might also recall those names such as mercury (*suigin*, from the characters for "water" and "silver"), zinc (*aen;* "next to lead"), and platinum (*hakkin;* "white gold"), which are formed by combining two ideograms on the basis of their literal or analogical meanings. Then there are those names written solely in phonetic symbols, such as aluminum, natrium (i.e., sodium), tungsten, etc. Finally, there are also names [written in the Japanese phonetic alphabet] whose original language was Latin, German, and English, or that mingle these together. All of which, moreover, constitutes a noted contrast to Chinese, in which the titles of the elements today are all expressed with a single character. (Sugawara and Itakura 1990, 193)

In other words, the periodic table in Japanese represents a zone of cultural and linguistic collision.

Should we not declare an equal complexity for element titles in English? Haven't terms here come from the Latin, Greek, Arabic, German, Russian, the names of scientists, and so forth? Certainly they have. But in the Japanese case, there is a striking difference. To conceive an adequate parallel, one would need to imagine a wholly novel form of scientific English, in which terms were written in their original Greek or Arabic sign systems, in titles dating back to English's origins, in the symbols of Anglo-Saxon or Frisian. Moreover, one would have to accept the simultaneous use of medieval and modern forms of certain words, each with its own graphical integrity preserved. In short, all derivations would need to be *kept visible.* Clearly, this is not the periodic table as we know it. It is something else, for which we have no name.

The Mobility of Knowledge

Knowledge, whatever its contents, has always been a mobile form of culture. However one cares to define it—as a body of fact and hypothesis, the product of a specific labor, or an instrument of domination—human understanding, literary or scientific, has undergone enormous passages between peoples and places over the span of history. Its movement has come on the heels of war and conquest, commerce and trade, religious conversion, immigration, discovery. It has resulted, no less significantly, from the travels of itinerant scholars, pilgrims, and adventurers, either in the service of wealthy patrons or of their own curiosity and ambition. The mobilization of knowledge has taken place suddenly, during brief historical periods. It has occurred, more quietly and perhaps more profoundly, across the creep of millennia, as an elemental feature of daily life along national and linguistic borders, both within and between cultures.

Beyond any doubt, the transfer of learning has been critical to the building of societies, those we call "modern" most of all. Time and again, the introduction of new concepts and methods—Roman law, the system of Arabic numerals, the sonnet, Newtonian physics, linear perspective—has proved the source of new capacities for ordering, directing, and expanding human existence. Placing the knowledge of one people into the hands of another involves the transfer of certain powers: powers of expression in the case of literary or artistic knowledge; powers over the patterns and organization of life in the case of political, legal, or religious ideas; and, in the case of science, powers of imagination and practice with regard to the material world and uses of it.

Such transfer therefore defines a critical historical process: it is what scholars are really concerned with when they speak (and they do so often) of "influence" between different periods or societies.

Thus a question: how is knowledge rendered mobile? What makes it able to cross boundaries of time, place, and language? The answer would appear simple. But perhaps not so simple after all.

"Translation," one soon realizes, is not a word that describes any single activity. As the second oldest profession on the streets of authorship, it is generally conceived in fairly obvious terms, as a matter of rendering the words of one language into those of another, hopefully with little or no spillage of meaning. Yet this is hardly a definition; it is more in the manner of a description. It deals not at all with the enormous variety and complexity of the transfer itself.

Even as employed in everyday speech, "translation" is more like a canopy under which gather great crowds of phenomena. What has been designated "translation" in the strict sense, for example, has involved both oral and written exchange, sometimes simultaneously. It has been performed by individuals, by teams, by groups or entire communities, even by machines. Its materials have included original texts, copied and altered texts, reconstructed and stolen texts, fabricated, falsified, and even imaginary texts. Such texts, moreover, have comprised every conceivable type of prose and poetic document known to human society, as well as songs, tales, memories, and visions. In its methods, meanwhile, translation has indulged in unchecked *libertas utilis* (utilitarian paraphrase), labored under the yoke of *verbum pro verbo* (word-for-word faithfulness), and explored the vast plains between. It has therefore yielded works that have been completely rewritten, beautifully adapted, ideologically reconstituted, or so literally decanted as to be incomprehensible and useless.

As if this were not enough, there is much diversity beyond such questions of matériel and method. It is a simple matter to show how translations, in terms of their selectivity and their specific castings, have served the causes of ideology in innumerable and often conflicting ways, keeping faith not only with such concepts as Imperial Rome or "the white man's burden" but also with orthodoxies surrounding notions such as High Culture, Poetic Form, or "proper scientific style." Then, too, there are the realities of context—the who, where, why, and when of translation, as well as the what and the how. In looking into the matter a bit, one realizes that there have been historical periods in which massive amounts of transfer occurred as part of a distinct, sometimes centrally directed cultural policy, carried out by designated experts for the sake of particular ideas of "progress." On the other hand, translation movements at other times have depended upon a scattering of individuals or teams, working without any obvious support, seemingly for the cause of unpromised fame. In either case, meanwhile, the relevant episode may have been several decades long and particular to a single nation (sixteenth-century England; eighteenth-century Japan; early nineteenth-century Germany). Or it may have constituted a true epoch, lasting more than a century, involving hundreds upon hundreds of texts, and engaging, eventually, an entire continent in its struggle to procure the "wisdom" of one culture for the advancement of another (the so-called Twelfth-century Renaissance of Europe).

Who have been the translators to make these episodes or epochs possible? Here again one encounters wholesale diversity. In the case of science, no less than in literature, these mediators have included: monks, scholars, mercenaries, students, explorers, soldiers, ship's captains, commercial yeomen, diplomats, and scribes, to name but a few. What about the linguistic abilities of these culture-makers? Today, we tend to think of translators as requiring profound expertise in a foreign language. History shows this to have been the case less than half the time over the past two millennia. Some of the most astonishingly productive translators in all of Western history—Gerard of Cremona, for example, who brought into twelfth-century Europe dozens upon dozens of the most difficult scientific texts from the Arabic—appear to have used intermediaries on a fairly regular basis.

It should be made clear that what we are discussing here is not "translation" in any metaphorical sense. It does not seem wholly productive, historically speaking, to stretch this term out into generous shapes, such that it stands in for the negotiation of any and all differences across cultures. This has been done, in various ways and with varied advantage. But the effect is to divert attention away from the richness of linguistic transfer itself, something that has received far too little interest to date. For the purposes of the present study, I wish to define "translation" in a manner capable of including all the varied activities and intentions noted above. Generally, then, it is *the process of transforming a specific piece of one language (commonly a text of some sort) into another language.* If a bit inelegant, this definition nonetheless has the advantage of underlining the creation of a true cultural product, and of posing the important question: What happens to knowledge when it is given a wholly new voice and context?

Much more will be said and demonstrated along these lines in the chapters that follow. But a fundamental starting point, I hope, has been made: "translation" defines a process of communication every bit as varied as writing itself and no less central to what we commonly call "civilization," built as it is by movements of knowledge from one people to another.

A Tale of Sound and Fury

In the middle of the third century B.C., one of Aristotle's best-known students, Demetrius Phalereus, prior governor of Athens, escaped that city for Alexandria, there to become head of the great library. His departure was precipitated by political intrigue and adversity, which had affected the Peripatetic School in its wake. Notwithstanding the details of relevant events, it was now Ptolemy Soter I of Egypt, drawing descent directly from Antipater, successor to Alexander the Great, who became the conscious patron of Aristotelian thought and the assemblage of knowledge in all known fields it propounded. Aristotle, reputedly, had been the first in antiquity to amass a true library, and it was on the basis of his central thesis that true knowledge came from collection—a gathering of the facts of the world and the wisdom of others—that the great library of Alexandria itself was founded. This had been put in no uncertain terms: the central mission of the "universal" library was to bring to Alexandria "the books of all the peoples of the world," in order that this city, true to its namesake, would become a center for one of the greatest empires known. Thus it was fitting that Aristotle's own student became administrator over the vast project in its early years.

At some point during his regency, Demetrius hit upon a central theme for using the library to serve his new patron. This theme was to harvest from various peoples the scrolls they had written on the arts of kingship and the uses of power in building a strong and unified state. Part of this idea had its origin in Demetrius' belief that the books of Jewish history, law, and philosophy in particular would provide special insight, due to the lessons they might offer on constancy of belief and fortitude in the face of the vicissitudes of time. It seems that this belief was the result of persuasion by one Aristeas, a Hellenized Jew of disguised origins who nonetheless sought, under cover as it were, to secure greater benefits for his people. Ptolemy agreed to Demetrius' request for books from Judea and ordered it done. The relevant works, however, required translation: they were not written in Coptic or Greek or Phoenician, nor in Syriac, as commonly believed, but in Hebrew. As such, they could not be of any use until transformed.

Translation was not a problem for the great library. Indeed, it had been part of the vision for this institution endorsed by scholars employed to study, annotate, and categorize the collections as they grew. All the world's books were to be transformed into Greek—the library was to carry out Alexander's own incomplete conquest in another form. In fact, all the major Hellenistic capitals built or improved by the Macedonians had been given their own libraries, as repositories of textual resources. Put in place by Alexander, the rulers of these cities had been made to understand that, as victors, they were hopelessly provincial, and therefore had a deep need to know about the peoples over which they governed, from India to the Balkans. The way to do this was to possess and translate their books. In short, translation—not merely political triumph or library building—was the weapon with which they would solidify their own, elevated destiny.

The story, however, does not end here. Having agreed to the importance of the sacred Hebrew texts and the need for their translation, Ptolemy was pointedly reminded of the many thousands of Jews who languished in Egyptian prisons and slave camps. Most of these people had been taken hostage during earlier campaigns against Syria pursued by Ptolemy's own father. Now, to ensure that the desired translations would be done with maximum trust while his own image of authority remained untarnished, the younger Ptolemy agreed to free them all—more than a hundred thousand—and, what was more, to integrate them fully into his beneficent regime, making many of them soldiers and administrators and giving some individuals positions of high responsibility. In a relevant proclamation sent to Jerusalem, Ptolemy declared these things to have been achieved and, in response, received from the high priest of that city a delegation of seventy-two scholars, six from each of Israel's twelve tribes. These men came west to the small island of Pharos, where, in elegantly furnished and protected isolation, they completed their task in seventy-two days. It is unlikely they understood how their actions served the cause of intellectual empire, first set out by the vast writings of Aristotle, the greatest collector of knowledge in the ancient world. Yet history records that their single effort of translation helped save tens of thousands of lives.

The Once and Future Aristotle

A good deal of Western intellectual history, including science, would be unthinkable without the influence of Aristotle. This is a truism. For nearly two thousand years, Aristotelian thought ruled huge territories of the sciences in the more advanced civilizations of the West and Middle East, alternately serving the causes of epistemological revolution, orthodoxy, and even stasis in societies as different as those of Hellenistic Athens, imperial Rome, early Islam, and late medieval through Renaissance Europe. Only by toppling such unparalleled influence, along ragged chronological and conceptual edges, could space be made for a truly modern science. Thus runs the standard view, and there is much in it that continues to draw the attention of the contemporary scholarly eye.

This grand monument, Aristotle, thus appears among the most solid and enduring fixtures of occidental intellectual sensibility. Perhaps, then, it should be asked, simply: who was "Aristotle" in terms of the writings attributed to this most famous of names? What might be the origin and history of the many books that have given this monument a seeming lithic stability? One ordinarily assumes they were produced, in whole or large part, by the hand of a single man, living in the fourth century B.C. But what if this turns out to be untrue, a myth as old as that of Caesar's divinity?

Demetrius, as noted, had been a student of Aristotle, even then the most renowned teacher of "useful books" throughout the Greek-influenced world. Within a few years following the translation of the Jewish texts, Ptolemy I Soter died and his son ascended the throne. The new king had strong dispositions against Demetrius, who had openly sought favor for another of the father's children. This Greek, who had done so much for Jewish history and for Aristotle's teachings, was now summarily exiled to a remote village and there assassinated with a snake.

It was just about at this time, too, that Aristotle's own chosen successor, Theophrastus, died, leaving all his books—including Aristotle's library, with the only original collection of his works—to one Neleus, an old and trusted friend of Aristotle, whom Theophrastus assumed would be elected the new head of the Lyceum. This, however, did not happen. Once again, political realities determined the fate of scholarly history. Neleus, that is, had been linked to Demetrius, who had fled Athens to avoid trouble with a tyrannical regime he had plotted against. This association meant that Neleus was viewed as a liability by the administrators of the school, concerned as they were with their own survival. Neleus was therefore passed over in favor of Strabo, another well-known scholar. Neleus left Athens for his native Scepsis and took with him the all-important books, never to relinquish them again.

What were these works? It is something of a shock, perhaps, to discover that "Aristotle," as an author, probably never existed. The evidence is very strong in this regard. It indicates that the works placed under his name were, at the earliest stage of their history, compilations of notes, recordings, collections of facts, and other fragments, mainly from his lectures at the Lyceum, which were assembled, amended, and very often written by his students. They were, in short, communal creations. Moreover, their contents were apparently never looked upon as final in any sense, but were instead continually updated and replaced during Aristotle's lifetime and thereafter, in accordance with the changing levels of discussion, insight, student participation, and so forth. This gave them an evolving, organic type of reality, one rather at odds with the modern concept of an authorial end product, intended for a receiving but nonparticipatory audience. At some point, no doubt, these communal works, never set down for an outside readership, were effectively frozen by the death of Aristotle or a succeeding teacher, whose own direct participation came to an end. With such a death, the book came to life. Indeed, it is only through this loss of the "father" that such works entered their fate into the greater journal of outside history.[1]

According to the detailed account provided by Strabo, the scrolls of "Aristotle," perhaps due to their obvious value, were buried by Neleus's heirs, who therefore condemned them to the effects of decay and decomposition. From here, they survived later discovery, then sale to a Roman bibliophile who attempted to repair or edit them back to a form that approached their earlier wholeness. They then passed through a succession of owners who hired various scholars to work new corrections and interpretive fillings-in (especially of the holes resulting from burial), and still later, to reorganization and abridgement by Roman scholars of Greek. The detailed account provided by Strabo relates that the Aristotle corpus was eventually sold to an Athenian antiquarian known as Apellicon, who had the text of the decayed scrolls repaired in "abhorrent fashion." No settlement was yet in store for them, however, since political and military winds next placed them, as personal booty, in the hands of Sulla in 86 B.C., when he conquered Athens. Taken back

1. These and related points can be found in more detail in Canfora 1990 and in two works by Paul Moraux (1951, 1973). Similar questions, particularly dealing with "authenticity," are also taken up in Moraux and Wiesner 1983 and, more recently, in Delia 1992. For general factual information of relevance to the preservation and transfer of classical literature, there is no better general source than Reynolds and Wilson 1991. The major limitation to this work is its traditional focus on literary works (many pages devoted to Livy or Juvenal and almost none to Hipparchus or Archimedes). In the end, one cannot help but feel that all this recent scholarship ends up repeating, in newer and more well-considered words, the conclusion reached more than a century ago by Richard Shute:

> In a word, we shall try to get as near as we can to the earliest form of the teachings of the master, but shall not vainly and pedantically hope to restore his actual words . . . since we shall know well that the Aristotle we have can in no case be freed from the suspicion (or rather almost certainty) of filtration through other minds, and expression through other voices. (1888, 178)

to Rome, they were again worked on, this time by more competent hands, but also endlessly and poorly copied and sold to acquisitive aristocrats. Porphyry, in his *Life of Plotinus*, relates that the scrolls enjoyed the careful and skilled editorial ministrations of Andronicus of Rhodes, who set about dividing the entire corpus into specific books by gathering similar subject matter together under single headings. Copies of this edition by Andronicus seem to have survived into the second century A.D., when a revival of interest in Aristotle occurred. But the original, including the sacred scrolls, disappeared completely, and with them, any hope of true contact with "Aristotle."

The group of secondary Aristotelian works had its own fate. During the fifth and sixth centuries, it moved east, as a result of persecutions Nestorian scholars endured under the Byzantine emperor Theodosius. Nestorian communities in Syria and elsewhere became the sites of translation into Syriac, and later into Arabic and Persian—and, as a result, absorption into Islamic intellectual culture in the eighth and ninth centuries. From here, "Aristotle" underwent centuries of copying, further editing, and probable reorganization, before finally, along with the major portion of Greek philosophy and science, entering Latin Europe in the eleventh and twelfth centuries as part of yet another major era of translation. Aristotle thus arrived in the medieval West first through the Arabic language. It was at this point that these works became the foundation for Scholasticism and, therefore, large portions of modern science.

All along, survival of the Aristotelian corpus depended not upon mere preservation or upon any such thing as the innate "worth" of the relevant texts, but instead upon the very real actions of constant copying, editing, reorganizing, rewriting (often for pedagogic purposes), use, and translating—in short, the nativizing of these works to different historical epochs and to different cultural settings. Today, what qualify as the oldest extant manuscripts are scattered down the corridors of the last two millennia, some being in Greek, some in Latin, some in Arabic, but utterly nothing in Aristotle's own hand.

As an author central to the occidental canon, "Aristotle" is therefore, in concrete terms, a fiction, or rather a construct. The Aristotle we have today, the one that has existed since the beginning, is a classroom assembly rather than a textbook. He is a loose commonwealth of thought and writing that has changed, irretrievably, over time. And at the core of this change there has always been the process of passage—the transferring of textual matter between different locations, different peoples, and therefore different languages. Like the freeing of the Jews from

Egyptian prisons effected by his disciple, this same "Aristotle," in all his fertility of influence, is inconceivable without such transfer. It is the chariot on which he has ridden, in complex triumph, to the present.

The Great Library Returns

The history of Aristotle on the one hand, and on the other the fate of the great library at Alexandria, which, as we know, suffered eventual destruction, reveal two sides of a single truth: the written word is a fragile, changeable reality. In the end, however, we find that Aristotle is still with us, and the great library, with its dream of gathering the knowledge of the world, has never died. Indeed it has found a new, and perhaps inevitable, expression in the present era.

In July 1988, the late president of France François Mitterrand announced the launching of a new project in the history of text collection, a "très grande bibliothèque" whose mission would be to assemble in digital form written, audio, and visual material in every conceivable branch of knowledge from as many nations and cultures around the globe as possible. The design chosen for the main building is a spectacular evocation of the acquisitive vision itself: four huge towers in the shape of open books, each 80 meters in height, rising above an enormous hollow rectangular platform, suggestive of a closed cover. The scale of the project has been described as "*pharaonique*" and indeed seems worthy of such designation, encompassing 7.5 hectares in area, with an esplanade of 60,000 m^2; a garden spread over 12,000 m^2; 395 linear kilometers of shelf space; and a total usable floor surface of 2,900,000 m^2—in short, a pyramid dedicated to what some have called, ironically, the "death of the book."[2] Global in scale, the project is overlaid with a distinct nationalistic flavor, a fact made evident by its very title: Bibliothèque de France. The works of the world, in short, are to pass through the gates of *le patrimoine*, just as they once did through those of Alexandria.

Controversies that have raged since the inception of this project, however, have to do with other issues. They relate to factors both more concrete and more rarefied, factors that Demetrius would no doubt appreciate. The first of these regards the question of access. According to its mission statement, the aim of the project is to "open the future library to as wide a public as possible . . . [thereby becoming] a great

2. These and other facts about the new library, including historical information and an overview of the project, can be found on the World Wide Web, at the following address (English version): http://www.bnf.fr/institution/anglais/sommgb.htm

archival depository while at the same time serving as an agent of culture, stimulating the desire to read in the widest possible public" (Jamet and Waysbord 1995, 78). Use of the library was therefore to be free, equally open to all. Yet very quickly, certain realities, economic and political, set in. As of mid-1995, with the nation in the midst of recession and a new conservative government in charge, it was decided to impose an annual fee of 80*f* for the general public, 250*f* for researchers. A total of 1,650 seats on the upper level were now planned for use by general readers, who would be able to access about 382,000 volumes and periodicals, selected by a panel of experts. Professional scholars and researchers, on the other hand, were given 2,000 seats on the lower floor, an unimpeded view of the gardens, more sophisticated computer terminals, and complete access to the entire collection of more than 18 million documents. As a group, researchers are today aristocrats of the word. Their counterparts in the classical world were not the isolated thinkers such as Hipparchus or Ptolemy, nor the great teachers like Pythagoras and Plato, but instead the institutional scholars who roamed the great collections at Alexandria and Pergamum, endlessly annotating, editing, commenting upon, and therefore protecting the works of these thinkers and teachers.

As Demetrius knew, and as the story of the Jewish books makes plain, knowledge in textual form has been a type of raw capital ever since the dawn of literacy. Never fixed, linear, or hardened against time, written learning is amenable to manifold transformations and transfers. Like financial power itself, it has always been distributed on the basis of recognized class structures. The "largest and most modern library in the world" has become, like its archetypal predecessor, a container and divider of social space. The Bibliothèque de France seems poised, therefore, to act as a kind of archetype of its own, revealing how libraries of the future with their digitized contents, so often claimed to be "without walls," are likely to be nothing of the sort.

Does the digital library signal the disappearance of the book? Will the page meet its historical erasure in the primacy of the computer screen? Is the text about to be separated from any reality as an object? Do we live, in other words, "at a threshold moment in the history of libraries and the forms of knowledge they imply—a moment comparable to that of early antiquity when the clay tablets of the pre-Christian era were replaced by papyrus rolls" (Bloch and Hesse 1995, 1)? Such questions are no doubt important to definitions of the present, past, and future. But they are also, in a sense, irrelevant. The use of clay tablets lasted perhaps a thousand years; papyrus and pergamum (parchment) another eight hundred, before giving way to the codex in the fourth century A.D.; another millennium and printing in the West altered again, forever, the medium in which knowledge was held and transferred. Today, a mere five hundred years later, and we are perhaps in the infant stages of another such transformation. The book, as a type of literary technology, like the codex and the scroll before it, may well be in jeopardy. The resting place for our epistemological deposits appears to be in question.

Does this mean, however, that literacy is about to undergo a profound change? Does it signify an end to the text as we know it, a shift in the entire legitimacy of writing and reading? Is there something inherently oxymoronic, or simply wrong, about a current-day medievalist using a screen version of a fifteenth-century incunabula or tenth-century manuscript? Does, in fact, the movement from book to computer, "text" to "information," reveal to us a type of translation that marks a change in the nature and direction of Western culture as a whole? These questions all have an obvious answer.

The future of the book, as icon of knowledge, and of the library as a repository of print, are in doubt, true enough. But one should be clear about this: the attachments at issue are materialist ones, nothing more. The absolute dominance and centrality of the word can hardly be debated, even for a moment. Indeed, it is already evident that digitization *increases* this centrality, making the word that much more abundant, luxurious, exchangeable, ineluctable, alluring to censorship. The computer is a producer of many things—images, numbers, sounds, and so forth—but none of these, even in combination, compares with its manufacture of words. To a large degree, the reign of paper print has already fallen in defeat (where, for example, is the typewriter today? The printing press? What social corners does handwriting now occupy?). Yet instead of the final triumph by technology over "culture," we see something much closer to the opposite—the computer in service to language.

The book does seem something firm and solid, an anchor in time. It represents an unequivocal and direct link with the past. To recognize that, like the scrolls of Aristotle, it may be suffering decay and deformation means to many that the future is being cut loose from everything that has made it possible. And yet, for the vast majority of the world's literature, philosophy, history, and science, the concept of a reassuring solidity of the book is a fable—exactly in the same sense that "Aristotle," as a single author of a grouping of original texts now in our possession, is a fable. The greater output of literate society, even since the advent of print, has necessarily survived through processes of continual change: copying, editing, repackaging, republishing, and above all, translation. The book has never been a stable thing.

It is the translator, not the scholar, who has acted as the ultimate librarian. A chooser and a protector of texts, s/he has been a force for not merely preserving them but increasing their numbers, forms, and lives, for continually creating the possibility of new collections, in new settings. At once a destabilizer and a champion of the written word, the translator is also its depossessor and an enlarger of its dominance. It is impossible, in practical terms, to conceive of the universal library—as the memory of the world—without the central role of translation. Without it, there is no possibility of access to what has been written in so many different languages. This, too, was known at the beginning: all works accepted into the library at Alexandria, whether scientific or literary, were to be translated into Greek. This transformation of the world's memory would not only allow Greek society to profit by possession; it would also make this memory seem a native product. Indeed, the standard invisibility of the translator comes back to the fact that an erasure of origins—the reader's forgetting that s/he is holding a foreign object rendered into the familiar—defines one of the primal aims of translation. But this too is part of the migratory nature of knowledge.

Conclusion

The tale of Aristotle and the story of the Jews and the great library, past and present, suggest that in the history of knowledge, the power of translation is commensurate with the power of the word. Aggressive trade in books and ideas—the buying and selling of textual matter, whether in piecemeal, altered, or counterfeit fashion—has long been involved in the creating of knowledge systems and the saving of lives, well before the present-day notion of a "knowledge-based society." Literate culture has always been a floodplain for the fluid movements and deposits of learning. The creation of an Aristotle, by means of a meandering journey of texts through time, is an excellent allegory for what has happened to all important historical works in all areas of science.

One must take seriously, of course, in the history of science no less than in literature or art, the institutional presence of canonical works and authors. The idea of modern physics without Newton, or classical mechanics without Archimedes, is impossible, not to say nonsensical. No one would deny the intellectual achievements and influence of such thinkers. But, as the case of Aristotle suggests, there is very often a missing element in such worship. Many biographies exist of important thinkers in science, but not of their texts—the objects whose dissemination and passage define precisely the material reality of these individuals in the present. Newton is today a living textual result (so to speak), not an obituary or memorial. The movement of his work, which proceeded in many directions, through many languages—from Latin into the European vernaculars, and from there into other languages throughout the world—is a crucial frame in which to understand what scholars call "the impact of Newtonian thought."

Once published, any text has a "life" of its own, separate from that of its creator(s). This is an obvious and well-recognized fact, but, as I am suggesting, it implies that the conventional approach to understanding a work of importance—that is, in relation to its author's life (a distinctly literary approach)—ought to be augmented in many cases by an equally detailed history of this same work as it moved into the public domain, into other countries and cultures, there perhaps giving rise to new vocabularies while at the same time taking on a range of particular casts reflective of each particular cultural and historical setting. We have hundreds of books about the life and times of Sir Isaac Newton; it is unlikely that we have more than a few scattered volumes and essays about how his great *Principia* was translated into languages as disparate as German and Chinese, there to create in complex nativized form the nomenclature of modern physics. The past texts of science, like those of literature, seem relatively secure today. They have been rendered so by the invention of printing, digitization, vast libraries on every continent, the existence of universities. Yet, as implied repeatedly in this chapter, most of these works rest at the end of specific life stories that have resulted in the birth of a host of different versions, each produced for the people of a different linguistic community. These translations are not copies, simulations, or replacements: they are, in every case, true originals in each language. Newton in Latin or Archimedes in Greek had little meaning for Japanese scientists; nor did the algebra of al-Kwarizmi in Arabic for thirteenth-century Europe. Until such works were made available to these intellectual communities, there was little possibility of a functioning "original" within their borders. The textual heritage of Western science is many times larger, and many times more tenuous, than has been assumed.

PART I
The Heavens through Time and Space

A History of Translating Astronomy in the West

Since man, fragment of the universe, is governed by the same laws that preside over the heavens, it is by no means absurd to search there above for the themes of our lives, for those frigid sympathies that participate in our achievements as well as our blunderings.

M. Yourcenar, *Mémoires d'Hadrien*

1 The Era of Roman Translation

FROM GREEK SCIENCE TO MEDIEVAL MANUSCRIPT

The Outline of a Field

Astronomy, it is often said, comprises the oldest of the exact sciences, reaching back more than five millennia in the search for precise patterns in the skies and the power to predict them mathematically. Such venerability might seem to suggest a long, continuous movement toward the present, with various additions from different peoples flowing into the whole like roads into a great highway. Certainly astronomical knowledge—particularly in the West—has arrived from many quarters, passing through many hands and voices, representing today a vast inheritance of diverse origin. Indeed, only mathematics would seem to rival such diversity. It is, instead, the image of continuity that fails. When looked at with an unpeeled historical eye, the heavens have a much more complex story, or set of stories, to tell. Their passage to the present suggests a very different sort of analogy:

> What was brought [to America] has imprinted the New World with strange traces of prior origins. On an old road through the Santa Ynez Mountains in California, certain rock surfaces are scored with ruts spaced exactly as were the wheels of Roman chariots. The stage-coaches that marked them were built to Spanish measurements, and the wheels of Spanish coaches had been spaced to fit the ruts of Roman roads in Spain. (Kenner 1989, 68–69)

Each historical era, and each culture, through which astronomical knowledge has passed left its indelible imprint. Nowhere is this more evident, and in many ways significant, than in the deposits of language. The terminology and discourse of astronomy, including the names of the planets and constellations—the oldest nomenclature of this most ancient of sciences—together constitute a form of historical evidence that can be traced through time in a concrete manner. Reading and comparing texts with a view to such tracing provides a direct means for estimating what each successive people left engraved upon this knowledge, how they nativized what had come before, down (let us say) to the Renaissance in Europe, the gates of the modern era. Any such tracing, in order to be complete, would have to take account of three linguistic processes: coinage, compilation/standardization, and translation. As might be suspected, these are by no means easily separable. Translation, for example, has time and again resulted in the creation of new vocabularies in languages previously foreign to the relevant knowledge. In more than one case, multiple, competing, and imperfect vocabularies were produced by multiple translators, with true standardization coming only at a significantly later date, through retranslation by an acknowledged "master." The record is rich with intralingual complexity; committed study of it should not hope for formulaic patterns.

My discussion begins at a specific point: the transfer of Greek astronomy to Rome. I've chosen this point of departure because it is at this stage that sufficient evidence appears for the type of historical tracing noted above. But there is another reason for such a choice. Greek science has long been viewed by scholars as a necessary intellectual source from which modern Western science could be built. Attempts to chip away at this edifice of perception have had little effect overall—and rightly so, at least in the sense of historical influence. The textual evidence leaves no doubt as to the ultimate powers of what is commonly termed Greek thought. The question, rather, is this: which "Greek" do we mean? Or, to put it somewhat differently, how far can we realistically stretch this possessive in the face of what happened to the actual textual material involved, the manifold transformations that came to reconstitute the relevant textual corpus? The next three chapters will show how extensive an answer such questions demand.

As often noted, Greek science of the classical and Hellenistic periods incorporated elements from a wide range of cultural sources, such as Egypt, Babylonia, and other portions of the Near East. The names of the zodiac, for example, appear to have originated in Mesopotamia, as noted some time ago (Plunket 1903). Other constellations (the majority) and the planets have proven more difficult to trace. There are some obvious trends of shared imagination in their titles: Venus, for example, was known in Sumerian as *Niṇdaṛanna,* "bright lady of the heavens," and in Akkadian as *Ishtar,* "goddess of love," both of these being obvious predecessors to the Greek *Aphrodite.* Most literary remains, however, are confusing in their indications of influence.[1] Partly as a result, the true origins of most names and many terms in Greek astronomy remain obscure and may always be so. Much probably never existed in any traceable form, being derived from oral traditions handed down and modified since prehistoric times. "This much appears certain, that the image of the heavens in the old world formed through a long gestational process, one to which the people of the ancient orient provided not only the fundament but many of the details as well" (Kunitzsch 1974, 170).

Since a good part of the lineage of Greek astronomical discourse stands beyond reach, what matters is its passage: the hand-to-hand transfer of language and the gathering of it into systems. But this requires that one immediately recognize certain limitations. The great translators and systematizers of the classical and medieval eras lived and wrote within a manuscript culture. What this means is that there was no wholly stable, fixed "science," inasmuch as science existed in the written word. The indeterminancy of scribal culture, in which individual copyists regularly deformed their texts through errors, deletions, abridgments, misinterpretations, additions, and a host of other "editorial" changes, meant that few works were ever precisely repeated and therefore that a large and uncontrolled number of versions of any one text existed at any one time. Another aspect of scribal culture was the lack of any recognized standards for translation itself: translators, too, were prone to mistakes and omissions (for example, when they didn't understand a passage)—but even more significantly, they were free to consciously alter a work, to add new examples, reorganize or create chapters, insert commentary, change wording, whenever they felt the need, and they did so often.

Fictional Considerations: The Case of Ptolemy

This being said, it should also be stressed that no classical author has come down to us in anything resembling an original form (Reynolds and Wilson 1991). In the case of the greatest compiler of antiquity, we have nothing even approaching an original manuscript whatsoever, no Ptolemy in any true sense. What exists instead, and what existed in medieval Europe as well, is a textual institution, a collection of later versions and commentaries of highly varied vintage. Better said, "Ptolemy"

1. For example, Lloyd (1991, 279) makes the following observation: "it is clear that just how particular ideas, knowledge, myths, techniques travelled, and why some did easily and others only very slowly, are questions that have to be resolved case by case. There are indeed as many interesting problems relating to the *failure* of transmission, when cultural contact existed, as there are to successful transmission. . . ."

denotes a "textual community" (Stock 1983), made up of Greek, Byzantine, Arabic, and European translators, scribes, editors, and commentators, all of whom were involved in the interpretation and use of Ptolemaic writings. Even within half a century of Ptolemy's death (A.D. 170), his works on mathematics and astronomy were being copied, revised, and epitomized (Jones 1990). This was probably necessary. Ptolemy wrote on rolled, highly perishable material (probably papyrus). He lived in the twilight age of the scroll, just before the appearance of the codex (ca. third-fourth century A.D.). To survive, his works had to be continually reproduced, rendered contemporary for each new era, eventually to be legitimized by transfer to the codex form. This was done, with both spectacular and unfortunate results. "The eminence of these works, in particular the *Almagest* . . . caused an almost total obliteration of the prehistory of the Ptolemaic astronomy" (Neugebauer 1975, 5). In physical terms, this means that the writings of earlier Hellenistic astronomers (and there were many) either did not make the transition to the codex or did so only temporarily, in sporadic fashion, at least partly due to the fame of Ptolemy's achievement. Either way, their books were thus doomed to perish, while the Ptolemaic community grew. Between the fourth and eighth centuries, this community came to include Syrian, Hindu, Persian, and Arabic writers. Many of these copyists and translators tried to remain, at least at first, faithful to the Greek versions they received. But which versions were these? How might they have differed from the originals? Without the latter, we will never know. What finally arrived in late medieval Europe was many times removed from Ptolemy, the living author of second-century Alexandria.

On the other hand, "Ptolemy" has been often made a fiction in another, important sense that bears directly on our discussion here. Nearly every routine history of science has him reigning supreme in astronomical thought from the second century A.D. until the publication of Copernicus' *De revolutionibus* (1543)—that is, from imperial Rome to the High Renaissance. The "Ptolemaic universe," we are often told, as the dominant theoretical frame for "saving the phenomena," ruled Western astronomy "for over a thousand years." Notwithstanding the frequent (and, admittedly, necessary) simplifications of historical surveys, this is completely untrue. For the first millennium after it was written, the Ptolemaic corpus didn't even exist in Europe. It moved eastward to Byzantium, to Nestorian and Monophysite communities in Syria, to Persian and even Hindu culture, and to Islam, as innumerable scholars have pointed out. But it had little or no direct effect on European thought. The *Almagest*—the Arabic title to the *Syntaxis Mathematica,*

Ptolemy's great systematizing astronomical work—was not translated into Latin until the middle of the twelfth century and was not widely known until a hundred years later. At the most, the Ptolemaic universe had dominant influence in Europe for little more than two hundred years, and even then it had significant competitors. North of the Mediterranean, such works as Aratus of Soli's *Phaenomena,* Pliny's *Natural History,* and Plato's *Timaeus* defined the major content of astronomical tradition, adopted in large part from late Roman culture.

Ptolemy's astronomy was vastly more "modern" than that of Aratus or Plato—that is, it was geometrical, mathematical, and theoretical, highly abstract in other words, much more in keeping with the image astronomers (who, until recently, have been the ones to write the history of their discipline) of the past century have preferred. What actually existed in medieval Europe, before the thirteenth century, was something quite different. It consisted mainly of various "handbook" summaries describing the motions of the heavens in simple terms, increasingly accompanied by arithmetic calculations intended to help estimate the occurrence of various feast days, refine or replace existing calendars, determine the proper times for sowing and harvesting, and the like. "From beginning to end," writes Olaf Pedersen, "the Middle Ages were imbued with the idea that astronomy, more than any other science, was of immediate relevance to the human situation" (1978, 303). From Rome, medieval Europe inherited the idea of the heavens as a great realm of signs, whose reading was by no means confined to predicting the future but also included explaining many things of the present and past, both public and private. For over a thousand years, therefore, the stars and planets in Europe remained innocent of theoretical suppositions: they were instead a realm where ancient notions of divine presence, astrological destiny, and arithmetic regularity collided.

The "Ptolemaic universe," when it arrived, represented a distinctly new type of astronomy. The textual community that came to be "Ptolemy" was the pivotal influence that helped transform the skies into a truly mathematical phenomenon, soon to be possessed by science. Copernicus, coming a few centuries afterward, rearranged the pieces. Prior to the twelfth century, the lands that came to be Europe drew on the heavens of Rome most of all. It is therefore to these heavens, and their origin in Hellenistic science, that we now turn.

Background: The Character of Greek Science in the Hellenistic Age

Roman astronomy, as it came to exist in the late Republic and throughout the Imperial era, was almost entirely a result of translation from, and rewriting of, Greek works that emerged from Hellenistic intellectual culture. In a sense, the texts Roman authors took for their models represented a momentous choice. The absorption of these works into Roman culture ensured that a division in the textual cosmos of Greek science would endure for many centuries thereafter, with one strand (by far the more difficult and complex) being transmitted to the Middle East and Islam, the other to Latin Europe. These two strands would merge—not without struggles for dominance—only in the thirteenth and fourteenth centuries, in the wake of yet another episode of massive translation.

The two currents of Hellenistic science were represented by sophisticated technical works on the one hand and their selective, and at times highly literary, popularizations on the other. The first category includes mathematical treatises produced by a small elite that included such authors as Eudoxus, Eratosthenes, Hipparchus, and Aristarchus of Samos, as well as Ptolemy. These men possessed private incomes or worked under the nurturing shade of wealthy patrons, and the intellectual freedom helped them devote time and effort to producing complex treatments of various astronomical problems (e.g., planetary motion and size). Their works were written for like investigators, employing specialized vocabulary, difficult calculations, and advanced geometric reasoning.[2] It has been surmised that the impressive expansion of this scientific work, beginning in the later fourth century B.C., was tied to a range of social factors that combined to encourage royal patronage among the competing autocrats who took control over various pieces of Alexander the Great's empire following his death in 323 B.C. and who sought to acquire something of the stature and worldliness associated with cosmopolitan learning in an age when the library at Alexandria had become throughout the eastern Mediterranean a symbol of high civilization (Green 1990). Another important influence may well have been the rapid spread of literacy and book buying among the population. The sheer number of texts contained in the libraries at Alexandria and Pergamum, as well as their more local imitations in many cities, suggests that text production and the market for books were highly developed. Indeed, one scholar of the age has recently proposed that

> Writing was an exceptionally popular pursuit in the Hellenistic age, and . . . the written word by then dominated legal, political, economic and private life. . . . The levels of knowledge in all spheres of science, technology, craftsmanship, finance, and military science were set down

2. Though somewhat dated, an excellent source in this area remains Clagett 1957. For a general survey of astronomical thought during this period, the best up-to-date resource is North 1995.

> in specialist textbooks, and scholarly discussion was similarly conducted in book publications. . . . Large copying workshops in which armies of scribes reproduced texts from dictation, combined with a well-organized book retailing trade, ensured a dissemination of the written word that was unparalleled in the days of medieval Latin. (Dihle 1994, 281)

This is likely to sound a bit too modern, especially given that so little direct evidence remains of this vast literary output. Levels of public literacy, though greatly expanded from classical times, remained somewhat modest and confined to certain classes of individuals (Harris 1989). But it is clear that some types of literature grew considerably in the Hellenistic era, and primary among them are technical manuals or handbooks, whose "variety was enormous, going well beyond the medical, rhetorical, and other handbooks of the classical age" (fig. 1; see Harris 1989, 126).

For thinkers such as Eudoxus or Hipparchus whose work was especially advanced in mathematical treatment, widespread literacy came to have its negative, as well as positive, impacts. This was not merely because a majority of the Greek-speaking populace was interested in astronomy mainly for the sake of astrology. It also had to do with the contours of literacy itself, how it was taught, encouraged, directed. Education in the Hellenistic age was dominated by the study and teaching of rhetoric, whose great champion had been Isocrates, famed Athenian orator of the early fourth century (Marrou 1956; Bowen 1973). Isocrates' own school had included a broad array of subjects (far broader than his main rivals, the Sophists), but became, after his death, increasingly devoted to textual analysis, mainly of literary sources. In Hellenistic hands, this trend deepened and was given new practical attachments. The post-Alexandrian states were huge endeavors compared with the *polis,* built upon a new class of professional civil servants, career soldiers and diplomats, a less educated workforce, and royal extravagance. Increasing urbanization, civil strife, expanding trade, and cosmopolitanism demanded that leaders refine their ability to form and sway opinion, while also impelling a literate society capable of employing words to its own advantage. In such a climate, the techniques and tools of persuasion—specifically embodied in literary and philosophical works—soon became the requirements for advancement, even as they underwrote worldly success in a lavish court society.[3] By the second century B.C., when contact was first made with Roman culture, Hellenistic education had come to settle on types of analysis

3. For an extended discussion of this and related topics, see Jaeger 1939–1945 (esp. volume 3) and Peters 1973 (185–221).

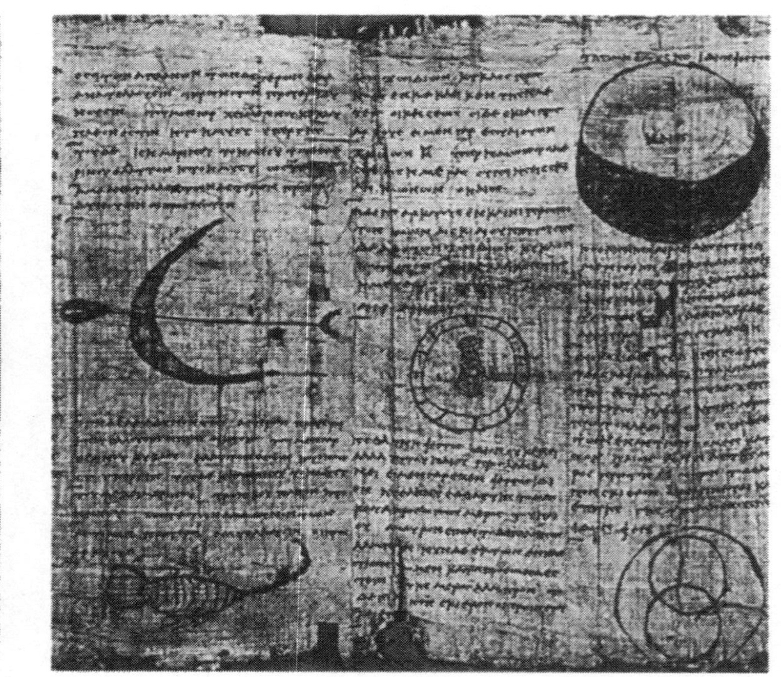

Figure 1. Portion of the oldest existing Greek papyrus (early second century B.C.), an elementary astronomical text, possibly a handbook of some sort, bearing crude images whose meaning is not always clear. Known as the "Eudoxus Papyrus," the work contains an acrostic bearing the famous astronomer's name. Museé du Louvre, Astronomic papyrus, Louvre E 2325.

that considered stylistic and rhetorical merit more important than any other content (Dihle 1994, 283). The combined effect was to exclude technical scientific writings from the curriculum and from most people's literary experience.

This did not mean, however, that people had no interest in scientific-type knowledge. On the contrary, the need to appear erudite in the age of the great libraries was a strong influence among the literate populace, and while some fields such as medicine were believed to be endemically specialist in their purview, the stars and planets existed for all to see, ponder, and, presumably, comprehend. Required, then, were simplified forms of difficult knowledge.

The Handbook Tradition

Hellenistic intellectual society yielded two answers to the demand for popularization. One response involved the writing of "handbooks," collected summaries of knowledge in accessible prose, spanning anywhere from one or two to literally dozens of areas of knowledge. In some respect, the great model for this type of writing was Aristotle himself, whose encyclopedic treatment of everything from poetics to meteorology remained for centuries an embodiment of the ambitions of the great libraries themselves. Aristotle's immediate successor, Theophrastus (c. 370–287 B.C.), wrote handbooks on botany, arithmetic, geometry, and astronomy, and the success of these writings rendered them rich ore for later compilers, who frequently mined entire passages for their own works. Indeed, so common was the practice of (what would be termed today) plagiarism that successive generations of handbooks almost completely replaced their forebears, even when the earlier texts were more accurate and comprehensive. The greatest of the early compilers, Posidonius (135–51 B.C.), was so widely read, dismembered, recast, and absorbed by subsequent authors that not a single work remains from his legendary output, possibly numbering in the hundreds of texts. Only by attempting to reverse this process of historical dissolution—extracting and reassembling the various purported fragments, or fragments of fragments, later writers appropriated from him—can we gain some estimate of his lost books (Stahl 1962, 45–53). Yet if Posidonius was erased by the very success of his work, this was due to the nature of the literary tradition he helped nucleate.

It has been common to refer to handbook writing as the effort of "professional hacks" or "dabblers."[4] One well-known scholar of hand-

4. See, for example, Stahl 1962 (55) and Green 1990 (642), as well as the more extended discussions in Laffranque 1964 and Edelstein and Kidd 1972.

book literature even reminds us that "our word *compiler* comes from the Latin verb meaning 'to plunder'" (Stahl 1962, 55). These sentiments, while perhaps understandable from the standpoint of the meticulous modern investigator, seem a bit inappropriate to the larger historical view. Rather, it makes more sense to speak of those like Posidonius, Theon of Smyrna, and their Roman progeny, Varro and Pliny the Elder, as writers in a different mode—that is, not as originators or creators or even critics and commentators, but instead as transmitters, carriers, messengers, combining the functions of "author" and "scribe." Such writers served a very real and important function within the greater social realm of knowledge production: they regenerated knowledge during a long era when texts were especially fragile physical objects (see fig. 1), when copyists (many of whom were slaves) could not always be trusted to preserve what had gone before, when changing literary taste demanded the past be retold in the rhetorical guise of the present, at least in part, yet also when famous authors of earlier times were considered the vessels of timeless wisdom, of a type that, once expressed in excellent form, should be allowed to speak itself thereafter lest the middling talents of various transmitters yield a succession of contaminations that would, in turn, diminish the contribution of future "borrowers." It is clear from the writings of compilers like Theophrastus, Strabo, and even Ptolemy (though not his astronomical work) that employing the past as a magnanimous lender of both word and idea served the purposes of private and collective gain.

This is not to excuse entirely the styles of plagiarism that so plainly existed. Nor can it explain away the outright theft that went on among contemporaries. Writing was most definitely linked to ambition, status, worldly success. Yet the great deal of "stealing" that took place, in fact, argues exactly for this, for an ever-increasing statutory power of the written word at a time when the technology of the writing medium itself (papyrus, vellum, etc.) remained extremely weak, vulnerable. Indeed, one finds that the compiler-transmitter was a central figure in European intellectual culture not only in antiquity, but throughout the Middle Ages, the Renaissance, and even beyond—right up to the late seventeenth century, to a Counter-Reformation figure such as Giambattista Riccioli, by whose time the printed book had finally become the sole means of legitimizing authorship. This continual, flawed, even egotistic rewriting of the already written, whatever its drawbacks for late modern scholarship, nonetheless defines a crucial process in the history of Western knowledge. It can hardly be a surprise that Roman authors, with their wealth of pragmatism, found it especially congenial to their needs when they eventually made contact with Greek literature.

Aratus of Soli: *Phaenomena*

It is unlikely that any handbook on Hellenistic astronomy ever matched in influence a single popularizing account of a very different sort. This work was not the effort of an accomplished amateur or scientist, but instead a poet of considerable merit: Aratus of Soli (ca. 315–240 B.C.), whose epic of the heavens marked the peak in a revival of didactic poetry initiated centuries earlier by Hesiod (Hutchinson 1988, 214–36; Green 1990, 183–86). A colorful, at times playful verse rendering in twelve hundred hexameter lines, the *Phaenomena* represents a rewriting of two prose works, the first being a popular account of the constellations and their movements by the astronomer Eudoxus, the second a treatise on weather signs by Aristotle's successor, Theophrastus. Most of the poem, as well as its title, were taken from Eudoxus' book. But it includes a number of errors and misstatements that were quickly pointed out by astronomers, including Hipparchus, perhaps the greatest observer-theoretician before Ptolemy. Aratus, however, was hardly concerned with precision, accuracy, or with mathematical reasoning of any sort. To his pen, the constellations were a source of magnificent imagery, a storehouse of poetic opportunity. He makes reference to practical uses of the celestial signs by sailors, farmers, soldiers, and the ordinary man. He does not speak of the planets—the primary focus of mathematical astronomy at the time—turning away from them with an intimidated, slightly self-justifying air:

> Of quite a different class are those five other orbs that intermingle with [the constellations] and wheel wandering on every side. . . . No longer with the others as thy guide couldst thou mark where lies the path of those, since all pursue a shifty course and long are the periods of their revolution. . . . When I come to them, my daring fails, but mine by the power to tell of the orbits of the fixed stars and signs in heaven. (Aratus 1955, 243)

Aratus speaks only of his "power to tell" the stars, not to explain or document them. His text offers an unending, glittering narration of physical (spatial) relationships. These are transformed into the effects of a life force manifold in movement, which is what the literary form of didactic poetry demanded:

> Her two feet [Andromeda's] will guide thee to her bridegroom, Perseus, over whose shoulder they are for ever carried. But he moves in the north a taller form than the others. His right hand is stretched toward the throne of the mother of his bride [Cassiopeia], and, as if pursuing that which lies before his feet, he greatly strides, dust-stained, in the heaven of Zeus. (p. 227)

Figure 2. Farnese globe, the earliest surviving map of the constellations from Greece and Rome, represented as described in the poem *Phaenomena* by Aratus of Soli. The globe is a Roman copy (second century A.D.) of a Hellenistic original, held overhead by a statue of Atlas, added in the Renaissance. British Library, Manilius 685.h.2.

Again, this is not astronomy, but an elegant use of astronomical material. It is not about the stars or their description, but rather their visual arrangement and the imagery this once called forth. In truth, Aratus has written the *Phaenomena* simultaneously in the form of a star chart, by which the skies might be memorized, drawn (fig. 2), and rendered part of cultural memory. The old-style hexameter line, with a venerable history reaching back to oral times, is the ideal poetic chariot to bear these ancient, communal stories of tragic determinism into a present full of change and insecurity. It is not the skies that Aratus revitalizes, after all, but the earthly (Stoic) vision of fixity in the midst of chaos (Sale 1966).

To call the *Phaenomena* a work of "popularization" is therefore, in some sense, misleading. It represents instead the replacement of science by literature in Hellenistic educated culture. It teaches little or nothing of the astronomical thought and advances of its time and doesn't seek to do so—only to narrate "the stars." The enormous and lasting fame this poem came to have, attested to by many preserved commentaries and multiple translations into Latin (see Aratus 1955, 190–94), proves all the more the considerable distance between technical astronomy and popular taste and interest. One of the most striking (and oft-mentioned) ironies attached to the *Phaenomena*, in fact, concerns Hipparchus, whom many scholars consider to be the greatest astronomer in antiquity and one of the most prolific (Neugebauer 1975). Of all his theoretical and observational writings, including those concerning his many discoveries and his refinement of the epicycle concept, not a single phrase remains: the only text to survive is Hipparchus' brief, hurriedly written commentary on Aratus.

Translating the Latin Tradition: Rome and the Imperial Heavens

Roman Views of Greek Culture

There is a famous statement attributable to Horace, but probably older in origin, to the effect that once conquered, Greece made a slave of savage Rome through its intellectual culture.[5] Often quoted, this statement seems at first irrefutable: for over a century, Roman poets and philosophers followed Greek models almost to the absolute letter. This was so much the case that the idea of "invention" (*inventio*) came to refer to only slight rewritings of existing translated material, while "contamination" (*contaminare*) denoted the impropriety of juxtaposing two or more pieces (e.g., scenes) out of order or inserting into them a section from a different Greek work (D'Alton 1962). But if Horace's claim for Greek influence appears true *in medias res,* it fails significantly to depict the larger circumstance of Roman-Greek relations *in toto literatis.*

The more precise attitude of Roman intellectuals, seen over the span of several centuries, involved a mixture of admiration and rivalry. The exact ingredients of this mixture, as well as their relative proportions, varied considerably between authors, and sometimes even within the work of a single writer. But its traces are everywhere evident and give rise to the deeper conclusion that the greater effort of Roman culture was to emulate, incorporate, and finally replace Greek learning. Such was the sometimes stated goal of translation itself (as we shall see). The power to recreate in Latin the best the Greeks had to offer meant the power to produce new models that would come to supersede, or at least substitute for, the Greek originals (Copeland 1991, 29–32). It is a significant truth that Roman authors rarely spoke of "translation" except with regard to Greek sources: even in the age of empire, when Rome controlled nearly the whole of the central and eastern Mediterranean, including North Africa, little or no mention was ever made of bringing into Latin works originating in such languages as Coptic (Egypt), Phoenician, Persian, or even, for the most part, Hebrew. Hellenistic Greek culture was itself intensely, complexly cosmopolitan, an amal-

5. *Epistles,* II.1.157–59. The Latin text is: *Graecia capta ferum victorem cepit et artis intulit agresti Latio.*

gam of many influences. But Rome, even in the midst of its own manifold urbanity, did not view it as such in the realms of intellectual pursuit. Translation, in large part, thus retained this power: it proved to be the medium by which Roman dependence on a relatively monolithic "Greek achievement" could be both expressed and overcome.

This is not to impose upon the whole of Roman culture a single, dull gray garb. Astronomy, as understood and practiced within the greater empire, was a diverse affair. Alexander Jones (1994) has recently documented, for example, that a more sophisticated level of astronomical activity existed in Roman Egypt than elsewhere, from the end of the Hellenistic period to as late as the third century A.D. The overwhelming majority of this activity, based on examined documents, would appear to have astrological prediction as its purpose: numerical tables, horoscopes, and other computational exercises were focused on forecasting celestial events and drawing earthly meanings. Moreover, in the later texts of the second and third centuries A.D., the methods used clearly brought together certain ancient Babylonian arithmetic techniques with newer means of calculation influenced by Ptolemy's *Handy Tables* (essentially a tabular section of the *Almagest*). All of which is significantly more advanced than what was commonly practiced and understood in Rome itself. Egypt, after all, with Alexandria as its intellectual jewel, had been a center of Hellenistic thought. Astrology had a very ancient and sacred role within Egyptian religious practice, and it seems evident that various methods were adapted to this venerable need in sophisticated fashion. In most other portions of the empire, Rome included, astrologers acted like rhetoricians in lowercase, casting nativities as forms of topical persuasion (Barton 1995). Such men were involved in the economy of symbolic capital; they wrote few or no texts and were interested in using the stars to contract favorable alliances with powerful and wealthy families who might then become their long-term patrons. It is interesting to see that they were often shunned by authors such as Pliny, who did write on astronomical matters and whose work had wide influence in the medieval period as well.

In the realm of written astronomy, meanwhile, Rome visited upon Greek textuality a very particular type of inadvertent vengeance. This is most apparent in the domain of science, astronomy included. The weapon employed by Roman authors is precisely what scholars have so often lamented as a profound limitation—narrowness of selection. Indeed, in their choice of books to translate, Roman writers effectively silenced the vast majority of Greek technical thought. They took in only what they felt was needed, what suited their immediate cultural tastes and abilities, and this meant the substance and vocabulary of handbooks and popular accounts like that of Aratus. The larger Greek achievement in mathematical science found shelter under other roofs, while a tradition of summaries, nativized to Roman proclivities, was passed on to medieval Europe, where it effectively remained dominant for nearly eight hundred years. The power of replacement succeeded entirely in this sphere, far more than the eventual effort by Roman authors to rival and supplant what they did adopt from the Greeks. The historical consequences of this were not merely momentous for medieval Europe. They proved equally profound for those peoples to the east, into whose caring and transformational possession the full textual corpus of Greek science found its way.

The educated Roman of the late Republic and imperial eras, meanwhile, was required to have a significant smattering of knowledge regarding the Greek "classics" and frequently sought much more than that. Posidonius, the great epitomizer, commanded a towering height among the hills of the Roman intelligentsia. The young Cicero eagerly traveled to Rhodes to become his pupil, while the most influential Latin scholars of later times—Varro and Pliny the Elder, for example—used the Posidonian texts as an infinite storehouse from which to draw literary, structural, and factual grain, not to mention large chunks of material on any number of subjects. Indeed, Varro's *Nine Books of the Disciplines,* which helped set the tone for Roman education from the first century B.C. onward, seems likely to have been derived from a Posidonian model (Laffranque 1964). Among the disciplines of Varro's title was astronomy, with which every literate Roman gentleman was expected to have some acquaintance, mainly through translations of Aratus. Where, however, did these translations come from? Who performed them and what were they like? Such questions deserve more than a simple answer.

Roman Theories of Translation: The Larger Context

Both as idea and as practice, translation in the late Republic and imperial periods took place within a larger domain of intellectual debate. This debate continually returned to questions about rhetoric generally and the role of "imitation" in particular. From the beginning, Roman authors were highly conscious of their dependence on Greek literary models. One expression of this was a complex range of attitude devoted to the notion of *imitatio*—the term given to using Greek sources as a basis for creating works in Latin. Aside from which specific models should be chosen, debate surrounded the problem of "servile" vs. "legitimate" imitation. Lower forms involved lifeless copying of an original, either Greek or Latin, resulting in a skimming of surface effects: "Nothing is easier than to imitate a man's style of dress, pose or gait," Cicero noted in this context.[6] If done from a Latin source, this counted as plagiarism; if from translation of a Greek text, slavish mimicry. At an early stage in Roman letters, for example in the late third and early second century B.C. when Plautus and Terence were active, word-for-word Latin renditions of Greek works, either in whole or in part, affixed with different titles and given Roman authorship, were counted honorable "creations," the rightful bringing to Latin audiences of an otherwise inaccessible treasure (D'Alton 1962; Gärtner 1988; Fuhrmann 1973). But as time went on, this became less and less palatable. Constant use and reuse of Greek material made writers ever more aware of their dependence, but also their need for such dependence. Horace, in the *Ars Poetica,* expresses very well the mixture of attitudes involved:

> It's difficult to write
> On common material in an original way.
> You would do better to tell the tale of Troy on stage
> Than to be the first to bring forth something new and unknown
> Into the world. In the public domain you'll have private rights
> If you avoid most commonplaces and do not linger
> Along the easy path, and if you neither translate
> Word for word, in a fawning manner, nor, as an imitator, hurl yourself
> into a well from which shame or poetic law prevent any escape.[7]

To the post-Romantic eye, there seems something contradictory in this injunction to venture but a few tender steps from the beaten path. But in the great age of Roman letters, this was the very essence of the literary "golden mean," and in any case must be understood within the context of Roman self-consciousness stemming from centuries of "faithful imitation" (Kenney 1982).

Part of the discomfort involved a sense that the Latin language was impoverished compared to the Greek, particularly when it came to Greek ideas. Lucretius, in *De rerum natura* (On the Nature of Things) makes this explicit:

> Nor does the thought escape my mind how hard
> A task it is to expound in Latin verse
> The dark discoveries of the Greeks; for first

6. Cicero, *De Oratore* II.xxii.91. The Latin is as follows: *Nihil est facilius quam amictum imitari alicuius, aut statum, aut motum.*

7. *Ars Poetica,* 128–35. My translation represents a merging of several renditions, including those by H. R. Fairclough (in Horace 1926); Smith Palmer Bovie (in Horace 1959); and Copeland (1991, 29).

We must treat of many hidden things
In novel words, since our own native tongue
Is poor, the subject new. . . . (I.136–41)

Lucretius is cleverly serving two masters, proclaiming obedience to Greek models yet also the need for originality. The difficulty of the task is at once the guarantor of its conventionality and the promise of authorial achievement. Yet this was realistic: Roman translators did indeed greatly enrich the Latin tongue through their efforts, while "novel words" meant either Latinizations of Greek terms or creative deformations of existing words to approximate Greek usages (e.g., producing new verbs and modifiers from established nouns). Roman writers thus used the Greek tongue as a means to fertilize and gain mastery over their own.

The greatest volume of new coinings came in the second and first centuries B.C. This coincided with a time when the dominant view of rhetoric emphasized public utility and service (Kennedy 1972). As expressed by Cicero, one of the leaders of the "new oratory," "the very worst fault is to depart from ordinary usage and meanings agreed upon by the community" (*De Oratore,* I.12.). Translation thus became a means to both obey and go beyond such limits. By adapting Greek works to Roman usage, it directed the sense of linguistic difference toward discovery of opportunities for innovation.

Viewed as a subset of rhetoric, translation was commonly performed as a type of exercise. Cicero and Quintilian, the most influential writers on the subject in the late Republic and Imperial periods, respectively, argued for such practice in clear terms and with the definite aim of encouraging an enrichment of Latin:

> Thus I saw that to employ the same expressions [as previous orators] profited me nothing, while to employ others [less refined] was a positive hindrance. . . . Afterwards I resolved . . . to translate freely Greek speeches of the most eminent orators. The result of reading these was that, in rendering into Latin what I had read in Greek, I not only found myself using the best words—and yet quite familiar ones—but also coining by imitation certain words new to us, provided they were suitable. (Cicero, *De Oratore,* I.155)

And,

> Our earlier orators thought highly of translation from Greek to Latin. . . . The purpose of this form of exercise is obvious. Greek authors are conspicuous for the abundance and variety of their material, and there is much art in their eloquence, so that translating them gives us li-

> cense to use the best words we can find, since anything we might use is naturally our own. (Quintilian, *Institutio Oratoria,* X.iv.3)

Servile imitation was rather strongly discouraged by these writers, who were most responsible for formalizing Roman ideas on translation. The translator, it was stated, must be more than a mere device, clicking out word for word, phrase for phrase; he must be actively involved in the choice of language, dancing the border between imitation and innovation.

This describes very well the translating career of Cicero himself, whose versions of Aratus and of Plato's *Timaeus* had enormous influence. On many occasions Cicero repeats his ambition to translate "in such measure and in such manner as shall suit my purpose" (*De Officiis,* I.6.). These statements, offered time and again, contain the seeds of a later attitude toward Greek sources that came to the fore, fittingly, in the imperial age. Quintilian, for example, whose *Institutio Oratoria* appeared in A.D. 93, is especially forceful on this point: "I would not have paraphrase restrict itself to the bare interpretation of the original: its duty is rather to rival and vie with the original in the expression of the same thoughts" (X.v.5). But it is Quintilian's student and contemporary, Pliny the Younger, who gives the new sentiment its fullest and most famous formulation in recommending translation as a form of literary training:

> You desire my opinion as to what method of study you should pursue during your current state of retirement. The most useful approach, which is constantly recommended, is to translate from Greek into Latin or Latin into Greek. By this type of exercise, one acquires propriety and richness of vocabulary, diverse powers of figuration and exposition, and a similar faculty for original composition arising from imitation of the best models. . . . From this, one's understanding and judgment are enlarged. It may not be amiss, after you have read an author, to become, as it were, his rival and attempt something of your own upon the same subject, then to make careful comparison in order to determine whether you or he is the better. Congratulations are well deserved should you find yourself superior; humiliation will be your reward should he always win out. (*Epistulae,* VII.9.1–4)

By this time, Greek sources are no longer sacred material that must be handled with adoring and faithful hands. In some part, they have come to represent a type of trial, a burden of example that must be matched and superseded. The greatness that was Greece has evolved from a model of literary virtue and beneficence into an adversary and an antagonist. Translation thus itself becomes more than a mere exercise; as a process, it embodies a "political agenda in which forcibly substituting Rome for Greece is a condition of acknowledging the foundational status of Greek eloquence for *Latinitas*" (Copeland 1991, 31).

The core of this agenda lay in the moral precepts attached to the study of rhetoric, within which translation gained meaning. Hellenistic influence in this area, focused in the work of Cicero, Varro, and Quintilian, helped place rhetoric above all other subjects in importance, but not for academic reasons. Oratory according to these authors was the means by which all worthwhile knowledge could be put to use, given a public presence, and devoted to civic benefit. Having the power of spoken brilliance meant having power over others: power to convince, to see through deception, to move the populace, make policy, frame and impose laws, keep order, guide others toward virtue. In Rome, the *artes* of speaking and, to a lesser extent, writing, came first, because (as Cicero put it) they were the essence of *humanitas*. The aim was to produce "good and wise men," meaning public men, for the purpose of a strong and wise Rome. The art of rhetoric, once brought to a high level, thus constituted a branch of political science, since it possessed the power to make any branch of knowledge persuasive, but more, to move men's minds and hearts toward virtue in the midst of the most changeable and challenging conditions.[8] The superiority of oratory also resided in the fact that it represented a type of social action, a praxis.[9] Ethical activity, these writers agreed, was the final goal. Mathematics, dialectics, law, astronomy—"all these professions are occupied with the search after truth," wrote Cicero, "but to be drawn by study away from active life is contrary to moral duty. For the whole glory of virtue is in activity" (*De Officiis*, I.vi.19). Or, still more to the point: "service is better than mere theoretical knowledge, for the study and knowledge of the universe would be in some part lame and defective were no practical results to follow" (*De Officiis*, I.xliii.153).

It is difficult to determine whether this helps explain, or simply justifies, the lack of interest among Roman intellectuals in the more advanced works of Hellenistic science. At the same time, there is no better means of showing how these same writers sought to replace Greek sensibilities with their own than by comparing the above comments by Cicero with those of Ptolemy in his preface to *The Almagest*, the most sophisticated scientific work of Greco-Roman antiquity:

> For that special mathematical theory would most readily prepare the way to the theological, since it alone could take good aim at that unchangeable and separate [substance]. . . . And indeed this same discipline would

8. E.g., Cicero, *De Oratore*, I.xi–xvi. For a more recent discussion, see Kenney 1982.
9. Cf. Quintilian, *Institutio Oratoria*, II.xviii.

> more than any other prepare understanding persons with respect to nobleness of actions and character by means of the sameness, good order, due proportion, and simple directness contemplated in divine things, making its followers lovers of that divine beauty, and making habitual in them, and as it were natural, a like condition for the soul. (I.i)

Ptolemy's universe, no less than Cicero's, contains an axis of moral precept, but it turns in the opposite direction. Mathematics and its application to the heavens are a means of initiation into the higher order of things and thus harmony between human and divine. In contrast, Cicero views such an idea as hopelessly detached from ordinary abilities and experience. In the *Dream of Scipio,* the one work where the Roman translator of Plato and Aratus takes time to contemplate the heavens and the place of human beings within them, the narrator is borne upward into the celestial realm not to gaze on the stars and planets but instead to look back upon the Earth, and to see there in magnified form the strivings of men. This movement is all-important: even for the most intellectual of Romans, the stars and planets remained a stage for visualizing daily needs and demands.

Translations from the Greek: Texts and Patterns in Astronomy

Roman astronomy was derived primarily from two traditions of translation. The first of these, and possibly the more important with regard to general knowledge, involved Latin versions of Aratus. This tradition was inaugurated by the young Cicero, whose rendition of the *Phaenomena* appears to have done more than any other work to establish a new vocabulary for the heavens. At least two later versions were produced, meanwhile, one by Germanicus Caesar (15 B.C.–A.D. 19), who incorporated corrections advised by the commentary of Hipparchus, and another by Avienus (fourth century A.D.). Judging from the huge popularity of the *Phaenomena,* other translations probably existed as well. In any case, Cicero's had by far the greatest influence, and though not slavishly obeyed by Germanicus and Avienus nonetheless gave these later translators the bulk of their nomenclature.

A second source of authority for Roman astronomy, more difficult to trace in specific contour, stemmed from the handbook tradition (see discussion above), whose remains are best preserved in Pliny the Elder's vast, rambling compilation, *Naturalis Historia.* Here, problems immediately arise, for Pliny arrives relatively late on the scene (the work was completed about A.D. 77), and it seems clear that he and his Latin predecessors closely imitated their Greek models in appropriating already existing secondhand material while disguising such "theft" by citing the foremost authorities of various eras (Thales, Plato, Eudoxus, Hipparchus, etc.—Borst 1994). Certainly the mathematical sophistication required to comprehend the writings of Eudoxus or Hipparchus was well beyond the reach of encyclopedists such as Pliny, judging from all available evidence.[10] One also finds mention of the greatest Greek astronomers in the work of other Roman authors who wrote on natural philosophy, for instance Lucretius, Seneca (his *Naturales Quaestiones*), and Vitruvius (book 9 of *De Architectura*). But again, the lack of mathematical treatment leads one to suspect that these authors too were familiar with Greek science through the intermediary of handbooks or the perusal of selected, less technical passages.[11] If their intended audience consisted not of specialists but of educated laypersons (Borst 1994), the same had been true of the authors and educators who preceded and taught them in turn.

Pliny's own immediate model, meanwhile, and the most influential Roman writer in the handbook tradition, was Varro (116–27 B.C.), who compiled an encyclopedia of existing knowledge (the *Disciplinae*), along lines earlier established by Posidonius, laying out in extended summaries the nine liberal arts every educated Roman should be required to study (grammar, dialectic, rhetoric, geometry, arithmetic, astronomy, music, medicine, and architecture). If Pliny's Greek was middling or weak, Varro's was excellent, and he undoubtedly borrowed directly from Greek sources.[12] Though his work is now lost, significant portions of the *Disciplinae* were in all probability unacknowledged

10. Pliny's attitude toward the subject of mathematics is actually rather complex. When it comes to astronomy, however, he is apt to heap dismissive scorn upon the efforts of mathematicians, even in the midst of discussing their results, for their apparent hubris. See *Naturalis Historia* II.xxi.85–88.

11. Given the status of Greek works in general and the common (if no doubt imperfect) knowledge of Greek among authors of the late Republic and early imperial era, it seems unnecessary to assume (as modern scholars often have) that Roman authors had no direct familiarity with the texts of Hellenistic astronomy. Large private libraries that included Greek texts were a common phenomenon among the educated aristocracy, and writers of such fame as Lucretius and Seneca must have owned copies of the most important Greek works. The more important truth here comes back to the fact that these works never appear to have been translated, and were thus reduced, in effect, to a glitter of citations.

12. Varro's knowledge of Greek, and his readiness to use Greek works for his own writings, are confirmed by the remains of his *Saturae Menippeae,* which adapt to Latin the prose-verse satires of Mennipus (third century B.C.). More generally, Varro and Cicero were educated in the late second century B.C., when the learning of Greek was commonplace, even required, among the intellectual elite. For a general summary of the changing place of the Greek language in Roman culture, including its decline under the empire, see Marrou 1956 (255–64).

translations of various Greek handbooks (Stahl 1962). Pliny's astronomy, as rudimentary and purely descriptive as it is, must have accepted and passed down such translations.

Aratus and Pliny were not, to be sure, the only authorities affecting Roman discourse on the stars and planets. Cicero also appears to have translated portions of Plato's *Timaeus* and may also have been responsible for rendering into Latin pieces of Aristotle's *De Caelo* (On the Heavens). Neither of these works have been preserved, and their influence on astronomical discourse does not appear to be crucial. A lost work by Eratosthenes, polymath and head of the library at Alexandria, entitled *Catasterisms,* seems to have helped establish—or confirm—a tradition of telling mythological tales about the stars and planets that was passed on by means of such writings as Ovid's *Metamorphoses* and Hyginus' *Fabulae.* In terms of actual nomenclature, however, this tradition can be considered a subset of Aratean discourse.[13]

Aratus in Latin: The Legacy of Cicero

The renown of the *Phaenomena* in Rome was even greater than it had been in Greece, judging by its literary influence. Authors of every stripe and color used this text, mainly in Cicero's rendition, as a basis or aid for commentary, for incorporation, for theft, or for selective translations of their own.[14] What type of translation is this? What types of choices did the great orator of the late Republic make when bringing to his countrymen this Homeric poem on the heavens? As our discussion above on the context of Roman translation would suggest, Cicero's version is nothing less than a transformation that sought, ultimately, to both imitate and replace its source. A recent detailed study of astronomical nomenclature in classical Rome, as well as a number of other readings of the Ciceronian Aratus, concur in this general impression: "Above all, the process of translation favored by Latin authors [especially Cicero] consisted in finding in their language common words that might act as an equivalent to the Greek model, and then endowing them with a meaning appropriate to the science of astronomy" (Le Boeuffle 1987, 19). This type of "patriotic" adaptation or "lexical chauvinism" can be seen in all aspects of astronomical discourse, including general

13. A few additional renditions from Greek sources are included in the useful list of important Latin translations offered by Clagett (1957, 155–56).

14. A detailed sampling of such authors would have to begin with Ovid, Vergil, Catullus, Hyginus, Vitruvius, Quintilian, Manilius, Pliny the Elder, and such later writers as Martianus Capella and Macrobius, who acted as important intermediaries to the early Middle Ages. For a listing of sources, see Le Boeuffle 1987 (13) and Kenney 1982 (2.2.2–3 and 72–73).

terms used for celestial phenomena, technical terms created for movements and positions of the stars and planets, and even in the names of the constellations. The effect was to do away with Greek abstraction by decanting the Hellenistic nomenclature—with its definite precision, often touched with aristocratic suggestiveness—into a more prosaic imagery that might be immediately accessible to the average educated Roman. In astronomy, perhaps even more than in other subject areas, the guiding ethics of translation meant producing a discourse not for specialists but for the entire *res publica.* Decanting Greek science was a moral activity, whose end lay in popularization.

For the Roman *civitas,* astronomy meant, above all, three things: moral philosophy about the origins of the universe; simple descriptions of the planets and constellations, whether for agricultural use or for entertainment; and, on an everyday level, astrology. There can be little doubt that when the average, educated Roman looked at the skies, it was most of all the tales of Aratus and the hazy shapes of destiny that he or she saw. In addition to the *Phaenomena,* books on various forms of divination had great authority in the Roman world. Astrology was practiced as a regular part of daily life, both privately and publicly (Barton 1995). The Romans took over with appetite the Greek innovation of applying astrological predictions to the individual. As citizens of a military empire, in which battles, harvests, and personal position in the hierarchy were paramount, they looked to the heavens for the signs of present direction and future necessity. In such a frame, abstract theories about movement and position of the planets did not matter. Instead, the most important thing was the ability to selectively adopt from such theories what might prove most useful to certain styles of prediction.

The process of this decanting, meanwhile, had a number of different aspects. As indicated by Cicero's version of Aratus, these included: (1) adapting for a Greek term one or more words from the vocabulary of Roman religion (where it spoke of the heavens); (2) using existing Latin words for simple celestial phenomena; (3) forming composite terms from existing Latin words; (4) poetic adaptations of ordinary Latin words; (5) direct phonetic borrowing from the original Greek, nested by explanations told in ordinary language and often introduced by a phrase such as "Whereas the Greeks say, we use the term . . ." (*Graeci vocant/dixerunt, nostri . . .*), suggesting an independence of discovery and coinage (Le Boeuffle 1987, 19 n.55). A few examples of these categories might be given. Cicero's use of the verb *lustrare* to denote the "shining" of the heavenly bodies, for instance, had a venerable religious origin, and came replete with suggestions of purification through movement (Le Boeuffle 1987, 170). A term such as *plenilunum,* for the Greek *πληροσεληνοζ* (*pleroselenos*), was a composite of existing Latin words, *plenus* and *luna,* with the first of these bearing the suggestion of "large with child" in association with a female entity, thus embodying the popular idea of the moon's power over fertility. The mixture of ordinary and poetic adaptations of Greek technical terms, meanwhile, are particularly interesting. In this regard, even a word like "zodiac"—which the Greeks had taken from the Babylonians—though cited by Cicero in several places, was more generally refused in favor of *signifer,* sign-bearer, which was also sometimes used to designate "constellation." For "eclipse," on the other hand, two ordinary terms were employed with metaphorical flourish: *labor,* from the verb *laborare* meaning to "be in distress," and *defectus,* as in "failure" or "diminution," suggesting that Cicero borrowed ancient religious terms for this celestial event. The concept of opposition (two bodies occurring 180° from each other with regard to the celestial sphere), which had its specific Greek technical term (*αντιοζ, antios*), became in Latin *adversus,* a common word combining the sense of opposed entities with adversity and omen.

"Planet," which in Greek (*πλανηται*) signified "wanderer" with connotations of the "illusory," of "seducing the eye," was a term that also differentiated this type of body from the fixed stars, "astron" (*αστρον*). No such distinctions were made in the Latin of Cicero's Aratus: here, instead, "planet" was rendered with a number of different terms and phrases. These included (but were not limited to): *stella,* also used for "star," "comet," and any other bright object in the heavens; *stella errantes,* or "star gone astray," sometimes used in opposition to *stellae inerrantes; stella vagae,* "star that roams"; *sidera* or *sidus,* also employed generally for any heavenly body (stars, planets, sun, moon, constellation, etc.), with *errantia sidera* sometimes used; *astrum,* a Ciceronian adaptation from the Greek but again used (especially in its plural form, *astris*) to designate more than one phenomenon.[15] Such abundance of vocabulary, one notes, provided sufficient linguistic rationale to refuse direct adoption of "planetos," which had to wait until

15. Compare also Cicero's *Dream of Scipio,* found in *De Re Republica,* xxii. This brief, elegiac treatment of man's place in the universe, forming the final section to the *De Re Republica* (ix–xxvi)—a work intended as a Roman version of Plato's famous *Republic*—was frequently excerpted and read as an individual work during the medieval and Renaissance periods, and even after. It was written by the mature Cicero and uses a selection of astronomical terms to speak of the heavens, to which the younger Scipio (Scipio Africanus the Younger) was introduced by his adopted grandfather, Scipio the Elder. The *Dream* contains a number of astronomical terms mainly in accordance with what is found in Cicero's Aratus. One exception, however, is his abundant use of *orbis* to speak of celestial motion, e.g., planetary "circles" or Eudoxian "spheres," but also the Milky Way (*orbem lacteum*). Again, that is, the stability and precision of Greek terminology has been abandoned for more flexible, even vernacular uses.

Martianus Capella (fifth century A.D.). More, it completely erased Aratus' own disclaimer that his abilities "failed" when it came to these "shifty"—that is, overly complex—bodies.

Many more examples of this sort might be given, both from Cicero himself and from his near contemporaries, influenced by the great orator's choices and the spirit behind them.[16] Fortunately, much of the detailed scholarly work has been done in excellent fashion (Le Boeuffle 1977, 1987). A dominant perception that emerges from even a brief examination of this work is that, far from suffering an impoverishment of vocabulary (as Lucretius would have us believe), Roman astronomy was flooded with an impressive excess of terms and names, and consequently, more to the point, with a resulting superabundance of choice and inexactitude.[17] Cicero was only the beginning in this general process. Subsequent authors tended to add their own coinings to his initial nomenclature of the skies, so that the total range of available terminology continually expanded rather than gaining stability through standardization. As Le Boeuffle (1977) notes, there is some evidence of more standardized usage during the imperial era, as seen in comparisons between such authors as Manilius, Pliny, Germanicus, Macrobius, and Martianus Capella. But this applies to a limited number of terms (e.g., *signifer* for "zodiac"; *orbis* for planetary motion) and in any case was always tentative. Subsequent authors could reintroduce older terminology to suit their particular narrative purpose. In this regard, we might note that Roman authors never settled upon a single term for "translate" either, employing a whole spectrum of verbs, such as *verto* (to turn, appropriate), *converto* (to change, transform, pervert),

16. To cite a single example, Vitruvius, who devoted all of book 9 in *De Architectura* to astronomy (on the basis of its "constructed" and "designed" forms), used a great deal of Cicero's vocabulary, with some alterations and additions of his own inspired by the same style of intention. When speaking of the planets, for instance, he uses *astris* (ix.15) in addition to *stella*. For celestial motion, he employs the Ciceronian *volvo* and *verso*, but introduces an even more inexact and ordinary verb, *pervagor* (to range over, rove about), to refer not only to the planets but to the fixed stars, constellations, and zodiac as well (e.g., ix.3, 4).

17. There was an element of need involved in this, too, stemming from innate differences between Greek and Latin. Greek astronomers, such as Eudoxus and Hipparchus, had developed a highly specialized language for astronomical phenomena. Such an early "jargon" was possible because of an almost unlimited ability to use existing Greek suffixes, prefixes, and verbal roots to produce new compound nouns, each specific to a given mathematical or observational context. This language was difficult, self-referential, and precise: even today, scientific vocabulary is sometimes less exacting than that of ancient Greek. Most terms were neither metaphoric nor part of ordinary, vernacular speech (Lloyd 1987). They were instead coined constructions, very often derived from mathematical descriptions of the heavens. Some examples include: apogee, perigee, parallax, horizon, epicycle, synod, all crucial terms in the vocabulary of Greek astronomy.

transfero (to carry over, transfer, interpret), *interpretatio* (to explain, expound upon), *explico* (to unfold, set forth, express), and *translatio* (to transport, carry across), which in the study of rhetoric also had the specific sense of "metaphor," a "transfer of meaning" from one object to another.

The reasons for this have everything to do with the nature of authorship regarding astronomical works. Rome metabolized its portion of Greek science through the writings of orators and poets, authors concerned with eloquence and richness of language, but also with virtue, public recognition, and influence. These writers were not interested in epistemological adventure; their task was not to overturn or augment past knowledge on the basis of new researches. They viewed Greek science through the filter of an intellectual ethics that had its distinct patriotic side, in the sense that suitability of material and fame for its translator should result from a mixture of its moral and aesthetic uplift for the *res publica.* Linguistic abundance, meanwhile, supported this moral purpose through its power to make technical knowledge appear less fixed, more amenable to paraphrase, capable of being explicated in a range of different narrative formats familiar to Roman readers. Roman translators did not change the basic sensibility of Greek names for the planets, stars, and constellations. Emperors, famous senators, or other Roman dignitaries were never put in the place of gods, mythic heroes, and other figures. A respect for the past, as a deserved inheritance of excellent example, was kept. Thus "Aeneas" was never used to replace "Herakles" or "Perseus"; "Romulus" was never put in the place of "Cepheus." Catullus, and later Pliny the Elder, accepted and passed on the image of "Bernice's hair" (*Crines Berenices*) from an old Greek story told by the third-century B.C. poet Callimachus, whom many Roman authors had used as a model. Translators preferred to accept the famous "Cassiopeia"—possibly because the name was already well known from Greek tragedy—even though other titles in Latin existed (e.g., *Mulier Sedis,* "woman of the chair").

If Cicero's Aratus was a foundational text in Roman astronomy, and thus in European astronomy at least down to the time of the Carolingian *renovatio*, it is important to point out that what remains to us of this text is not the original translation. Instead, what we possess are pieces of it that appear in a scattering of manuscript texts. One of these is an extraordinary ninth-century illustrated version of the constellations (fig. 3). This work includes a drawing of each constellation filled with the descriptive text of Hyginus, a second-century A.D. fabulist, and captioned by the relevant passage from the *Phaenomena* in Cicero's version. One is struck by the degree to which the heavens have here been completely textualized, entirely given over to literature even in a literal sense. Such would appear to show the Aratean tradition brought to a particular apex. Yet there is also an antiquarian motive, typical of Carolingian intellectualism, clarified by the writing of Hyginus' words in a late imperial *capitalis rustica* script. This is not merely astronomy in the service of literature, as was true of Aratus' own time. A thousand years later, the effort has become one of preservation—using the stars as a means to hold on to a distant literary past, which includes translation.

Another version of Cicero's Aratus that remains in existence is a series of selected quotations that appear in a work of the author's later years, *De Natura Deorum* (On the Nature of the Gods).[18] This text, inspired by the death of Cicero's daughter and by the impending collapse of the Republic, his cherished political idea, is a critical survey by the author of the whole of Roman philosophy, including natural philosophy—a final *summa philosophia et moralia* intended as an eloquent plea for Roman adherence to Stoic views: "In the light of current affairs, I considered it to be my highest duty to the commonwealth to set forth philosophy to our people, since I believe this will contribute greatly to the honor and glory of the state" (*De Natura Deorum,* I. iv. 7). Cicero then immediately argues against those (e.g., Lucretius) who bemoan the presumed inadequacy of the Latin language to accomplish such a task: "On this subject . . . I would profess that we have made sufficient progress that even in abundance of vocabulary the Greeks no longer exceed us" (*De Natura Deorum,* I. iv. 8). Given these sentiments, it is very interesting that Aratus is chosen for an extended appearance. But the fact is that the Aratean heavens, particularly in Cicero's hands, were the very embodiment of Stoic theology projected onto the skies, a theology of eternal divine order and the need for human beings to gaze upon, comprehend, and submit to such order.

In its translated form, therefore, the *Phaenomena* served many purposes and demanded many forms of duplication and preservation. The two works just mentioned reveal the wide range such forms inevitably took, as a result of the passage of this work from Roman to medieval society and beyond. Before recovery in the Renaissance of the Greek Aratus, the oldest manuscript of which dates from the eleventh century,

18. Larger portions of Cicero's text are preserved in various early medieval manuscripts of the Latin Aratus, particularly those of the Carolingian era. See, for example, Dorey 1965.

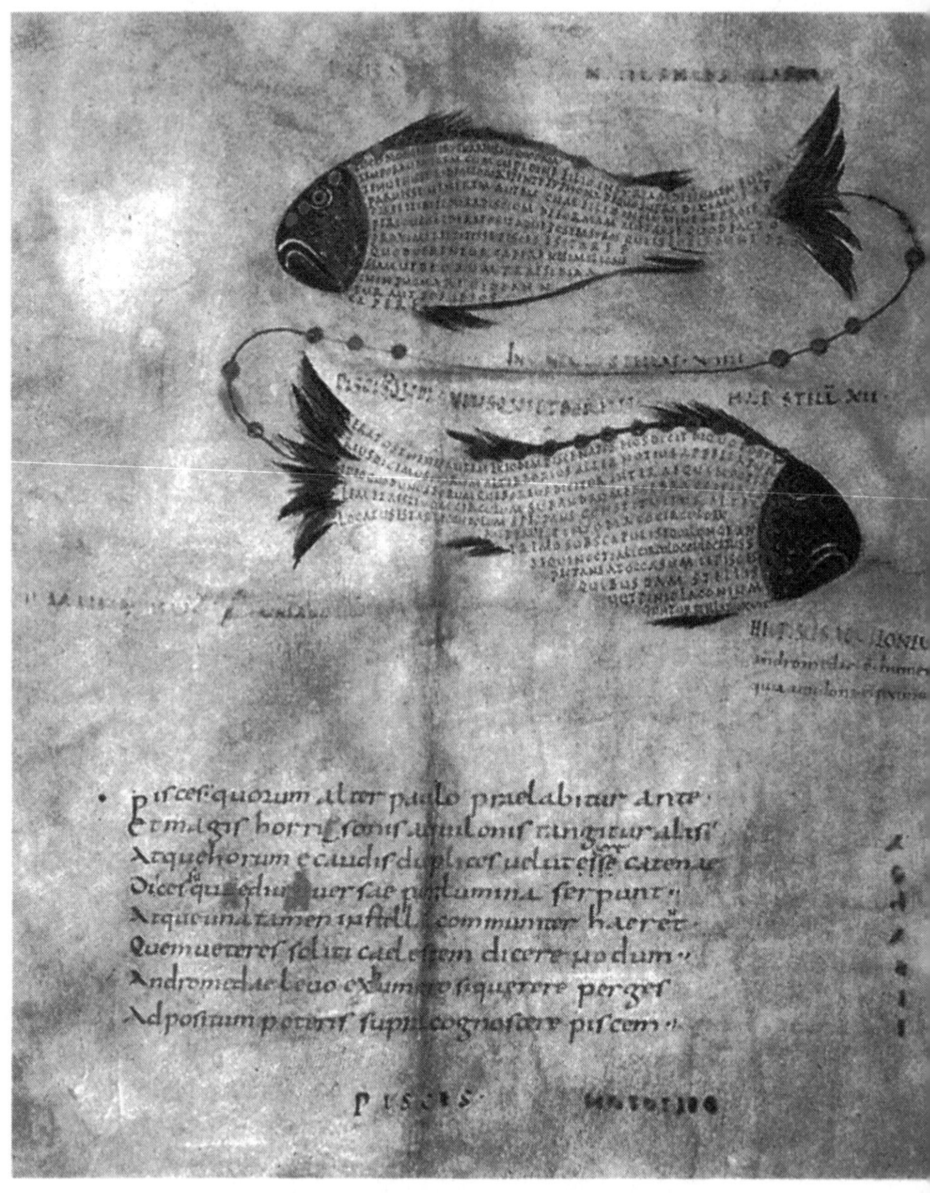

Figure 3. The constellation Pisces as represented in a ninth-century Carolingian manuscript, remarkable for its orthographic complexity. Within the two fishes, in an ancient *capitalis rustica* (late Roman) script, is the brief description ("scholia") of the constellation by the second-century A.D. writer Hyginus. The image is captioned by Cicero's verse translation of the *Phaenomena* by Aratus. British Library, MS Harley 647, fol. 3v.

Latin Europe had used, and added to, an extensive Aratean corpus. Passage of this corpus to the present has met with a mixture of success and failure, with the survival of a few texts and the extinction of many others.

At every step, however, the urge toward continued use and longevity was an effect of translation. Without his Latin garb, Aratus would have perished very early from the scene of European interest.

Pliny the Elder: Storehouse of Roman Knowledge

At both ends of the Roman transformation of Greek astronomy stood the work of translators, such as Cicero, but also compilers, the inheritors of the handbook tradition. In the beginning, these were men of massive learning, like Varro (said to have written more than six hundred books), who undoubtedly salted his works with selected translations from previous epitomes. But as time went on, they became much less so, much more rewriters of the already written and rewritten. What remains of this tradition in Rome is Pliny the Elder's *Naturalis Historia,* composed of thirty-six books, offering a vast, sometimes rambling assemblage of Roman understanding on nearly every topic under (and including) the sun. Pliny's work, even more than that of Cicero's Aratus, has been preserved because of its abundant use throughout the medieval period as a central text of study. Pliny was not a translator, and, from the point of view of Greek knowledge, his work is aptly characterized as "written by a learned amateur for the benefit of unlearned amateurs" (Goodyear 1982, 174). Yet because of this very fact, it also represents the general state of astronomical discourse in Rome under the empire.

Pliny claims, regarding his own effort, that "there is not one among us who has made the same venture, nor any among the Greeks who have alone tackled the whole of the subject" of natural history (Prefatio 14). He is obviously avoiding mention of Aristotle, but this was to be expected in a work written by a longtime imperial procurator who was presenting his labor to the emperor Titus himself. Putting the face of originality on the body of imitation was, by this time, a rhetorical commonplace. In Pliny's case, it was simultaneously a claim for Roman superiority over Greek intellectualism and an acknowledgment that, under imperial oversight, true originality could be seen as a threat. The author, therefore, must reorient his bravado: "It is a difficult task to give novelty to what is old . . . brilliance to the common-place . . . attraction to the stale." And for those who have gone before but sought only personal fame:

> It would have been a greater merit . . . to have rendered this service to the Roman nation and not to himself. As Domitus Piso says, it is not books

> but store-houses that are needed; consequently by perusing about 2000 volumes, very few of which, owing to the abstruseness of their contents, are ever handled by students, we have collected in 36 volumes 20,000 noteworthy facts obtained from one hundred authors that we have explored, with a great number of other facts in addition. (Prefatio, 11–12)

This author also chastises his literary forebears, with the notable exceptions of Cicero and Virgil, for having copied word for word and without acknowledgment their sources. To support his own claim for an amending honesty, he provides in Book I page upon page of the names of his presumed sources (which, in the case of astronomy, curiously include the emperor himself), divided into Roman *auctoritates,* given first, and "foreign writers," or *externis,* occurring in a disordered jumble revealing of how the author is here merely piling up support rather than citing works he had actually read.

It is exactly this grand effort at compilation, clearly intended as a reference work, that makes Pliny so worthwhile, an indicator of where Roman intellectualism stood in its period of final efflorescence. In this respect, he should be contrasted with Cicero, whose purpose was to introduce new material and language under guise of the familiar; a century and a half later, Pliny (let us take him at his word) desires to be a "storehouse" of common usage.

Pliny's "nature" is that of the Stoics, some great, cosmic machine (Beagon 1995). It is an enormous gathering, united by divine plan, perfectly treated by a single author of a similarly vast compendium of existing knowledge. Such scale and intention meant a rapt obedience to tradition. Pliny is both more extensive and more constrained in his astronomical terminology than Cicero, indicating an overall simplification of this discourse, not expansion. For example, among Cicero's several designations for the zodiac, Pliny declines the Greekism *zodiacus* in favor of *signifer.* For the idea of "revolution" or axial turning, he refuses the great orator's *revolvo* and instead employs a host of more inexact, prosaic nouns such as *vertigo* (whirling around) and *conversio* (turning around). The planets are again dubbed *errantia* and, despite Pliny's general disavowal of astrology,[19] are said to have "stations," *stationes.* For their orbits, Pliny chooses *circulus* (circle) or *ambitus* (going

19. The side of astrology unacceptable to Stoic philosophy concerned personal destiny, the assignment of individual planets and constellations and their influences to people's lives. This ran against Stoic beliefs on the importance of free will, the nature of moral conduct, and other ethical points. For Pliny's attitudes on this topic, see Book II.v.28. (pp. 187–89 in the Rackam translation, Loeb Classical Library).

around), which Cicero had alternated with *orbis*, a term that bore connotations of cyclic movement.[20] On the subject of comets, meanwhile, Pliny says, "The Greeks call them 'cometas,' but we call them 'crinitas,' or 'long-haired stars' because they have a terrible blood-red shock of hair at their top" (II.xxi.89). He then goes on to discuss a vivid nomenclature for these phenomena that includes such terms as "Javelin stars" (*barbae longae*), "daggers" (*xiphias*, a Greek borrowing that also meant "sword fish"), "glowing torches" (*lampadias ardentes*), and so forth (II.xxii.89–90). The moon is said to have spots that are "nothing other than dirt snatched up from the Earth along with moisture" (*non aliud esse quam terrae raptas cum humore sordes;* II.vi.46). Only in one section, where he discusses the geometry of planetary motion, does Pliny break down and admit "it will be necessary to employ Greek terms" (II.xii.63). This lasts for a single, brief paragraph, in which we find the words *apsides, polos*, and *orbes* (here used as "orbit") in close proximity, suggesting that the passage represents a semidigested chunk from another work, possibly by Varro.

Many other terms Pliny accepts from Cicero and other early author-translators, such as Seneca (*Naturales Quaestiones*) and Lucretius. For the Greek word *συνοδοζ* (synodus), denoting a conjunction, Pliny employed *coitio*, "coming together," with obvious relation to *coitus*, sexual union. He also continues use of *aequinoctium*, ("equal nights," presumably meaning equal to the length of the day) for equinox, which, according to Varro (*Lingua Latina*, vi.8), had an ancient origin among the Romans and contrasts interestingly with the Greek *ισημερια* (isomeria) "equal days." Among the most telling of choices, however, is one Pliny goes out of his way to discuss in linguistic terms: "I am inspired by agreement among our nations. The Greeks have named the heavens with a word that means 'ornament' [*κοσμοζ, kosmos*], while we have called it *mundus* [neat, elegant] for its absolute perfection and elegance" (*Nat. Hist.* II.iii.8). Nothing could better express the Roman proclivity to view nature through the lens of a simplified divine rhetoric than this reduction of a complex aesthetics to a brief, aristocratic possession. Pliny's huge opus, with its chatty survey of the heavens, is thus an excellent indication of the state of the handbook tradition at the time. Astronomy, in Roman hands, comprised a series of descriptions and names, with minimal addition of mathematics. These descriptions and titles, moreover, bore immediate links to astrology. Pliny rejects any

20. At the same time, *orbis* had come to be used by poets such as Virgil and Ovid to indicate the Earth or universe, and it may be in view of such usage that Pliny chose the more mundane and semantically restricted *circulus* and *ambitus*.

such connections: "the stars are part of the universe, not assigned to each of us in the way the vulgar believe" (*Nat. Hist.* II.v.28). And yet: "Saturn is a star with a cold and frozen nature (II.vi.34) . . . [Venus] is the cause of the birth of all things . . . scattering a genital dew with which it . . . fills the conceptive organs of the Earth" (II.vi.38). Indeed, Pliny's names for the planets also include those of older Roman usage, derived in part from adoption of Hellenistic astrological usage, e.g., "Hercules" for Mars; "Lucifer," and "Vesper" for Venus; "Apollo" for Mercury. Pliny reflects very much the attitude of serious writers of his time who sought to avoid any direct association with such popular cults, while simultaneously including elements of their larger legacy. His fidelity was thus, in Cicero's terms, to the greater "service" of the skies in Roman culture, as a source of both knowledge and omen. He is the best example of how Greece, in terms of its science, had come to be replaced by its own filtered remains.

The *Naturalis Historia,* like nearly all works of antiquity, exists today as a shatter of different pieces assembled from a range of historical eras. The oldest manuscript segments appear to be from the fifth-sixth century, with others from the eighth-ninth centuries (associated with the Carolingian revival of classical learning) and eleventh-twelfth centuries.[21] A recent translator of Pliny has commented that "the mass of scientific detail and terminology and the quantity of curious and unfamiliar erudition . . . necessarily afforded numerous opportunities for copyists' errors and for the conjectural emendation of the learned [such that] many of the textual problems raised are manifestly insoluble."[22] The problems, of course, are not merely "textual." As a crucial work in European science for nearly a thousand years, excerpted and plundered during the whole of the medieval period and the early Renaissance as well, Pliny's *Naturalis Historia* constitutes a major cultural influence in the history of occidental thought, and the biography of its text, including its eventual translation into the European vernaculars in the sixteenth century, must be seen as an integral part of that influence. Detailed comparison of versions from different periods suggests that textual adaptation, including the reordering of various elements, addition of new material and illustrations, and changes in language, was commonplace throughout this long period and took place in accordance with the needs and demands of each era. The standard

21. The most often-used "standard" edition is by K. Mayhoff, who discusses his choice of manuscripts in some detail. See Mayhoff 1933.

22. H. Rackam, "Introduction" (in Pliny 1942, 1.xii).

editions of today therefore comprise a historical-cultural amalgam; Pliny's larger fate as an author might be said to neatly match his own original process of composition. The modern scholarly desire to produce a "definitive text"—always an admitted impossibility in such a case—defines an urge to erase the traces of history and thus to produce a kind of false final version for use in the present epoch of "standard editions."

Such versions, however, always betray themselves in translation. No better example of this can be found, in fact, than in modern English renderings of Pliny, and to some degree Cicero and other authors too. Whereas these Roman authors tended to simplify the language of their Greek source texts, translators of the last two hundred years have done exactly the opposite, giving these authors something of a contemporary scientific (Newtonian) vocabulary. Thus, we read in *Naturalis Historia* of the "revolutions," "orbits," and "velocity" of the planets, of water "evaporating" from the Earth's surface, of the "acceleration" and "deceleration" of certain bodies, and so forth. In one notable case concerning Cicero, the great orator is posed as the sturdy shoulder on which Newton must have stood:

> Hence if the world [*mundus*] is round and therefore all its parts are held together by and with each other in universal equilibrium [*undique aequabiles*], the same must be the case with the Earth, so that all its parts must converge towards the center . . . without anything to break the continuity and so threaten its vast complex of gravitational forces [*gravitatis*] and masses with dissolution.[23]

Inasmuch as the Roman translation effort was to adapt Greek learning to its own cultural present, so has this exercise been practiced inevitably upon the result.

Martianus Capella: A Textbook for the Ages

Versions of Aratus and the *Naturalis Historia,* however, were by no means the final word in Rome's astronomical legacy to the Latin West. Pliny, in particular, was himself succeeded by a small group of encyclopedists of the fourth, fifth, and sixth centuries, who tried to survey all important knowledge of the classical liberal arts in far more condensed fashion. Some of these authors knew Greek and introduced new words into the discourse of astronomy; others worked strictly from translations, but found in these works terms not present in Pliny. Thus, astronomical discourse was broadened, in significant ways. The most influ-

23. *De Natura Deorum,* as translated by H. Rackam, II.116.233.

ential of these authors in the medieval period were those who produced summaries of existing summaries, specifically for pedagogic purposes, transforming the handbook into a textbook. In one estimation, they came at the end of a "long line of compilers and commentators who had long since lost contact with the classical origins. In many cases, the late encyclopedists were removed from classical Latin authors by five or six, and from Greek authors . . . by ten intermediate sources" (Stahl 1962, 9–10).

This paring down and reconstituting of older material thus represents a crucial process in the history of Western knowledge, one that deserves greater study. As with translation, it is a process that reflects changing needs and uses with regard to authorship, readership, and education. The Hellenistic handbook authors were appealing to an expanded literate public; they raided the precincts of specialized knowledge to sell to a ready market for the trappings of intellectualism. The Roman compilers, on the other hand, made huge, imperial collections that perfectly mirrored the self-image of the *civitas* and its presumed destiny—in Ovid's famous phrase, *Romane spatium est urbis et orbis idem,* "the extent of Rome is that of the city and the world." The later encyclopedists, on the other hand, wrote much shorter works in a time of increasing political instability and social uncertainty. Their purpose seems to be more one of a conscious passing on, preservation rather than collection or popularization. In them, the pedagogical impulse stands out most of all.

Among these rewriters of the classical liberal arts, there is one whose influence came to outshine all others. This was Martianus Capella, whose highly allegorical work, *The Marriage of Mercury and Philology* (ca. A.D. 450), was more often used than any other to frame the medieval curriculum of the *artes liberales* and whose ultimate effects therefore extend down to the present. Varro had originally delineated nine *artes* altogether, two of which, architecture and medicine, were deleted by Martianus for unknown reasons, possibly because they were seen to be too practical in nature (*artes labores*). Martianus sought a curriculum free from immediate worldly concerns: it is this "freedom"—*artes liberales* means something like "skills of knowledge free from mercenary pursuits"—which has been so long associated with Western learning.

Yet, if true, there is a certain ruse in such a characterization. Like Varro, Cicero, Quintilian, and other "teachers to Rome," Martianus accepted the primacy of the verbal arts—rhetoric, grammar, and dialectic, the *trivium* as it came to be known, the higher portion of the medieval curriculum (Stahl, Johnson, and Burge 1972, 1977)—and thus carried forward the Roman emphasis in this area, with its notions of political power and moral service. In his discussion of rhetoric, Martianus uses a greater variety of source material and is more knowledgeable than in any other section of the *Marriage,* suggesting that he himself was a teacher of this subject. Indeed, the work is an extended allegory on the union between Mercury, posed as the god of eloquence, and knowledge, in the person of the bride *Philologia.* This same *philologia,* one should note, was largely commensurate with literary culture, learning embodied, conveyed, and studied through attention to language and expression. Martianus accepts the notion of rhetoric as a means to *humanitas;* to a lesser extent, for him as for earlier Roman authors, no such thing as total, ivory tower "freedom" existed. Certainly the Romans would have considered such an idea not merely ludicrous but dangerous. In adopting most of Varro's scheme, therefore, Martianus Capella adopted an imperial hierarchy of learning, and because Varro's encyclopedic work appears to have been lost soon after, Martianus was left as the main conveyor of the liberal arts tradition to early medieval Europe. Augustine, whose influence could have well supplanted that of any other author, began his own survey of deserving knowledge at roughly the same time, but never finished it.

The *Marriage* is a unique and narratively complex work. The eight *artes* appear in allegorical form, as female slaves comprising part of the bride's dowry. Following a highly elaborate poetic prologue, each art is given a separate chapter in which she steps forward to offer a more sober prose *summa* of her respective field. Astronomy is in Book VIII, following arithmetic and geometry and preceding music (these four subjects made up the *quadrivium*). Martianus' *De astronomia,* "despite its gross errors, [is] the most orderly and comprehensive treatment of the subject by any Latin manuscript extant" (Stahl 1962, 53), and was also one of the most popular portions of the entire work. It is a brief survey of much late Roman astronomy, somewhat inferior to Pliny's and infinitely below Ptolemy's synthesis of Greek astronomy, but a summary nonetheless that helped standardize the zodiac, the major constellations, terms for celestial motion, and other important elements of vocabulary.

Martianus begins, interestingly, with a swipe at Aratus: "whereas the Greeks have filled the sky with mythological figures, I prefer to discuss the precepts of the discipline itself" (Stahl, Johnson, and Burge 1972, 1977, 2.320). This is his claim to authority in the tradition of Eratosthenes, Ptolemy, and Hipparchus, whose names he has cited slightly earlier. Though it is extremely unlikely that Martianus had read (or could read) any of these authors, he appears to have been influenced by certain handbooks more recently translated from the Greek. His terminology (the major part) comes from a mixture of poetic convention, Pliny, and Cicero, and also from other, unknown sources in which latinated Greek words were abundant. In his fanciful prologue (the first two books), Martianus speaks of the planets in highly imagistic fashion. Mars, for example, is the source of a river flowing "to the lower regions." Saturn is "stiff with cold," wearing "now the face of a dragon, now the gaping jaws of a lion, now a crest made of the teeth of a boar." In Book VIII, however, the author shifts to a more technical vocabulary that is an alloy of mostly Latin terms, with significant Greek additions. For example, while he passes on *vertigo* for rotating and *signifer* or *sidera* for constellation, he uses the phrase *signis zodiaci cycli* (signs of the zodiac), as well as words like *axes, polos,* and *revolvere,* the last of which was re-added from Cicero. For the moon's motion, he employs *helicoides* (spiral-shaped), and for its various phases he provides descriptions in Latin but terminology in Greek: *menoeides* (crescent-shaped), *dichotomos* (halved), *panselenos* (full Moon). The term for "orbit," meanwhile, is given several common equivalents—*circumcurrens circulari* and *tractum circuli* being the most frequent. The author seems to have sifted through existing manuals, simplified what he could, and tried to standardize the rest of what appeared to him crucial. Possibly in imitation of an earlier model, Martianus is deeply conscious of the language he is using:

> Not because of their errant motions (*errantes*)—for their courses are defined in the same way as the Sun's, and they do not admit of any error—rather, because their peculiar behavior confounds mortals' minds, I shall call them not *planetae,* "errant bodies," but *planontes,* "confusing bodies," as Aratus declares. They have their proper names and they are also called by other names. Saturn is called 'the Shiner' (*Phaenon*), and Jupiter 'the Blazer' (*Phaëthon*), and Mars 'the Fiery' (*Pyrois*), Venus 'the Light-Bringer' (*Phosphoros*), and Mercury 'the Twinkler' (*Stilbon*). (Stahl, Johnson, and Burge 1972, 1977, 2.331)

The term *planontes* does not appear very much in later centuries, but the neologism, even if adopted from another, uncited work, is significant because of its attempt to coin a new term from the Greek. At this stage in the history of Roman culture, at the verge of its political, military, and social collapse, the desire to rival or outdo Hellenistic Greek culture is no longer an important factor. Indeed, if Martianus is any indication, Greek language and knowledge had gained new prestige by virtue of being reduced to the work of "sages" from the distant past (more than six hundred years earlier, in the case of Hipparchus and Eratosthenes). Greek works, especially on astronomy, increased in fame with their disappearance from the scene of use, acquiring the glow of touchstone wisdom.

The *Marriage of Mercury and Philology*, in modern terms, is many things: a textbook, a reference, a literary compilation, a satire, a romance, a nomenclatural catalogue, a neoplatonic vision of a harmonious universe. Perhaps because of this, as well as its brevity, it remained a powerful standard down to the thirteenth century. Though used in the early monastic schools, Book VIII on astronomy became a crucial manual on the subject in the ninth century, as part of the Carolingian *renovatio* in classical learning. Judging by the number of existing manuscripts today, it was possibly more commented upon before 1400 than any other secular work in existence (Eastwood 1993). In a real-world sense—that of institutional influence—the *Marriage of Mercury and Philologia* must certainly be admitted among the "great books" of Western history.

Other Crucial Sources: The Building of an Astronomical "Legacy"

In addition to *The Marriage of Mercury and Philologia*, several other handbook-type works, more technical in content, were significant in transmitting astronomical discourse to medieval Europe. These included two manuals in particular: one by Chalcidius, his *Commentary* on Plato's *Timaeus* (early fourth century A.D.); the other, Macrobius' *Commentary* on Cicero's *Dream of Scipio* (early fifth century A.D.). That both works were commentaries indicates their derivative scope. It is significant that the astronomical sections of both works, like that of Martianus, were frequently copied and bound separately as textbook treatments of this subject.

Chalcidius was a late translator of the *Timaeus* into Latin. Plato's work had long since penetrated Roman sensibility by being absorbed into the handbook tradition at an early date. The other portion of Chalcidius' writing was itself a loose translation of a handbook by a Greek popularizer, either Theon of Smyrna or Adrastus of Aphrodisias (both second century A.D.—Kren 1983). Chalcidius and Macrobius were concerned with Neoplatonic cosmography and included diagrams on such things as the division of the zodiac, eclipses, and climatic zones. Such diagrams had been drawn before, by Ptolemy, Hipparchus, and others. Chalcidius and Macrobius, however, were responsible for transmitting this tradition to early Latin Europe. They thus helped guarantee a visual rhetoric would be in place. Both authors are more difficult, and therefore proved more challenging to medieval readers, than Martianus. Their writings eventually added weight to a vocabulary that went beyond the mundanity of astronomical nomenclature in classical Latin and that had its effects from the tenth and eleventh centuries onward.[24]

With regard to discourse, Macrobius is especially interesting due to his frequent discussion of terminology and proper usage. What makes this significant, in part, is the fact that he is using as a basic source Cicero, who stood near the origin of Roman astronomy through translation. Macrobius peppers his various treatments of astronomical subjects in the *Commentary* with mention of Greek terms, suggesting a desire to demonstrate erudition. This double aspect of the work—returning to the most famed translator of Greek astronomy and to Greek terminology itself—shows that the displacement of Greece by Rome attempted by earlier Latin writers was now, in a sense, complete: no longer was there a need even to pretend to consult Greek texts directly. In his section on terminology regarding the stars and planets, in fact, Macrobius focuses on both Latin and Greek nomenclature (I.xiv.21–26). He differentiates, for example, *stellae* and *sidera,* saying that the first "refers to the solitary planets [*errantes*]" and any stars not included in the known constellations, i.e., those "borne along by themselves," while *sidera* includes only constellated stars. Similarly, "among the Greeks *aster* and *astron* do not have the same meaning, *aster* signifying a lone star and *astron* a group of stars, which form a constellation or *sidus.*"

What seems notable about this passage is that little of it corresponds to common usage, among either Roman or Greek authors. Macrobius is attempting to impose order upon what probably seemed an unruly and overly prolix, even confusing vocabulary (see Le Boeuffle 1977). The rationale behind his choices is not entirely clear, unless one considers that it roughly corresponds to the usage of Aratus as translated by Cicero and Germanicus more than to other authors (Vitruvius, Lucretius, Pliny, Manilius).

Other terms discussed by Macrobius include *circus* and *orbis,* with the latter referring to "one complete revolution of a star [planet]." He also employs the word *sphaera,* Greek for "sphere," to refer to the planets, which he otherwise calls *stellae errantes.* Elsewhere, he speaks about the names of the planets: "We must remember that the names Saturn, Jupiter, and Mars have nothing to do with the nature of these planets but are fictions of the human mind" (I.xix.18). And yet, "when he [Cicero] calls Jupiter 'that brilliant orb, so propitious and helpful'

24. There has been some debate on the subject of when the astronomies of Macrobius and Chalcidius became influential. Probably the best source remains Duhem 1913–1959 (esp. 3.44–162).

. . . and Mars 'the ruddy one, dreaded on earth,' the words 'brilliant' and 'ruddy' are appropriate, for Jupiter glistens and Mars has a ruddy glow" (I.xix.19). These two passages are immediately juxtaposed in the text. What they reveal is that the planetary titles had become established on the basis of a simple logic that could overturn more thoughtful musings. Names, that is, might well be "fictions," but they were also, at this point in history, loyal to the evidence of the eyes—to "observation" mediated by culture.

Pliny Revisited: Adaptation to Medieval Contexts

Scholars have often judged Macrobius' *Commentary* the second most influential work on astronomy in the Middle Ages (e.g., Kren 1983). This is due to the large number of existing manuscripts and the quantity of marginal scribblings (glosses) in them. Yet almost half this work is concerned not with astronomy but with Neoplatonic views of life, and it seems that the *Commentary* had much of its impact in this more spiritual realm. The astronomical portion became popular later on, after the twelfth century, when Ptolemy and Aristotle were in circulation. Before this, the most influential author and, especially from the Carolingian period onward, the one most often turned to for the learning of astronomy after Martianus Capella was surely Pliny.[25] Indeed, the *Naturalis Historia* became one of the most authoritative and widely available works during the spread of education that took place in Carolingian times, inspired by Alcuin of York and his students. Portions of Book 2 (planetary astronomy) and Book 18 (stellar astronomy) were excerpted time and time again for various texts dealing with practical computation and time-keeping. These, in turn, had been important

25. This is not to overlook the fame of certain other authors and works, most notably the encyclopedists Cassiodorus (*Institutiones divinarum et saeculorium litterarum,* sixth century A.D.) and Isidore of Seville (*De Rerum Natura* and *Etymologiae sive Origines,* seventh century A.D.). But the sway of these authors appears to have been much stronger in the subjects of the *trivium* (rhetoric, grammar, dialectic): nearly half of Isidore's *Etymologiae,* for example, is devoted to grammar alone. More to the point, the treatment of astronomy in these works was less sophisticated and useful (e.g., to timekeeping) than that of Martianus Capella and in any case was largely derived from similar sources. Isidore's nomenclature, it is true, was a complex mixture of Plinian terminology and various Christianizing revisions introduced by Gregory of Tours in *De cursu stellarum* (late sixth century A.D.), whose astronomy, rudimentary as it was, found its own inspiration largely in Martianus. But Isidore's astronomical lexicon, and thus in some part Gregory's, as interesting as it is as an example of historical adaptation, did not take hold. The Christianized heavens continued as a minor strand within Western astronomy down to the Counter-Reformation, with Julius Schiller's magnificent 1627 celestial atlas, *Coelum stellatum christianum,* in which all the constellations are stripped of their pagan titles and given biblical names.

from the eighth century onward, particularly for solving the crisis that involved calculating the date of Easter (McCluskey 1998). For this purpose, Pliny's simple descriptions of the lunar cycle were helpful. But for the expanded interest in astronomy that followed, especially planetary astronomy, the *Naturalis Historia* was like a feast.

Pliny, however, was now set before a different host, at a different table. His writings on astronomy, authoritative as they might be, were significantly altered and adapted to a historical era and culture far removed from his own. At the most basic level, the *Naturalis Historia* was carved up into selected excerpts, rearranged, and either made to stand alone or included within new compendia made up of works from various authorities. This had the effect of making Pliny into a specific expert on astronomy. Lifted out of their narrative contexts, often stitched together into a new whole, Pliny's writings on the heavens were transformed from a rambling collection of observations, facts, superstitions, and suppositions, all admitted to have been gleaned from earlier authors, into a single, concentrated work that appeared to be the result of a single, trained individual. It is as much this loss of narrative context, and the adaptive reuse it reveals, that made Pliny into an *auctoritate*.

Moreover, from about the eighth century onward, the Plinian excerpts on astronomy were changed, in both a narrative and a visual sense, by being studded with illustrations. Most often, these were diagrams showing the universe as a series of concentric circles, variably spaced, sometimes slightly elliptical in shape (Eastwood 1987, 1993). The planets are given their respective geocentric orbits and are labeled, with their Roman names intact. Often, the orbital circles are themselves inscribed with quotations from Pliny's text, from other authorities, or with simple descriptions relating to each body. These diagrams represent a profound change to the Plinian corpus and, more broadly, to the larger discourse of astronomy as well. Not only do they add a new dimension of order and regularity; they also act to fix visually both theoretic relationships and nomenclature in a manner almost instantly available to the eye. Pedagogically, they represent an enormous advance, a new branch of discourse, and their use for teaching purposes guaranteed a new type of standard. This, in large part, is what helped make them far more useful than other late medieval texts, such as the *De Rerum Natura* by Isidore of Seville, which also sought to compile existing knowledge of the heavens. Plinian astronomy, as it came to exist, offered a map of the Roman heavens adapted to medieval use.

But Pliny's work had been altered in a number of other ways too. As Bruce Eastwood writes,

> Anyone who has not looked at these planetary excerpts must understand that they are not simply unmodified paragraphs from Pliny's [original]. Not only is there strategic ellipsis, there is also rearrangement of the sentences to create a more orderly presentation of the material. . . . The Carolingian composers of these texts . . . [created] a relatively unified picture of a planetary system. Not just the brief and essentially descriptive account of Bede's *De Natura Rerum* but the causal picture of interrelated phenomena is what the Carolingian excerptors set forth. (1993, 165)

The users of Pliny, and also Martianus, were obeying needs and demands of their own time. These exigencies included a concern with precise and concrete information related to regularity of motion—ritual time-keeping, calendrical reform, agriculture, navigation. But they also involved placing this information in a theoretical framework that made theological sense, and this meant a God-decreed regularity to whose signs human beings must in some part accord their own daily and yearly movements. A need to expand astronomy beyond its previous role led Carolingian writers to systematize the primary written sources they possessed.

In a textual sense, Pliny had to be made a member of the Church and of feudal society. What this meant was that Pliny's work at this stage—no doubt like Martianus Capella's—was therefore no longer the work of Pliny himself. What we said earlier of Ptolemy was now true of these Roman encyclopedists. In a very concrete way, they were no longer "authors"; they were instead textual communities, whose expanding presence was itself the sign of this change.

Plinian astronomy thus offers an excellent example of the fate of Roman science, as translated residue from Greek sources, in the European Middle Ages. It is perhaps fitting to end this discussion with a final example that sums up much of the process involved. During the early medieval period, there was no lack of spectacular phenomena in the skies and, indeed, it seems to have been the assigned work of monks to record such phenomena as comets, meteors, aurorae, and novae. Representing and, in part, helping set the sensibility of the age, Gregory of Tours in *De cursu stellarum* rejected the precepts of astrology but spoke of such celestial happenings as divine signs of coming events. Yet he offered very little vocabulary for describing these phenomena. Instead, the source most often employed for this terminology was once again Pliny, the only one among the encyclopedists to treat these phenomena in any real detail, especially with regard to titles. His descriptions of comets, as noted previously, are highly evocative:

> there are stars suddenly born into these very heavens and they include several kinds. Where the Greeks call them *cometas,* we say "covered with hair" [*crinitas*]. . . . The Greeks also give the name of *pogonias* [bearded] to those from whose lower part spreads a mane resembling a long bear. "Javelins" quiver like a dart and portend terrible events. . . . The same stars when shorter and tapering to a point are sometimes called "daggers." . . . The "tub-star" presents the shape of a cask, with a smoky light all around it. Another has the shape of a horn, like the one that appeared when Greece fought the decisive battle of Salamis. The "torch-star" resembles glowing torches, the "horse-star" horses' manes in very rapid motion and turning in a circle. (II.xxii.89–90)

Here, in a nutshell, are revealed the priorities of Roman translation. No real debt is admitted to Greek sources, only parallel, rival coinage (in fact, *cometa* is derived from *kometes,* meaning to "wear long hair"). The Latin terms are adopted from ordinary speech; Pliny is not interested in technical vocabulary, and, indeed, cites a Greek word only where it is similarly vernacular to the Latin. For meteors, meanwhile, he names two kinds, *lampades,* meaning "torches," and *bolides,* or "missiles." Aurorae he calls *soles nocturnos,* "night suns." Again, such titles obviously reflect Roman experience, particularly in the period of its history when military events were paramount and celestial occurrences were linked to ideas of violence or sudden change, to "fate" as a kind of trembling vulnerability.

What became of this vocabulary in the Middle Ages? To a large degree, its basic sensibility remained intact, but its specific language did not. The medieval period did not make much use of the specific titles classical authors had coined but instead invented those reflective of its own experience and preoccupation. This might seem obvious and inevitable: yet it is fascinating to see how Plinian nomenclature was transformed, while retaining every drop of metaphorical charge. Instead of javelins or daggers, for example, comets are recorded as fiery staffs (*baculus igneus*) or as death-bringing stars (*stella mortis*). They are also called "signs shaped like a bowl" (*signum velut contum*) and a "writing instrument" (*tenaculum*), names which bear a more immediate relation to a monk's livelihood. But for aurorae, which were more common in Europe than in Rome, a great number of terms concentrated on violence and power were used: sulphurous crosses; crucifixes; swords; bishops; castles; armies of horsemen. And for meteors, one finds: fiery stick; flying torch; flying angels; split-in-the-sky; heavenly flame (Dall'Olmo 1982).

Those monks who recorded such terms were engaged in setting

down God's wrath and warnings. Like the Romans, they read the skies as an offering of signs (holy, terrible, punitive), with a similar general image of "fate." And like the Roman translators well before them, they adapted an already existing, indeed ancient, discourse to their own time and place. This context did not, any more than did imperial Rome, demand a highly technical, scientific vocabulary but instead a language of making sense, of attributed meaning in a daily idiom. It projected earthly realities and relationships upward through the lens of a terminology densely expressive of medieval dispositions. It adapted Rome to a new set of everyday realities.

2 Astronomy in the East

The Syriac and Persian-Indian Conversions

The Issue of "Transmission" Reconsidered

A great deal of modern astronomy would be unthinkable apart from the inheritance of what has long been called "Greek thought." Rome transmitted but a very minor part of this to Christian Europe, a part that did not include the great systems of Hipparchus or Ptolemy. The majority of Hellenistic astronomy remained, for a time at least, in Greek, not Latin—and therefore in cities, libraries, and in the hands of scholars to the east, first in Byzantium and later in multilingual Christian communities in Syria, where Syriac and Greek were spoken, thus ever more removed from Rome both in distance and in language. A point to be considered is that at this stage, Greek astronomy is not European, or "Western," in any consistent sense. While it might certainly be argued that such terms are at least partly irrelevant or ideological when applied to a time period before Europe existed as a political and cultural entity, this is not the argument I wish to make. Admittedly, Plato is no more a European than Hegel is an Athenian. But there are other matters, more to the issue, for consideration.

Two realities of history, in particular, make "Greek thought" problematic as an origin of the "European." One of these involves the question of context and thus of repeated beginnings. The other, equally profound, concerns transmission. Taking astronomy as an example, it can be said that knowledge of the skies came to Europe via four major episodes of translation. We have spoken already about the first of these. The second episode, about which very little is known, involved the transfer of Greek texts into Syriac (a form of Aramaic), on their way (so it is often said) into Arabic. These texts, in considerable numbers, had been gradually transferred eastward during the fifth and sixth centuries A.D., in part under the harassing influence of the orthodox Byzantine Church against Nestorian and, to a lesser extent, Monophysite teach-

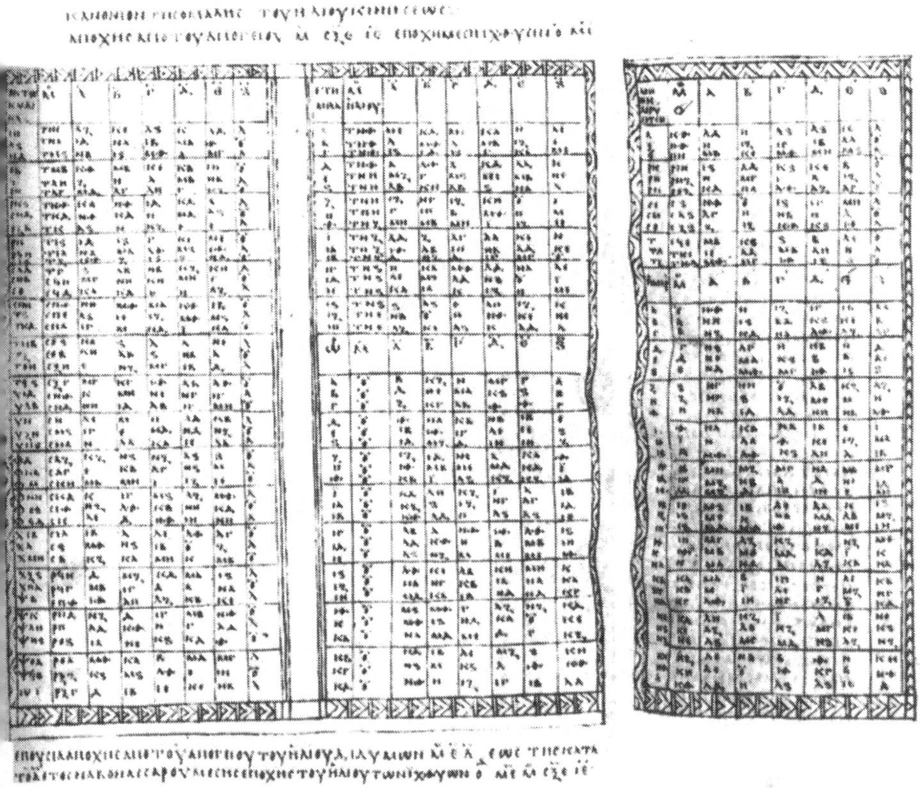

Figure 4. Tables on the mean motion of the Sun, from a ninth-century Greek manuscript of Ptolemy's *Almagest* (Chapter 2, Book III). The manuscript is one of the earliest versions of this work and represents nearly seven hundred years of transmission by means of scribal copying. Bibliothèque Nationale, Paris, graec 2889, fols. 68v–69r.

ers and intellectuals. In particular, systematic purges by the emperors Zeno and Justinian in the late fifth and early sixth centuries urged members of these communities to migrate to the fringes of the Byzantine empire and beyond, into Persia (Syria, Iraq), and there to set up schools for studying, copying, commenting upon, and eventually translating the texts of Hellenistic knowledge. While a brief and significant revitalization of higher learning did take place in Byzantium, with important scribal effort aimed at preserving crucial works, including those by Ptolemy (fig. 4), this "renaissance" was too short-lived, dependent on only a few individuals, to have any permanent effects. Compared with the work of Syriac and, later, Arabic scholars, it was paltry and tentative. Indeed, the more successful efforts by Nestorian scholars took place in the midst of a great flowering in Syriac language and literature, which had begun in the small city of Osrhoene, capital of the local kingdom of Edessa, and then had branched out to many parts of the Near East with the continued spread of Christianity (Wright 1966; Brock 1992). It continued for several centuries, within the limits of a vibrant, changing, cosmopolitan culture, well removed from the walls of Rome, Athens, and Alexandria, eventually to be absorbed into Islam, at which point the next great episode of translation began.

Regarding the transmission of scientific works, this period of translation into Syriac has been widely ignored by scholars in the history of science as a separate episode worthy of attention.[1] Studies have tended to focus almost entirely on the ninth century, when the great translation efforts supported by the ʿAbbasid caliphate began. In the familiar scenario, Syriac is posed as a linguistic "halfway house" for the movement of Greek learning into Arabic. Yet it is well known that the translators of this later period frequently made use of Syriac versions without the Greek, at least at first. In the case of medical writings, for example, Sergius of Reshaina (fifth century A.D.) had rendered a number of crucial works by Galen into Syriac (Hugonnard Roche 1989), and this corpus became an important touchstone for Arabic versions. Moreover, the very fact that Syriac was often used in the ninth century as an intermediary begs closer consideration of the evidence. It was, after all, these Syriac versions that determined the transmission of the relevant knowledge—but why and how was this so? The Syriac language was used for specific historical, cultural, and linguistic reasons. A substantial momentum of intercourse between the Greek and Syriac languages, spanning the whole of late antiquity, lay behind such use (Brock 1984b, 1992). To pretend there was no such reality as "science in Syriac" prior to the episode of translation into Arabic would obviously be wrong.

The matter fairly demands that one therefore also distinguish between "Greek science," with its temporal, spatial meanings, and "science in Greek," a much broader, culturally diverse, and historically significant corpus. Similarly, it may be fruitful to consider questioning a standard concept like "Hellenization," as in such phrases as "Hellenization of Syriac/Arabic culture." Such phrases, after all, speak of im-

1. Note, for example, the absence of any such study in three recent volumes that otherwise deal extensively and in excellent fashion with the translation of Greek works: Ragep and Ragep 1996, Nasr and Leaman 1996 (esp. vol. 1), and Butzer and Lohrmann 1993.

manent power, particularly when it comes to textual knowledge, and distract one's attention from problems related to the changes the relevant knowledge inevitably underwent when transferred to new cultural and linguistic contexts. When Justinian closed the doors to the schools in Athens and Alexandria in A.D. 529, he helped guarantee a new era in the movement of knowledge, one that involved more than either the simple transfer or the radioactive influence of "Greek science."

This era, as stated, had already begun, but gained increasing momentum thereafter. Instead of suffering interruption following the Islamic conquests after 632 (the year of Muḥammad's death), translation of Greek works into Syriac increased, culminating in the ninth century, when the ʿAbbasid caliphs and their court patronized the transfer of secular Greek learning into Arabic on a massive scale, using Syriac as a fertile intermediary. Syriac had been excellently prepared for this historical task: as in the case of Latin, part of the maturation of Syriac from a local dialect of Aramaic to a widespread language of religious, literary, and educated expression came from an evolution in vocabulary, syntax, and grammar partially inspired by Greek models (Brock 1975, 1977). At the same time, however, influences clearly moved in the other direction as well. As Syriac grew in influence and sophistication, many theological and poetical writings were translated into Greek. The situation is perhaps best exemplified by the fourth-century poet Ephrem, whose work was instrumental in elevating Syriac to new heights of expressive power and prestige, and who plainly rejected Greek culture as a required touchstone. Yet Ephrem nonetheless used a Syriac that had already absorbed significant influences from the Greek and his writings, in turn, were quickly translated into Greek itself, there to become a constant source of study and imitation (Brock 1992). Indeed, Ephrem was educated, lived, and wrote in Edessa, the center of early Syriac Christian culture and a nexus for Greek learning, especially philosophy and rhetoric, and for translations from the Greek as well (Wright 1966, 61–63). The procreant interplay between these two languages no doubt yielded changes in both, particularly with the continual rise of Syriac as a language of popular and literary-ecclesiastical expression.

The complexity of the Near East in late antiquity, in terms of linguistic diversity and interchange alone, is indeed daunting. In addition to Syriac, there existed several other dialects of Aramaic, as well as such widespread languages as Hebrew, early Arabic, Armenian, and Georgian, to name only the most well known. Such complexity resulted in many hybrid creations, e.g., "while an Arab ḍynasty ruled Edessa . . .

yet the same city has produced a third-century mosaic of Orpheus complete with Syriac inscription" (Cameron 1993, 185). The relative universality of Greek as the language of political administration and secular learning, as well as its crucial importance as a reservoir of ecclesiastical writings (including the New Testament), mean that any simple differentiations made between so-called "Syriac culture" and "Hellenistic culture" are to be distrusted. Rather, it makes more sense to confine the discussion of translation to specific writers, texts, patterns of method, and identified linguistic influence. Even this may not be an easy matter, however.

For example, while some analysis has been devoted to the changes wrought in Syriac (Schall 1960; Brock 1977, 1992) by its use of Greek linguistic models, little has been written on the evolution of Greek during this period, or changes produced in Greek texts, particularly with regard to secular works. Yet, as mentioned at the beginning of the previous chapter, even so great a work as Ptolemy's *Syntaxis Mathematica* (the *Almagest*) was being altered within a half-century of its author's death, apparently for reasons of clarification and teaching. What further changes might have been effected upon so influential a work as this during the next five or six centuries? Such changes can only be imagined, as the evidence is lacking. The thick corpus of Greek philosophical and scientific texts translated into Syriac and Arabic in the ninth century cannot be accurately judged in these terms. In too many cases, the earlier versions have been lost. Indeed, the oldest manuscripts of certain Greek works (even of Aristotle) are preserved only in Syriac or Arabic translations.

What *is* clear is that the body of secular knowledge chosen for use was entirely different from that which had been taken up in Rome and later transmitted to early medieval Europe. The style of "humanism" so long prized in European intellectual circles as the essence of the classical tradition—the writings of poets, playwrights, orators, and so on—was largely absent from the body of learning translated into Syriac and Arabic. Rather, it was philosophy (especially Aristotelian), history, and science that composed the chosen realms of knowledge in the east. The specific reasons for this remain unclear, but no doubt have much to do with the fact that both Syriac and Arabic had developed rich literary cultures of their own before the main episode of translation began. In the case of Arabic, there were religious reasons too; learning in Greek was kept separate from the "Islamic sciences," which were focused upon study of the Qu'rān and its application to literature, law, politics, and other areas. Whatever the case, however, the fact remains that the bulk of technical learning in Greek was transferred eastward in late antiquity, away from Europe and thus from Latin.

Translation into Syriac

Intellectual and Cultural Background

If a significant amount of science in Greek passed into Syriac during the eighth and especially the ninth centuries, this took place within a milieu well prepared for such an effort. From the conquests of Alexander onward, Greek had been established as a widely spoken tongue in the region, and during the succeeding centuries when Hellenism flourished and thereafter when the Byzantine empire came to rule much of the region, it became the standard language of worldly power in the eastern Mediterranean (Bowersock 1990; Brock 1994). Traditions of intellectual discourse in Greek are clearly in evidence during Roman occupation of the Near East (e.g., Jones 1994). Many of the most cosmopolitan cities in the region, such as Palmyra, Damascus, Bostra, and Emessa, and even large portions of the non-urbanized countryside, were bilingual in Aramaic (one of several dialects, including Syriac) and Greek (Bowersock 1990, 1–13). The common presumption of fundamental animosity between early Christianity and "pagan" (Hellenistic) learning appears to be entirely mistaken: "In language, myth, and image [Greek culture] provided the means for a more articulate and a more universally comprehensible expression of local traditions. This became the precious mission and character of Hellenism in the Christian empire of late antiquity" (Bowersock 1990, 9).

The eastward spread of Christianity was thus also instrumental in supporting, indeed even broadening, the use of classical intellectual materials in Greek, just as it was necessary to the expansion of Syriac. This can be seen in some of the earliest influential works by Christian writers in Syriac, such as Bardaisân (or perhaps one of his students), whose *Book of the Laws of the Countries* (third century A.D.) contains much favorable discussion of Hellenistic astrology and is written in the obvious form of a Platonic dialogue, with the author in the role of Socrates (Brock 1982; Drijvers 1984).[2] Bardaisân came under attack later on, notably by Ephrem, who, in one of his hymns, wrote "happy is the man who has not tasted of the venom of the Greeks" (Brock 1982, 17). But,

2. Readers should note that the work of Sebastian Brock, as exemplified by the collected papers in the *Syriac Perspectives* volume, represents an invaluable source of information on many aspects of Syriac learning and translation in late antiquity.

as noted above, Ephrem himself, as an educated and worldly author, had drunk deeply of this same poison. And his own works were quickly decanted into its reservoir, there to become part of the larger literary legacy of the region. In the next two centuries, meanwhile, the establishment of Christianity in the region involved complex political and cultural divisions that aided the spread of learning in Greek. The eastward migration of Nestorian believers after the condemnation of 435 established a community in Edessa, home to a famous school of Greek study under the Roman occupation. Driven out in 457 beyond the border of the Byzantine empire, the Nestorians then settled in Persian Nisibis, which became an outlying center of Greek learning and Syriac-speaking intellectuals. Continued eastward migration of Nestorian missionaries founded similar communities elsewhere, notably in Jundishapur, where, from the sixth century onward, a school for the study of philosophy, medicine, and science in Greek thrived (Meyerhof 1930; Peters 1968, 1996). The influence of Greek studies only grew in the centuries after Ephrem. Indeed, it came to have transforming effects upon Syriac itself.

Ephrem represents an era when the Syriac language, as a vehicle of literary and religious expression, was rapidly coming into its own, among both urban and rural populations. The success of Syriac, particularly as a poetic language of great eloquence, did not cause Greek to suffer decline. On the contrary, it helped ensure that Greek would remain the language of material power—that is, the speech of politics, law (crucial Roman legal documents were translated into Greek), theology, and higher learning in general. Ephrem is the sign of a rising self-confidence among Syriac writers in the face of this already established power. He is evidence of a short-lived rivalry similar in spirit to that expressed by the Romans nearly four hundred years earlier. The phrase "short-lived" is accurate, since, within a century of his death, the writers and translators of Ephrem's time became themselves the subject of criticism. This arrived from a leader of the early Syriac church, Philoxenos of Mabbug (early sixth century), who called for, and undertook, a new translation of the Bible, taking his predecessors to task for their lack of exactitude, saying "our Syriac tongue is not accustomed to use the precise terms that are in currency with the Greeks" (Brock 1982, 20). A large part of Philoxenos' desire for a new Syriac Bible surely derived from recent theological revisionism and contestation among the recently established Monophysite and Nestorian sects, leading to a sundering of eastern and western churches. The western Syriac Church, with Philoxenos as one of its living patriarchs, was Monophysite in character and actively rejected Nestorianism. In 489, the Byzantine emperor Zeno closed the famous school of Greek studies in Edessa, which soon reestablished itself in Nisibis under the recently exiled Nestorians. The demands made by Philoxenos (himself later exiled by Justinian in the early sixth century) to modify or replace older translations should partly be seen in light of such religious politics. His was a strategy for updating the faith and securing its textual legacy. Yet this should not distract one from the fact that for both Monophysites and Nestorians, learning in Greek remained a crucial source of useful eloquence and knowledge. Indeed such learning is evident throughout Philoxenos' own writings (see Budge 1894).

As Sebastian Brock, our best guide to this era of translation, has indicated, Philoxenos is actually an early revelation of the ultimate prestige that would come to embrace Greek learning in the sixth and seventh centuries (Brock 1982, 1994). One sees that the concept of "the wisdom of the Greeks" now comes more fully to include secular and religious works both. But the new influence granted to Greek studies should not be misunderstood. It was not a sign of the superior vibrancy of Greek intellectual culture per se, but rather of the parallel literary cultures that had come to exist in Syriac and Greek, each with its own arena of subject matter and related distinction. Whereas the literary culture of Syriac was still in ascendancy, eager and absorbing of influences, the philosophical and technical body of Hellenistic learning had more and more been reduced to a collection of hallowed texts, a source of "wisdom" upon which to draw. This fate was secured by the Islamic conquests of the early seventh century, after which Greek underwent inevitable decline. Bilingual and bicultural communities in Syriac and Greek significantly decreased in number between A.D. 600 and 800 (Drijvers 1984; Kennedy and Liebeschuetz 1987).

Part of the reason for the gaining prestige of Greek textual (secular) "wisdom," which came to be studied by fewer, more elite members of the community, seems to be that it offered benefits of an elevated, practical kind, in such areas as medicine and astronomy, rhetoric, dialectic, and philosophical conception. Its powers had shifted in some sense, that is, from the more purely political-spiritual to include material-intellectual aspects as well. The work of translation remained loyal to ecclesiastical writings, to be sure, but now came to include more and more secular texts, particularly from the late seventh century on. Translation into Syriac meant, above all, the movement of an institutional knowledge, increasingly restricted to a textual existence, into a language of daily speech and use. Thus, it is not entirely surprising or ironic that

> Syriac, which had started out confined to Osrhoene [and] had spread west, across the Euphrates, as the literary language of Aramaic-speaking Christians by the early fifth century . . . [thereafter] should have continued to increase in this area over the course of the next two centuries precisely at a time when Syriac literature was itself becoming more and more phihellenic, adopting not only large numbers of new Greek loan-words, but also many features of Greek style. (Brock 1994, 159)[3]

It was Syriac, among all the various dialects of Aramaic, whose writers most avidly sought to incorporate and transform Greek models of learning, eloquence, and vocabulary. And it was exactly this effort to enrich the written tongue through selective transformation of the recognized language of authority, in matters both religious and secular, that provided one more element leading to the magnificent efflorescence of Syriac between the time of Ephrem and that of Severus Sebokht. Certainly, it made of this language an excellent, perhaps inevitable intermediary between Greek and Arabic, which were otherwise very dissimilar except for the respective components they shared with Syriac.

Patterns in Translation: From Paraphrase to Literalism and Beyond

These historical-cultural developments, involving as they do an evolution in attitude toward Greek learning, find immediate expression in the work of translation among Syriac authors. On the simplest level, one sees this in the very quantity and content of translations performed. During the formative period of Syriac expansion and establishment of Christianity (mainly the third through the early fifth centuries), translation was focused upon the Bible and patristic texts. From the fifth century onward, when Syriac had become widely established, with Greek attaining prestige as the language of political power and of textual "wisdom," efforts at translation increased markedly and began to include many secular works. The audience and appetite for Syriac versions of Greek works had grown enormously by this time; indeed, it must have become obvious that the overwhelming majority of Christian literature, from Jerome and Basil to more recent hagiographic (lives of saints) and homiletic writings, was in Greek and that these writers, whatever their statements against pagan learning, had been classicists themselves insofar as their training and rhetorical style were concerned.

3. Brock, it seems to me, wisely distinguishes within the larger Syriac-speaking community between those people who could still speak Greek and those who could read and write it, i.e., who had been educated in the "wisdom of the Greeks." "The latter group," says Brock, "will have been very much the smaller" (1994, 160).

By the seventh century, it appears that certain monasteries of the orthodox Syrian church, such as Qenneshrin (Ken-neshre) on the Euphrates or Ennaton near Alexandria, specialized in Greek studies and produced scholar-translators of significant accomplishment (Brock 1977, 9–10). The work of Jacob of Edessa, James of Nisibis, Severus Sebokht, Athanasios of Balad, and Paul of Tella are notable in this regard. It is at this time that a significant amount of Greek philosophy and science makes its way into Syriac, most of all in the form of Aristotle and related commentaries (e.g., Porphyry's *Eisagoge*), but also including astronomical works on the constellations, fixed stars, and the astrolabe.

In terms of translation technique, meanwhile, one sees a distinct evolution that can be characterized as follows:

> The significance, for translation technique, of a language's cultural prestige is in fact very well illustrated by the history of Syriac translations from Greek. The earliest versions of the fourth and fifth century are almost all very free (Basil's homilies, for example, are expanded by about fifty percent), and significantly the translators adapt the Greek biblical quotations to the wording familiar to their readers from the Syriac Bible; Aramaic was, after all, [in the standard claim of Syriac writers] the original language of mankind. The rapid hellenization of the Syriac church began in the mid-fifth century, and the precise wording of the Greek original now becomes all-important, and biblical quotations are translated exactly, even when they diverge from the text of the Syriac Bible. In other words, with the Greek language's new position of prestige, translation techniques change. (Brock 1984a, 75)[4]

Brock further notes that this shift of concern from the language of reception to that of origin—that is, shifting focus from the Syriac reader to the Greek text—spilled over into secular works and had to be overturned later by the great Arabic translators who employed Syriac versions. The literalism of Syriac translators came from a desire for "precision" (as in the case of Philoxenos of Mabbug, quoted above), but a precision based on textual fundamentalism.

In contrast to their Roman counterparts, then, these translators from the Greek seem to be engaged in double acts of rivalry. On the one hand, they did not at all revere and adopt the work of their predecessors (as, for example, Cicero and Varro were revered by Pliny); indeed, they perceived a need to delegitimize earlier translations, and thus clear space for newer and holier versions. On the other hand, the literalism of their

4. See also Brock's more recent evaluation in "A History of Syriac Translation Technique" (1983, 1–14).

translation technique, which might at first seem to represent a devaluation of Syriac at the hands of Greek, is perhaps more accurately viewed as evidence of a desire to replace, once and for all, these Greek originals by transferring to Syriac a complete ability to bear the holy word. Severus Sebokht (d. A.D. 666–67), for example, educated in Nisibis and later bishop of the monastery at Qenneshrin, and thus one of the most educated men of his time in Greek learning, argued against any assumption that scientific knowledge belonged to the Greek language, noting that the great Ptolemy himself, in the *Syntaxis Mathematica,* had depended significantly on Babylonian astronomy, and moreover that "nobody I think will dispute that the Babylonians are Syrians" (Brock 1982, 23–24). These are the words not of arch rejection but of desired possession. The move to effect a linguistic transfer without spilling a single semantic drop is thus not necessarily an indication that the Greek text had attained the rank of sacred object (if such were the case, translation itself would have been viewed as approaching blasphemy, as later happened with the Qu'rān). The new period of translation indicated a greater confidence among Syriac writers that God's word and the word of the material world could both reside entirely in their tongue as well as in the Greek.[5]

To achieve this, translators introduced many new Greekisms into Syriac. The number of loan words from Greek had grown steadily from the fourth century onward, and had entered Syriac through a number of different venues, popular and ecclesiastical alike (Brock 1975). Early on, most of these words were nouns, testifying to the more advanced

5. A different interpretation to this episode of translation, sometimes called upon in the scholarly literature, is stated by Wright as follows:

> The effects of [the Muslim] conquest soon begin to make themselves manifest in the literature of the country. The more the Arabic language comes into use, the more the Syriac wanes and wastes away; the more Muhammadan literature flourishes, the more purely Christian literature pines and dwindles; so that from this time on [early seventh century] it becomes necessary to compile grammars and dictionaries of the old Syriac tongue, and to note and record the correct reading and pronunciation of words in the Scriptures and other books, in order that the understanding of them may not be lost.(1966, 140–41)

The problem with such an interpretation is that it collapses centuries of complex linguistic evolution into the span of a single, simple formulation. The evidence hardly suggests a rapid deterioration of Syriac literary culture from 638 onward; indeed, more writing than ever takes place in this language, displacing Greek. It seems, rather, that the long-term effects of the Muslim conquest included the support of tendencies that had already come to exist within Syriac literary culture, the trend toward literalist translations being one of these. As to the composition of grammars and dictionaries, one should note that this same activity began in Arabic during the identical time period, i.e., the seventh and eighth centuries (See, for example, Fleisch 1994).

differentiating and technical aspects found in Greek. Later, however, borrowed and adapted words came to include prepositional phrases, particles, and adverbs, while complex lexical and syntactic calques became common as well, as a direct result of translators' efforts to reproduce original Greek religious texts word for word. Brock points out that many of these introductions are ingenious, imaginative, and thus required considerable creative effort, not merely static imitation. Again, translation proved to be the means by which Syriac became a true competitor to Greek as a language of knowledge embodiment and transfer. And it is exactly this maturation, whose total product was a language that retained its Semitic qualities while adopting many Greek elements, that made Syriac the very logical, indeed inevitable, intermediary between Greek and Arabic later on.

Early Astronomical Translations: The Crucial Role of Example

With regard to astronomy specifically, we have several examples of work performed by translators prior to the ninth-century episode of massive transfer. This work has considerable importance. It not only helped establish an early form of astronomical discourse in Syriac, but also provided essential models—both negative and positive—against which later translators could react. These models proved to be not merely examples for revision and nomenclature, but the stimuli for an increased awareness of technique and the honing of translation skills. Indeed, in more than one instance, the work of these earlier translators was used as a focal point for debate on the best methods to render Greek works into Arabic.

In this regard, three works in particular are worthy of mention. The earliest of these is by Sergius of Reshaina, chief physician and priest of that town, who appears to be the first great translator of Greek secular literature into Syriac, having rendered not only works of philosophy (Aristotle), but medicine (Galen), logic, botany, and cosmology as well.[6] His translations of Galen are especially impressive, numbering no

6. The literature on Sergius is significant. See, for example: H. Hugonnard Roche 1989 and references contained therein; Sherwood 1952; Baumstark 1894; Ryssel 1880–1881; and Furlani 1923. The list of specific works attributed to Sergius has been the subject of no small debate, exaggeration, and revision. Until very recently, it was believed he had translated dozens of major works in a host of fields, including writings by Aristotle, Porphyry, Isocrates, Themistius, Plutarch, Menander, Galen, Dionysius Thrax, and even Ptolemy (the *Syntaxis*). For a sampling of such attributions, see Peters 1968 (58 n.3); Baumstark 1894 and Baumstark's revised opinion, Baumstark 1922; and Wright 1966 (88–93). For a brief review and history of the debate, with appropriate corrections, see Hugonnard Roche 1989 (esp. 1–7).

less than thirty-seven works and encompassing nearly the whole of the so-called "Alexandrian curriculum," the fundamental corpus for the training of doctors in Alexandria and Byzantium generally, much attended to by the great Arabic translators of the ninth century. Sometime in the early sixth century, Sergius produced several brief treatises on astronomy-related subjects, including *On the Universe* (Peri Kosmou)[7] by the Pseudo-Aristotle, *On the Causes of the Universe, According to the Views of Aristotle, Showing How it is a Circle,*[8] and *On the Action and Influence of the Moon* (with brief appendix on the Sun), adapted from Book III of Galen's *On the Natural Faculties.*[9] At one time, modern scholarship also collected under Sergius' name a variety of other works, including those by Menander, Pythagoras, Isocrates, and even Ptolemy's *Syntaxis Mathematica.* These attributions are now known to be false. Yet it is extremely significant that such works exist in manuscripts dated to the seventh and possibly the eighth century, indicating that translation and therefore possible study of these writings was active in this period, when the level of astronomy in Byzantium was largely dominated by astrological concerns (Tithon 1993). Indeed, even at this level, the most sophisticated practitioner of eighth-century astrology in Constantinople was Theophilus (d. ca. 785), who had been born and educated in Edessa and was known as a translator of Greek works into Syriac (Tithon 1993; Pingree 1989).

In terms of influence, the most important of Sergius' astronomical translations was that of the Pseudo-Aristotle, which came to be studied by Islamic authors from the eighth century onward. Originally written in Greek sometime before the fifth century, the work was widely used as a philosophical reference text in late antiquity and the early medieval period, both in the Near East and in Christian Europe, where it came to be known as *De Mundo* (Lorimer 1924). A detailed comparison of the Greek and the Syriac in the version by Sergius suggests that the astronomical vocabulary is Aristotelian—fairly basic and comparable to Aristotle's *Peri Ouranos* (Latin, *De Caelo,* On the Heavens—Ryssel 1880–1881; Furlani 1923). This makes it significantly more advanced than the Roman Aratus, but far shy of the mathematical astronomy of Ptolemy. Such is very likely to be true for the other works translated by Sergius, as they are focused upon the ideas of Aristotle and Galen. The fundamental terms of spherical (circular, really) astronomy are included, as are Greek titles for the planets and vocabulary related to or-

7. Rendered partly into German, with a commentary by Ryssel (1880–1881).

8. Translated into Italian by Furlani (1923).

9. No modern translation appears to exist for this work of Sergius. See Sachau 1870 (101–24).

bital motion. A majority of these (and other) technical words were apparently adopted into Syriac either as phonetic transliterations (loan words) or as lexical cognates. This comports with what other scholars have identified as a typical pattern in Syriac translations (Brock 1975, 1977), for example concerning technical terminology in legal, ecclesiastical, and administrative matters.

What of the character of the translation itself? Sergius, we are told, makes the claim: "I have taken great care to remain entirely faithful to what I found in the manuscript, neither adding anything to what the philosopher wrote nor leaving anything out." The nineteenth-century Swiss scholar Ryssel, meanwhile, who offers the most detailed modern commentary on the text to date, is wholly effusive on this point: "the translation is a masterpiece of adaptive rendering of a source text, in which the author employs a nearly literal method of translation with utter precision to give the truest version of the Greek text, with all its details, a Syriac expression" (1880–1881, 10). As a modern commentator, Ryssel seems at pains to emphasize the literal exactness of Sergius' version over any "free disposition of Syrian vocabulary"; he thus appears to reveal a degree of reverence for the Greek original entirely characteristic of German scholarship in his own time. Sergius, meanwhile, was well known to the great Arabic translators of the ninth century, and, judging from the comments made by the most prominent among them (Ḥunayn ibn Isḥāq), his mature method was indeed a nuanced mixture of literal and free rendering that made him a formidable model of the translator's art. This is also attested to by modern criticism regarding his versions of Galen's works (Degen 1981).

Regarding the other group of astronomy-related translations, there is much less information available. These were performed in the mid to late seventh century by Severus Sebokht (Severus of Nisibis, d. 666–67), possibly while he was bishop of the Qennenshrin monastery and active as the teacher of succeeding Greek scholars and translators, such as Jacob of Edessa. Severus appears to have translated and partly adapted, without attribution, a treatise on the astrolabe originally written in the late fourth century A.D. by Theon of Alexandria (Nau 1910), whose commentaries on Ptolemy and whose handbook reworkings of Euclid were widely known in late antiquity and became important texts for later Arabic translators. A modern French translation exists (Nau 1910) and indicates that Severus was in full command of much Greek astronomical terminology, particularly as it related to the motions of the constellations, zodiac, and planets, as well as such phenomena as precession, and their measurement. The first portion of the treatise concerns the different parts of the astrolabe, how to assemble and disassemble the instrument, and its practical uses for determining time, season, planetary longitudes, the obliquity of the ecliptic, and so forth. This is supported by a basic theoretical discussion of the structure of the heavens, which the astrolabe is said to mimic in its own construction.

The "Treatise on the Constellations" by Severus Sebokht: A Unique Convergence

A second work by Severus that deals with astronomical subjects, far more interesting in terms of the history of translation, comprises a "treatise on the constellations" that mainly concerns the imagery of the Greek mythological heavens (Nau 1929–1932). What makes this work especially noteworthy is that, in no uncertain terms, it poses the "true astronomy" of Ptolemy and "the geometricians" against the "fictional" astronomy of "the poets and astrologers," among whom Aratus is paramount. Indeed, Severus quotes Aratus fairly extensively in this work, while drawing on other literary sources as well (e.g., Eratosthenes' *Catasterisms*), in a manner that makes it clear Severus considers the *Phaenomena* in a league with fables and fabrications, rather than with any useful text for the learning of astronomy per se. As a bishop of the Eastern Church, Severus was more than prepared theologically to view Aratus as a bearer of pagan myth, not at all commensurate with the "true philosophers" of science. But such a division, by which the *Phaenomena* was reduced to mere historical interest, marking a striking contrast both to Roman uses of Aratus and those of early medieval Europe, is based in this treatise not on religious grounds but instead on purely intellectual considerations.

In form, the "treatise on the constellations" contains eighteen chapters addressing a range of subjects announced in the title: "Whether the named figures that one speaks of in the heavens have been placed there by nature or by human convention; on their risings and settings; on the circles, that is zones, of the celestial sphere, and the positions of the poles and latitudes of terrestrial climates, the measure of the heavens and the Earth, and of the inhabited and uninhabited portions thereof" (Nau 1929–1932, 344). Above all, it is the first of these topics—whether the constellations are named and identified "by nature" or by human invention—that is most relevant to translation, and this for several reasons. Severus is fully aware that the question (which has apparently been posed to him by a second party) involves both epistemological and linguistic aspects. Indeed, it is worth quoting his statements along these lines in full:

> Above all, it should be understood, gentle reader, that everything for which we intend to provide instruction—that is, make understandable to others—would not be teachable without the use of names and words, whether it be a matter of teaching things that exist in nature or those that exist due to convention. Indeed, the *most eloquent philosopher* [Aristotle] notes four things that, being simple, one must begin with in order to [*sic*] make matters comprehensible: events, thoughts, words, and writings. . . . One sees here that the first two must be expressed, while the latter two serve to produce such expression. For example, the heavens and the Earth and related phenomena are of the first type and must be expressed; they are the same for everyone, since they exist in nature. On the other hand, names, words, and writings are of the second type and serve as the means for such expression; they are not the same for all men, for they are products of convention. One has neither the means nor the capability to teach or learn anything on the subject of what exists in nature without utilizing names and terminologies that are conventions and fictions. . . . (pp. 345–46)

Severus thus provides a basic theory for the local nature of knowledge. Due to its dependence on language, human understanding is defined largely in terms of conventions of speech, which include even "fictions." It is neither Platonic (a reflection of naturally occurring forms and ideas) nor wholly Aristotelian (involving, in part, a precise correspondence between word and object). Still more to the point, the author states that nothing of this knowledge can be transferred between individuals, groups, or entire peoples without movement through the medium of language—in this case, names and terminologies, which are implied to be the linguistic substance of science.

With this as a beginning, Severus proceeds to translate and then discuss the literary heavens of Aratus, to offer many Syriac titles for the constellations alongside those of Greek origin (e.g., "beard of the goat" for Corona Borealis), and to dismiss astrological interpretations as the effort of "the ignorant" who would deprive the world of free will and of God's power. It is interesting to see that, after designating them "conventions and fictions," Severus goes on to employ the Aratean constellations extensively, in his own explication of basic celestial motions. It is clear that he uses these titles (at times with their Syrian equivalents in tow) as a type of shorthand, a widely accepted nomenclature, implying therefore that the Greek heavens were fairly well standardized by this time among those with *both* astronomical and astrological leanings.

The Syriac Transition: Concluding Statement

In the late eighth or early ninth century, a northern Syrian scholar and poet, David bar Paulos, wrote in one of his letters, "every kind of wisdom derives from the Greeks." In another writing, he even composed a poem on the point:

> Above all the Greeks is the wise Porphyry held in honor,
> the master of all sciences, after the likeness of the godhead.
> In all fields of knowledge did the great Plato too shine out,
> and likewise subtle Democritus and the glorious Socrates,
> the astute Epicurus and Pythagoras the wise;
> so too Hippocrates the great, and the wise Galen,
> but exalted above these all is Aristotle,
> surpassing all in his knowledge, both predecessors and successors.[10]

If we can take this as a general sign of where "the wisdom of the Greeks" stood in this period, we can see how far things have come from the rivalry of earlier centuries. What is notable about this passage includes not merely its honorary tone, but the manner in which it places side by side authors of widely differing times, subjects, and achievements—a sign that this wisdom has indeed become frozen in a timeless domain, with the implicit job of the present confined to preservation and study. Preservation and study of what, exactly? The suggestion here (and elsewhere in his writings) is that David bar Paulos did not know Greek especially well, and did not need to.

There had been centuries of direct connection between Greek "wisdom" and biblical study, such that the more useful works had been translated and incorporated into Syriac intellectual culture. The scholastic method of Aristotle, focusing on textual exegesis, had proven itself of great worth and power during the centuries of sectarian struggle: the Nestorians, Monophysites, and Chalcedonians all employed the procedures of Aristotelian logic in the analysis of scripture to prove the rectitude of their specific beliefs regarding divine substance, the nature of Christ, the eternity of the universe, and so forth (Peters 1968). Aristotle, and the commentary by Porphyry, served as crucial weapons in these theological battles. By the seventh century, they were "thoroughly domesticated in Syriac [as] a hallmark of the education shared by the Christian exegetes and theologians who constituted the east Syrian intelligentsia" (Peters 1996, 50). By this time, too, the study of medicine, based on the translated curriculum of the schools in Alexandria, was fully established "and must already have been in use at what was emerg-

10. Quoted in Brock 1982 (25).

ing as the Nestorians' chief medical centre at Jundishapur in Khuzistan in Persia. The material was Hellenic and Hellenistic, but its study did not necessarily imply a knowledge of Greek" (Peters 1996, 50–51).

At the dawn of the episode of Arabic translation, a considerable portion of Greek learning already existed in a Semitic language of wide usage and elegant literary and religious expression. That the Arabs chose to employ Nestorian translators who had been raised in this language was a mark not only of historical practicality but of necessity. Indeed, it has been fairly said that the great period of Arabic translation was less a wholesale innovation than a more systematic continuation of a process begun long before (Corbin 1993). Thus, the fields initially chosen for particular attention by the Arabs were precisely those that had been selected by Syriac translators over the centuries—above all, disciplines relating to Aristotle's philosophy, to medicine, and, of course, to cosmology and astronomy.

Texts and Translations from Persia and India

Sources of "Wisdom" from the East

In its earliest phases, Islamic astronomy was like a vast reservoir into which flowed many rivers of influence with textual nutrients of wide and often overlapping provenance. This movement, or rather these movements, were made possible in part by Muslim conquests in the Mediterranean region in the early seventh century and in areas to the east thereafter. As often noted, the effect was to bring under one linguistic roof an array of different cultures: in the east alone, Syrians, Greeks, Persians, Jews, Hindus, Armenians, and Arabs now came to exist within the embrace of the Arabic tongue. Each of these groups, in the centuries after Alexander, had adopted portions of Greek science, modified and adapted it to their own native traditions, added to it according to their needs, and entered the result into a continual interchange of textual materials that took place between the second and seventh centuries A.D.

Indeed, when considered from the point of view of secular texts, those in the sciences most of all, the Near East appears as a region in almost constant motion. Embassies, refugees, armies, missionaries, merchants, and lone travelers all contributed toward the passage of texts among cities and peoples. Translation proved, time and again, to have a fertilizing power, and this effect was by no means limited to periods of conquest. In a manner of speaking, Islamic rulers and scholars of the eighth and ninth centuries sought to gain control over this process of textual passage and fertilization. To do so, moreover, they had important models to draw upon.

In its initial stages, the desire among Muslim thinkers for the textual wisdom of other cultures was very eclectic: Islam brought together once again under one dominion all the varied portions of Alexander the Great's empire in the east that had long since absorbed, interpreted, and rejected any original Hellenistic elements in terms of local, evolving cultural idioms. This may seem a misuse of emphasis; nearly a thousand years had passed since Alexander's death. Yet detailed studies on the written traditions of these areas reveal that what was true in the Near East (for Syriac-speaking people, for example), within and just beyond the reaches of Byzantium, was true of Persia and India as well: Hellenistic ideas and systems had early on provided opportunities both for intellectual stimulus and for new forms of expression among a diversity of peoples, each of whom subsequently transformed this influence to suit their own cultural precinct. By the time of Islam, the relevant intellectual forms had a truly venerable history; they could no longer be called "Greek" or "Hellenistic" without the impunity of historical erasure. But this cannot, in its turn, delete the complex origin of such forms, their long-term dependence on traditions of translation. Indeed, it might well have been partly a result of such perception that Islamic scholars themselves eventually came to favor the translation of Greek works at the expense of all others.[11] For the early formation of Islamic natural science, however, and for astronomy and mathematics above all, the use of Indian and Persian writings proved of crucial importance.

Study of these writings, the method and manner of their translation, and their exact routes into Islamic thought constitute the least developed chapter in the greater story of Arabic science. Many of the critical texts are now lost and are known only through later reports. The opportunity to compare original and translated versions of individual works is very small, in many cases nonexistent. According to one recent discussion, many or most of the remaining manuscripts in Pahlavi, Sanskrit, and Arabic lie scattered among various libraries and museums around the globe and have not yet been examined, let alone systematically analyzed (Haq 1996). As a result, the type of study offered above for Syriac and Greek works is not yet possible. It is thus the general contours of movement and influence that I will try to summarize below.[12]

11. This is a widely recognized fact among contemporary scholars of Islāmic intellectual history, including those who have recently offered valuable critiques of the previous over-celebration of Greek influences. See, for example, Saliba 1994 (esp. chapter 1) and Sabra 1987.

12. It should be noted that the works of the Oriental scholar David Pingree are absolutely essential to any study of this field. My own summary is largely a condensation of the following studies: Haq 1996 and Pingree 1973, 1976, 1978, 1981.

Jundishapur: Intellectual Cosmopolis of the East

Some idea of Persian and Indian influences upon what later became Arabic astronomy can be achieved by examining the intellectual activity that occurred in the city of Jundishapur, which had been a major center of translation and scientific learning for over five hundred years before its annexation by the Islamic empire. This activity had been initiated in the third century A.D. by the first Sassanian kings, Ardashīr I (ruled 226–241) and, especially, Shāpūr I (241–272). The latter had chosen the town as a place to settle Greek prisoners taken in the wars against the Roman emperor Valerian.[13] Very soon thereafter, following a long and destructive siege, Shāpūr I captured Antioch, then a cultural center of the eastern Roman empire, and many Greek-speaking Syrians were given refuge in Jundishapur. Thus, from the very beginning, the city acted as a nexus of Hellenism and cultural-linguistic mixing. This was further enhanced during the sectarian purges by Byzantine emperors: Zeno's closing of the Nestorian school in Edessa (A.D. 489) and Justinian's draconian measures against secular teaching throughout large portions of Byzantium led to a "general reduction of higher education in the cities of the empire" (Cameron 1993, 22) and a migration of Nestorian intellectuals eastward. Byzantium's loss became Jundishapur's gain. The schools established by the Nestorians in the late fifth century were elevated by Chosroes I Anūshirwān ("the wise," A.D. 531–579) into a full-scale center of translation, study, and teaching, patterned after similar institutions in Alexandria. Greco-Syrian medicine flourished, and mathematics, logic, and astronomy were all included, to varying degrees, within the curriculum. Elements from India were introduced, for example in the form of the famous Aesopian Fables of Bidpai, various pharmacological tables, and even several Indian physician-teachers, all brought by the physician-vizier Burzuyah of Anūshirwān on his return from a trip to India ordered by Chosroes (Rashed 1996, v. 3). By the time of the Arab takeover of the city, in A.D. 738, the fame of Jundishapur as an intellectual nucleus was canonical. Indeed, it is recorded that al-Hārith ibn Kalāda, a relative of the Prophet Muhammad, studied at the medical school there in the early seventh century (Dunlop 1971, 204–5).

As a site for intellectual activity, Jundishapur was a necessary locus of translation in late antiquity. Within its borders, movements of medical, cosmological, astronomical, and Aristotelian works took place at various times among Pahlavi, Syriac, Greek, Sanskrit, and later on, Ara-

13. This and other details about the early history of the city can be found in Nöldeke 1973 (32–34).

bic. Equally impressive is the fact that this linguistic transfer was directly supported and sponsored by the ruling kings of the city—a truth whose significance for the history and fame of the city did not escape the early ʿAbbasid caliphs. This type of "state-sponsored" translation work began very early, under Ardashīr I and Shāpūr I, who were particularly interested in Greek and Indian astrological texts. No less a work than Ptolemy's *Syntaxis* was rendered into Pahlavi in the third century A.D., less than a hundred years after its original composition (Pingree 1968, 7–13). Subsequent to the founding of the Nestorian school in Jundishapur, translation activity increased significantly. Later Sassanian rulers continued and expanded the interest in astronomical works and used the available linguistic expertise in the city to produce new texts that combined elements from various sources. They tended to focus on astrologically useful texts, in particular the so-called "zik" (Pahlavi, "zij" in Arabic), comprising various tables for predicting the motions of the planets and constellations, often accompanied by explanatory writings that could be quite extensive, incorporating portions of more theoretical works, such as by Ptolemy or his commentators. In this connection, it seems useful to cite a few specific examples giving some idea of how complex the related process could be:

> In the late fourth century the Era of Diocletian was employed in computing a horoscope inserted into the earlier Pahlavi translation of the Greek astrological poem of Dorotheus of Sidon [first or second century A.D.]; this may indicate the existence of a set of astronomical tables in Pahlavi using that era.
>
> But, in any case, it appears that in *ca.* 450 a *Royal Canon—Zik-i Shahriydran*—was composed; the one element that we know from it, the longitude of the Sun's apogee, is a parameter of the brahmapaksa of Indian astronomy. A century later, in 556, Khusrau Anushirwan ordered his astrologers to compare an Indian text called in Arabic the *Zij al-Arkand* (*arkand* being a corruption of the Sanskrit ahargana [the number of days elapsed since the beginning of a given epoch, used to calculate planetary longitudes] with Ptolemy's *Syntaxis;* the Indian text . . . was found to be superior and a new redaction of the *Zik-i Shahriydran* was based upon it. . . . A final version of the *Zik-i Shahriydran,* like its predecessor incorporating many ardharatrika parameters though employing others of unknown origin as well, was published during the reign of the last Sasanian monarch, Yazdijird III. In its Pahlavi form it was probably the zij used by the computer of a series of horoscopes illustrating the early history of Islam, computed shortly after 679 and of another series computed during the reign of [the caliph] Harun al-Rashid. (Pingree 1973, 35–36)

This type of tracing of sources and influences suggests several important conclusions. First, no individual work was viewed as sacred during this period, to be left unmodified. If there were sages of astronomy, prophets and magi from the past whose fame was unalterable, such was far from the case with regard to their actual writings. Pieces of various texts, phrases, words, titles, tables, calculations, and so forth might all be combined and recombined to produce a needed manuscript. Frequently, this involved the updating of older tables and retranslation of accompanying text, but it included many other aspects as well, such as the rearrangement and editing of one work and its insertion into another. Second, the great Ptolemy, whose *Syntaxis* had surely achieved no small degree of fame by this time, was also seen as simply one more useful source, though an important one. Indeed, Ptolemy could even be found wanting in certain comparative respects and subjected to redaction. Third, the city of Jundishapur, with its cosmopolitan institutions of learning, acted as a type of marketplace for textual goods. These products were constantly being transferred, adapted to current uses, and sent on their way again. There appears to have been, if not exactly a constant search, at least a standing interest in news of any overlooked or newly produced work that might aid in a particular purpose.

The Nature and Cause of Influence

If these sorts of conclusions seem exaggerated, it should be remembered that the remaining evidence is extremely meager, yet universally points in this direction. Indeed, this evidence is unequivocal in showing that the Muslims were not at all slow to comprehend the relevant process and to try to take charge of it:

> [One] of the astrologers consulted on the propitious moment for the foundation of Baghdad was Muhammad ibn Ibrahim al-Fazari, the scion of an ancient Arab family of al-Kufa. When an embassy sent to the court of al-Mansur from Sind in 771 or 773 included an Indian learned in astronomy, the caliph ordered al-Fazari to translate with his help a Sanskrit text related to the brahmapaksa and apparently entitled the *Mahasiddhanta:* this work seems to have been dependent on the *Brahmasphufasiddhanta* written by Brahmagupta in 628. The result of this collaboration was the *Zij al-Sindhind al-kabir,* whose elements, however, are derived not only from the *Mahasiddhanta,* but also from the aryapaksa (probably through the *Zij al-Harqan*), the *Zij al-Shah,* Ptolemy (perhaps the Pahlavi version), and a Persian geographical text ascribed to Hermes. (Pingree 1973, 38)

Not mentioned here is the fact that the *al-Sindhind* (as it came to be known) represents the first scientific work to be translated into Arabic. There is something magnificently appropriate in this, for Baghdād was to become, in the following century, the center of translation activity in Islam under al-Manṣūr's successors and thus one of the great centers of scientific learning in the history of the world.

The eclectic nature of the text, meanwhile, the most important of the early *zij* known to have entered Arabic and thus a forerunner of what would become a dominant type of astronomical work of Islam and, later, Latin Europe, is ample testimony to the style of "influence" to be found in important written works at this time. Since only the end products of the process remain, it is not possible to determine precisely the style of translation used, or, for that matter, just where "translation" leaves off and innovative adaptation begins. Yet, if the larger procedure indicated by the passage above can be taken as any indication of the local effort of transfer, it would seem that this, too, was a complex blending of literalism and paraphrase. The type of rather strict fidelity practiced by most Syriac translators of the seventh and eighth centuries does not at all seem likely in this case, given the wide array of source material and the readiness to combine it in ever new forms.

In the above passage, the author is essentially quoting from al-Fāzārī himself, particularly in the first few lines (Pingree 1970). Al-Fāzārī tells us that the Indian astronomer (Kanka may have been his name) was expert in many aspects of astronomy and speaks of a work that the caliph would do well to consult; he tells us further that al-Manṣūr became eager to see it and thus ordered the translation. Personal contact, word of mouth, offered at just the right historical moment, resulted in an effort of linguistic transfer of great importance in the history of astronomy. Al-Fāzārī was a Persian by birth, one of many astronomers from that region that al-Manṣūr brought to his court in the late eighth century. It is evident that the caliph recognized in al-Fāzārī an important intellectual and linguistic resource, and it may be that this was true of other Persian astronomers as well during this time, many of whom knew Arabic and Pahlavi, as well as some smattering of Syriac and possibly Greek and Sanskrit. How was the translation of the *al-Sindhind* produced? It appears that the Indian, who was at least conversant in Arabic, probably read a Sanskrit passage, translated it partly aloud for al-Fāzārī , who then gave it final written form—a process that would be repeated fairly often during the episode of Arabic translation and also, later on and more commonly, in the twelfth century, when the works of Arabic science were rendered into Latin (Dunlop 1971, 216–17). To pass between different languages, indeed different cultures, scientific literature had to

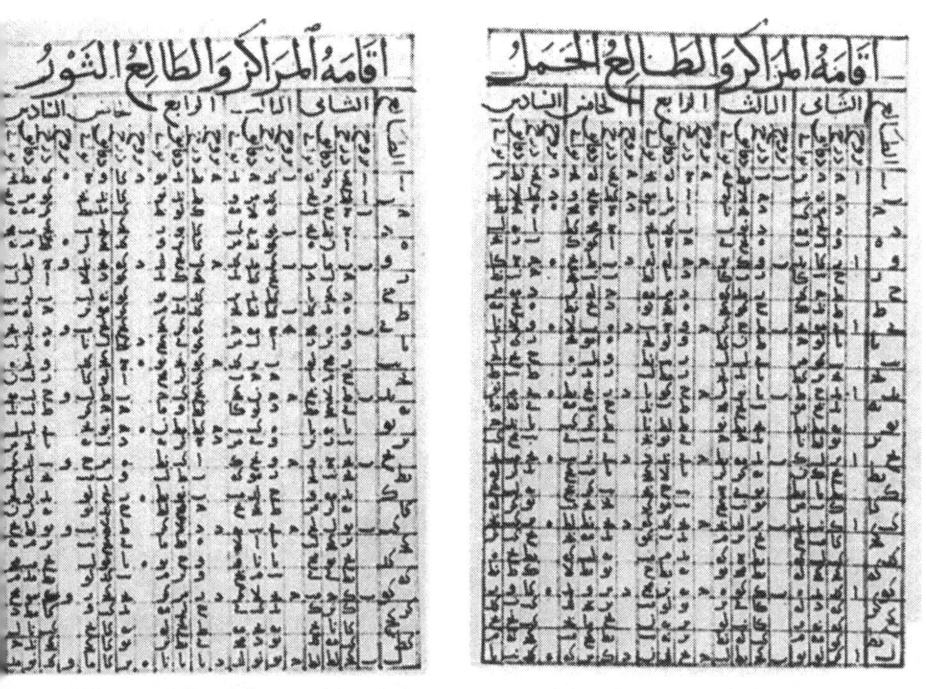

Figure 5. Pages from a thirteenth-century astrological/astronomical manuscript, showing a typical *zij* (astronomical table). In this case, the tables present positional information on various astrological "houses" for the latitude of Baghdād. Bibliothèque Nationale, Paris, arab 2426, fols. 173v–174r.

pass between individuals first of all. The "small" stories of such contact form the loci for exchanges sometimes of vast importance.

The *zij* tradition in Islamic astronomy is one of the most crucial in terms of both observational and theoretical work. Though often characterized as merely tabular information (fig. 5; the original Pahlavi term *zik* means "chord" or "thread," apparently referring to the woven-like appearance of the text itself), the *zij* was really a complex and sophisticated compendium of thought, method, and information, often including extended theoretical discussions of planetary motion. Its own origins seem to go back to Persian adaptations of Ptolemy's *Handy Tables* (a revised collection of the astronomical tables in the *Syntaxis* with instructions for their use) and to the *siddhānta* tradition in India, which comprised a series of treatises in verse form that sought to provide "rules" for arriving at mathematical solutions to various astronomical problems and which involved various systems of calculation (eventually including the sine function, which Indian mathematicians invented), geometric models, and diagrams representing celestial relationships and movements. This tradition had largely begun in the fourth and fifth centuries, reaching an early high point in the work of Āryabhaṭa I (b. a.d. 476), whose texts *Āryabhaṭīya* and (to a lesser extent) *Āryabhaṭasiddhānta* are generally thought to represent "the earliest extant mathematics-based astronomical work of great importance" in the history of Indian science (Subbarayappa and Sarma 1985, xxvii).

The *Āryabhaṭīya* is a poem laying out the structure and movements of the stars and planets, as well as methods of calculation, measurement, and prediction, in a total of 121 verses. The language, in places, is a highly evocative mixture of lyrical metaphor and technical terminology. The Earth, for instance, is referred to as the "hitching peg in the midst of space" and likened to a "*kadamba* flower covered all around by blossoms" (Subbarayappa and Sarma 1985, 31, 25). Yet, as in all *siddhāntas,* it is the technical discourse that eventually dominates, due to the very attempt to outline verbal and symbolic formulae for computing celestial positions, motions, and relationships. Several different systems of measurement, dealing with time in particular, were mastered and employed by Āryabhaṭa, setting a standard for future authors of these treatises. The difficulty of translating the *Āryabhaṭīya* is well illustrated by the following attempt to give a modern English rendering of a method for determining planetary positions:

> The corrections from the apogee (for the four anomalistic quadrants) are respectively minus, plus, plus, and minus. Those from *śīghrocca* are just the reverse. In the case of Saturn, Jupiter and Mars, first apply the *mandaphala* negatively or positively (as the case may be). Apply half the *mandaphala* and half the *śīghraphala* to the planet and to the planet's apogee negatively or positively. The mean planet corrected for the *mandaphala* (calculated afresh from the new *mandakendra*) is then called the true-mean planet and that corrected for the *śīghraphala* is known as the true planet. (Subbarayappa and Sarma 1985, 143)

Each of the italicized words represents a specific attribute of a planet's motion defined for purposes of positional computation: for example, *śīghrocca* refers to the apex of a planet's fastest motion, *mandaphala* indicates a correction related to the planet's apogee or aphelion, and *mandakendra* is defined as the longitude of the planet at any particular position minus its longitude at apogee. Such terms were apparently transliterated into Pahlavi or Arabic (or first into Pahlavi and then into Arabic), but many of the relevant details can only be guessed at, since the needed manuscripts are lost. It is more than clear, however, that the system of Indian terminology was no less highly developed and specific than that of Hellenistic Greek astronomy, and must have posed a formidable challenge to any translator of the time. This would seem to argue for a collaborative effort in the production of such translations.

Notwithstanding the question of Greek influence on early Indian astronomy, which appears considerable,[14] it seems beyond doubt that the *siddhānta* tradition was foundational to establishing a theoretical-computational—that is, mathematical—astronomy in Islam. Besides the two sources already mentioned, the *zij al-Sindhind* and the *Āryabhaṭīya,* the first generation of Islamic scientists made abundant use of a third text, the *Khandakhadyaka,* also written by Brahmagupta in the seventh century. This choice, like that of the *zij al-Sindhind* (also apparently derived from Brahmagupta's writings), was not in the least random. Āryabhaṭa I had essentially formalized the *siddhanta* into a set of verbal codes for detailed calculation, whose outline and progression were followed thereafter by later writers (Chattopadhyaya 1986). Brahmagupta, meanwhile, revised and corrected many of the rules in the *Āryabhaṭīya* and produced new methods as well, e.g., for calculating a planet's instantaneous motion, correctly determining parallax, and computing the precise progress of eclipses (Prakash 1968). These two authors—the one a hallowed "sage" of astronomical writing, the other a corrector of the first and the one whose works were composed during the years of the founding of Islam itself and that remained most in contemporary use—must have appeared to the early Islamic astronomers as the most canonical of Indian sources. It seems significant, in this respect, that Brahmagupta stated early in the *Khaṇḍakhādyaka:* "the methods given by Āryabhaṭa are generally impracticable for everyday calculation. . . . My statements in this work are more concise, yet give the same results" (Subbarayappa and Sarma 1985, 4). The two authors thus appear to stand at the beginning and the "conclusion" of the *siddhanta* tradition.

Conclusion: Choices and Meanings

Many first generation Islamic astronomers were Persian by birth, and it seems certain that a number of these men had fairly intimate contact with Indian astronomy, whether through travel, Pahlavi translations, or word of mouth. These same styles of contact appear to have existed from an early date in India as well, since Indian astronomers incorporated Babylonian elements (e.g., linear algebraic methods in planetary theory) even before being touched by Alexandrian astrology in the early

14. See, for example, Pingree 1971; 1981 (7–10). For a brief characterization of this influence, see Morelon 1996 (esp. 8).

centuries A.D. (Pedersen 1993, 152–53). One historian of this long period of obscure encounter makes the tantalizing and enigmatic statement that Āryabhaṭa I "mastered the advanced geometrical methods of the *Almagest* without any direct translation of it" (Pedersen 1993, 153).

Throughout this period, as the case of the Sassanian rulers makes plain, there existed between Alexandria in the west and Peshawar in the east a continual search for influence in the form of important and worthy texts with which to build a better royal astronomy/astrology. Knowledge and use of the stars offered specific powers: the power to comprehend and predict events, to order and manage time, to build great new cities, to touch God or the gods with human command. Such powers were absolutely integral to kingly authority; lacking any solid commerce with the heavens, royalty was bound to fail on Earth, for it meant that a king or queen was without divine imprint and sanction. Such beliefs were clearly part of early Islamic culture, too, as the notable case of al-Fāzārī and the caliph al-Manṣūr makes plain. Thus, the search for textual influence in astronomy was also, in large part, a search for political power, for an increased substance of terrestrial authority. Cities like Antioch, Edessa, Harrān, Nisibis, and Jundishapur acted as loci where the results of this search were converted into usable material through the process of translation. Whether in whole or piecemeal fashion, the writings of Ptolemy, Āryabhaṭa, Sergius of Reshaina, Severus Sebokht, Brahmagupta, Theon of Alexandria, and others, as well as a host of anonymous texts that likely represent redacted, plagiarized, and amended portions of known works, were all transferred time and again between cities, kingdoms, and languages. If Āryabhaṭa I was influenced in some manner by Ptolemy, it was probably by means of just such a recombinant text, whose utility and anonymity reflected the central place of translation in the making and remaking of cultural materials. In fact, it would be difficult, in such a case, to speak accurately of "Ptolemy's influence," because of the implied direct connection: "Ptolemaic influence" might be better. The reduction of authorship to adjectival status is precisely the point.

The translation activity of the Near East in late antiquity, and the texts involved, were far more complex and cosmopolitan than they ever were in Rome. No less important for astronomy than Indian and Persian texts were writings in Syriac that also encompassed full translations and pieces of adopted works originally in Greek. If Indian authors appear to be the most original and innovative, those who worked in Pahlavi and Syriac were no less significant in terms of the resource they supplied for future astronomical effort. This is because the baton of this translation activity in astronomy was taken by the new scholars of Islam—it is a mistake to say "the Arabs" in this case, since these scholars came from many cultural and religious backgrounds: Christian, Zoroastrian, Manichean, Muslim. During the earliest years of this transfer, under al-Manṣūr (reigned 754–775) and al-Mahdī (775–785), Indian and Persian works dominated: the *Fihrist* of al-Nadīm cites the writings of no fewer than fourteen different Indian astronomical authors translated into Arabic by the mid-tenth century (Nadīm 1970, 2.644–45). But starting with al-Mamun (813–833), and with the enormously influential writings of Yāḥyā ibn Māsawayh (known to the Latin middle ages as Mesue), a Nestorian physician, and the polymath al-Kindī (ca. 800–867), who became a powerful spokesman for Greek learning, the tide shifted toward Syriac versions of Greek texts, not only in astronomy but in medicine, natural history, and philosophy.

At the founding of Islamic astronomy, the chosen works were those that appeared to satisfy two major, overlapping conditions. The first of these was superior accuracy, and therefore usefulness. The *Zik-i Shahriydran,* as noted above, was compared with Ptolemy early on and found to be more precise in its determinations of planetary and stellar positions. Later, Ptolemy's *Almagest* came to be chosen in no small part because its methods allowed for updated corrections, as well as for continual refinement, commentary, and resistance in the name of improvement. The second major basis for textual choice seems related to a perceived need on the part of Islamic scholars to gain possession of the most renowned and comprehensive works—those that essentially encapsulated an entire tradition of effort. Thus we find both Āryabhaṭa I, the presumed originator of the *siddhanta* current in Indian astronomy (with its emphasis on exactness of method and result) and his most advanced commentator, Brahmagupta. And here, again, Ptolemy becomes a chosen source, since, as often remarked, it had long been clear that the fame of the *Almagest,* with its compendium and extension of Hellenistic ideas and methods, had largely obliterated interest in earlier sources.[15] These two conditions—accuracy and comprehensiveness—thus appear to define the vision of textual utility for early Islamic intel-

15. This sensibility has been recognized repeatedly, and in no uncertain terms, by modern historians:

> Any attempt to reconstruct the origin of Hellenistic mathematics and astronomy must face the fact that Euclid's "Elements" and Ptolemy's "Almagest" reduced all their predecessors to objects of mere "historical interest" with little chance of survival. As Hilbert once expressed it, the importance of a scientific work can be measured by the number of previous publications it makes superfluous to read. (Neugebauer 1969, 145)

The last sentence, with its rather abject projection of contemporary sensibilities, seems disingenuous with regard to the writing of history.

lectual culture. They give evidence for a significant level of discretion, for a decidedly rational program of translation whose foundations were subsequently built upon considerably during the second major episode of effort, encompassing the late ninth and much of the tenth centuries.

In the end, Islam had a true wealth of material to draw upon for its astronomical needs. The most advanced forms of ancient science were available to it because of centuries of translation, which, in the case of Syriac particularly, had proven not simply a preservative force, a calcifying process, but a multidimensional activity of cultural adaptation often "guilty" of deleting or deforming original contents and adding or adapting new ones. What Islam adopted on its own terms, then, was a wealth even beyond that of multiple coinage. Islamic scholars did not merely seek to mine cities and libraries from Alexandria to Peshawar for textual treasures; they sought (and largely succeeded) to gain control over, and to extend and expand, the refining process of translation itself. In this must be understood a good part of the civilizing effect of Muslim intellectual culture and its eventual effect on Latin Europe.

3 The Formation of Arabic Science, Eighth through Tenth Centuries

Translation and the Creation of Intellectual Traditions

The Problem of "Hellenization"

A reader who embarks today upon a study of this era of translation, arguably a major event in the history of world civilization, will soon find himself in both luxuriant and difficult waters. The seas here are indeed deep and rich in nutrients—no era concerning translation and science has been so extensively written about, particularly in recent years, and there is still much that remains to be said, even on a fundamental level—yet their expanse is continually crossed by contrary winds and currents, such that it is never easy to discern what shoreline lies ahead. Contemporary scholarship on this subject remains in the midst of energetic reconsideration and controversy over its own past, placing it alternately at odds and in (sometimes unadmitted or unconscious) sympathy with earlier interpretations and terminology. This rises to the surface in several main topics or "problem areas" that have an immediate and profound bearing on how one conceives of knowledge transmission during this and later periods, as well as its specific uptake within Muslim civilization.

The most central of these topics, and the one we shall discuss here, concerns the idea of Hellenism, or rather "Hellenization." This was briefly mentioned in chapter 2 with regard to Syriac translation, but it now comes to the fore with a great deal of added importance due to the much greater scale of the effort involved, its state-supported character, and its presumed transformational influence. The "Hellenizing of Muslim intellectuals" is the type of phrase one finds today both questioned and routinely employed as a scholarly commonplace.[1] The confusions

1. An example of the continued use of such terms can be found in the otherwise excellent essay by Goodman (1990).

involved, and the controversies implied, might easily be the topic of an extended study on the historiography dealing with this time and region. They are perhaps best exemplified for present purposes by a series of short passages from several important recent works, each of which makes a motion in the direction of such a study. The first of these works, and one of the most widely used and valuable sourcebooks on this era of translation, is Franz Rosenthal's *Classical Heritage in Islam.*[2] "The triumph of Hellenism," this author states apropos of the ancient Near East, "is a fact which is being confirmed again and again by fresh discoveries. However, at the beginning of the seventh century, this was a Hellenism which had largely dispensed with the Greek language, since Christian missionaries, and, in their wake, those of the Gnostic sects . . . insisted on using the indigenous languages for literary expression" (Rosenthal 1975, 2).

A few pages further on, in dealing with the translators themselves, Rosenthal makes this striking and entirely accurate statement: "The Greek of the works they translated was a dead language and known to be very different from the speech of contemporary Byzantines" (p. 16). We are also told that "almost all translators were Christians of various churches" and that "practical usefulness [for Islamic patrons] was the leading consideration" for what was translated (p. 3). As if this were not confusing enough, the reader also has the following to contend with:

> The works of ancient philosophy, medicine, and the exact sciences were taken over almost in their entirety, to the extent that they had survived into the late Hellenistic period, which means that our knowledge of Greek works in these fields does not substantially differ from that of the Arabs. . . . Works that had become classics were naturally the ones most studied in the schools. However, they had to be adapted to educational requirements, and changing times necessitated certain changes in emphasis. Furthermore, many a great author appeared long-winded, so that abridgments and paraphrases were deemed more suitable. . . . In the case of some authors, commentaries written on their works provided more information and had become more meaningful than the original text. (p. 10)

An alert reader is thus compelled to ask: to what degree, and in what specific ways, could "Hellenism" be said to exist as a force of influence

2. The original title in German, *Das Fortleben der Antike im Islam,* is far more suggestive than the flat English version and, indeed, expresses something of the difficulties at hand. More precisely translated as "the survival of classical antiquity in Islam," the word *fortleben* also carries the meaning of "afterlife."

when deprived utterly of its native language, when its primal texts had undergone centuries of change, and when these same texts, though seen as "classics," were rarely, if ever, accepted on their own terms, but instead as raw material to be adapted in whatever fashion seemed needed? In the case of Syriac and Persian cultures, for example, large portions of what had originally been scientific writing in Greek (particularly astronomy and medicine) had been translated, copied, amended, and adopted into both academic curricula and practical use for nearly five centuries before being transferred into Arabic. To what extent was this material still specifically "Hellenic?" What styles of legitimation are involved in taking this material away from Syriac and Pahlavi culture and giving it back to Greek?

I pose these questions not with a rhetorical answer in mind, but instead because they reveal the type of epistemological wetland that exists at every step once such terminology is brought into play. Indeed, the passages above should be enough to suggest that, even as a shorthand, "Hellenism" and "Hellenization" are at the very least problematical notions when applied to this era. It is clear from Rosenthal's characterizations, in other words, that what took place between the eighth and eleventh centuries in Muslim intellectual society was something quite different from "the survival of Greek culture." One might consider that historians have rarely, if ever, spoken of the "Arabization" of late medieval Latin culture, the "Romanization" of sixteenth- and seventeenth-century England, or the "Germanization" of late nineteenth-century European science. Yet all of these designations must at least be considered if such immanent force be granted the "Hellenistic element" during the era of Arabic translation and nativization.

The question of "origins," placed against the historical reality of intellectual use and change over time, can easily produce a slip or shove into ideological quicksand. It helps to keep the cultural dimension of Arabic translation specifically in mind:

> Scientific ideas move because people study books, compute with tables and use instruments, not simply because they translate books, transcribe tables or buy pretty artifacts. It suffices to recall that despite Byzantium's having direct access to whatever Greek manuscripts the Islamic world eventually came to possess—indeed to more of them and from an earlier date—the direct heirs of Hellenistic science showed themselves to be largely uninterested in their scientific heritage. (Berggren 1996, 265)

Yet even here one is confronted with the language of Islam-as-heir. Up until the late 1980s, it was common to encounter this discourse of "heritage" in a more raw form: scholars frequently spoke of this episode of translation as "preservation" and "safe-keeping," with Islamic civilization having "rescued" and "passed on" the "legacy" of Greek science. In these discussions there always hovered the image of Latin Europe as the true and final home of "recovery," the Ithaca (or throbbing Penelope) to a long-wandering, Odyssean knowledge. The West was thus posed as the rightful heir to an intellectual nexus that was somehow quintessentially its own. Greek knowledge was itself "lost" or entombed while in the hands of Islam, to whom it was always therefore foreign and indigestible. In this way, the role of Muslim society was degraded by the very terms of its most valued "contribution" to the West.

While a less Orientalist view has clearly been assembled during the past decade,[3] certain traces of this former discourse remain, like a flinty residue, particularly when the brush is broad:

> What we call the Greco-Arabic transmission of science and philosophy was a complex process in which translation was often much influenced by interpretations imparted through a prior scholastic tradition and, sometimes, by terms already in technical use in the newly formed disciplines concerned with Arabic language and with Islamic religion. . . . What the Muslims of the eighth and ninth centuries did was to seek out, take hold of, and finally make their own, a legacy which appeared to them laden with a variety of practical *and* spiritual benefits. (Sabra 1987, 225–26)

These points have had a resounding impact on the scholarship related to this era of translation and to other periods as well. The ideas of transmission, appropriation, and nativization Sabra adds into the scholarly mix have helped refocus attention away from the texts themselves to the realities of their reception, use, and transformation. Yet even here, problems reminiscent of the old discourse remain. Accepting the "Greco-Arabic transmission of science and philosophy" is one of these, for, as noted repeatedly, the translated corpus was far more diverse in content and history than such a term would suggest. Indeed, the true founding of the natural sciences within Muslim society came not from the uptake of Greek writings, but far more from material long appropriated and nativized to Syriac, Pahlavi, and Sanskrit-speaking communities—material, again, that is taken out of the hands and histories

3. The essay by Sabra (1987) has certainly played a central role in this new thinking—see, for example, the excellent essays based on his work and assembled in Ragep and Ragep 1996. There are also a number of other scholars whose writings have been important in this regard, and whose work should not be overlooked. See, for example, the essays and books by David A. King cited in the bibliography and also, even more to the point, the article by Rashed (1989).

of these communities by being called "Greek." However useful as a retooled shorthand, "Greco-Arabic" still essentially erases centuries of history—a history bereft of many crucial surviving materials and thus difficult to assemble, but history nonetheless.

Finally, too, we are left with the idea of "legacy." This, after all, implies an entrusted keeping, and a "rightful heir" somewhere down the road. Yet it may well be an entirely appropriate term, if given its own context. The ʿAbbasid caliphs, that is, whose support of the translation movement proved essential, appear to have envisioned for their shining new capital, Baghdād, a more magnificent version of the type of cosmopolitan center the legendary Alexandria, Antioch, and Jundishapur had come to represent. It seems this is the type of "legacy," related to the perceived model of a spiritual, political, and intellectual nucleus, inspired by divine motive and located through worldly knowledge, that can best be applied to the translation of Syriac, Pahlavi, Sanskrit, and Greek works into Arabic. Otherwise, in purely factual terms, one is left with the truth of David King's characterization that the transmission process "is to be viewed as an accident of Islamic history" (King 1996, 143).

The Promise and the Problem of "Foreign Learning"

Translation on the scale that occurred in early Islam must be rightly understood as a major historical movement. This means that it was really a complex series of events and relationships that did not begin and end with specific texts. Here, too, we are faced with confusion, particularly as to the significance for Islamic society. On the one hand, for example, in summary terms, the translation movement involved more than two centuries of planned and unplanned effort, hundreds of individual works, at least four major source languages (Syriac, Greek, Pahlavi, Sanskrit), efforts of manuscript collection, the setting up of translation "workshops," the building of libraries and observatories, all under the immediate and committed sponsorship of existing rulers and a variety of other patrons. The secular knowledge introduced in this manner eventually came to affect writing and thought in almost every conceivable discipline and was instrumental in making Islam into a "world civilization" of the highest order, despite political decay under the later ʿAbbasids. As such, it makes little sense to speak of the translation movement as hermetic. On the other hand, however, the entire movement and its fertile aftermath represent the work of a very small minority of scholars, most of whom were non-Muslims. While the resulting textual material and related knowledge came to be used by a very significant portion of educated Islamic society, far greater in actual number were those scholars and teachers engaged in the study, education, and exposition of the "Islamic sciences"—theology, law, language studies, exegesis—based on the Qu'rān. The "foreign sciences," or "sciences of the ancients," meanwhile, were sought out and encouraged at first, later nativized in variable fashion, taken up and advanced by princes, governors, courtiers, merchants, military leaders, and scholars, employed as part of the training of the civil service (especially mathematics), but always kept relatively separate from the most mainstream institutions of teaching and scholarship, the *madrasas.*

The modern debate surrounding whether the "foreign sciences" were central or peripheral to medieval Islamic society does not at all echo the arguments of that era. To a large degree, it is a debate that reflects a long-standing, but nonetheless fairly recent battle for the past, emphasizing as it does either the rationalistic or the theocentric side of early Islamic intellectualism, the power of "foreign" (scientific and philosophical) books versus those of the Qu'rānic tradition of writings. In the most reduced terms, the debate concerns the influence of *ʿilm* (knowledge, science) and *falsafah* (philosophy, especially relating to Aristotle) against that of *kalām* (theology, religious devotion through the intellect). What actually existed, however, was a much more complex series of interactions, in which a range of different cultural-intellectual groups battled for territories of influence over "the minds and hearts of the Muslim community" (Sabra 1987, 228). At the least, these groups included: (1) the newly risen *Sufi* mystics; (2) the *Ashab al-ḥadīth,* professing a faith based exclusively on the words and actions of the Prophet and the inscrutability of God and his creation; (3) the *fuqaha* believers in religious law as the basis of society; (4) the *mutakallimūn,* a series of theological schools specializing in *kalām;* and (5) the *Muʿtazilites,* a diverse group of theological thinkers who favored the powers of human reason (as a means to comprehend a rationally ordered universe), moral responsibility, and divine revelation. Relations between these different groups were sometimes combative, sometimes conciliatory, other times overlapping. The most consistent hostility existed between students of *ḥadīth* (sayings of the Prophet) and the *Muʿtazilites.* Yet even here, certain contacts can be seen to exist, for *ḥadīth* scholars were not averse to applying forms of logical exegesis to the Prophet's words, to using various types of rational ordering in *tafsīr* commentary, which came to include the collecting, organizing, and annotating of Qu'rānic passages for the sake of producing a canon of proper Muslim living.

Such rationalist "urges" in early Islamic society, as varied as they were, appear dependent on the full advent of a literate culture, based on the reading, reciting, and study of books (mainly the Quʾrān, of course), the consequent spread of literacy, and the ascent of intellectual discussion and controversy. The power of the Book in Islamic society brought with it a whole panoply of intellectual needs and demands that could not help but spill out in many directions.

Translation was a crucial part of this gathered and deeply founded interest in the written word. That the ʿAbbasids and their court came mainly from the Persian east, where a cosmopolitan culture had existed for centuries, should not be undervalued in terms of the stimulus given to the intellectual efflorescence that took place beginning in the late eighth century. The early ʿAbbasid caliphs, in particular al-Manṣūr (reigned A.D. 754–775), appear to have had a broad vision of the culture they wanted to build. This vision differed considerably from the type of society associated with the Umayyads, whose origins lay in the less cultured regions of Arabia. Indeed, it seems far more in line with the kind of cosmopolitanism and intellectualism pursued by the Sassanid kings of Persia, notably Anūshirwān, at Jundishapur. ʿAbbasid notions were founded on a complex merger between divine sanction, backed by military supremacy, and a vast accumulation of cultural wealth of every type. At the center of this vision lay a "heavenly city," overflowing with worldly knowledge, spiritual wisdom, and harmonious order, reflected (at times) in the display of material riches. Among other aspects, this city would be a nexus collecting within itself the highest achievements of the ancients—the greatness that had been Persia, Syria, India, Greece, Egypt. Following a century of conquest and of terrible civil war, culminating in the brutal extermination of the last Umayyad caliph in 750 along with ninety members of his family, Islam would transform itself into a realm of enlightened civilization. This meant employing the past as a resource, a kind of scattered metropoli of texts. Islam's own past had been troubled and crude, a time of *jāhiliyya*—ignorance, uncultured ways, and the over-reliance on powers of spirit and body at the expense of those of the mind. For the new caliphs, an original promise of ascendancy remained to be fulfilled:

> A literary and philosophic [scientific] culture was needed to present a vision of the universe as a whole, of the role of the state and the ruler in the divine plan and in the functioning of human society, and a concept of the nature of human beings and their destiny in this world and the next. In the complex and heterogeneous society of the Umayyad and ʿAbbasid

> empires this vision was expressed partly in Muslim terms, and partly in literary and artistic [and scientific] terms inherited from the ancient cultures of the Middle East.[4]

Al-Manṣūr was the first of the new caliphs to make this a reality, and there is every indication that he did so for reasons that were as much political and temporal as spiritual and elevated (Kennedy 1981). Early ʿAbbasid society was not merely heterogeneous; after the overthrow of the Umayyads, it was replete with conflict born of diverse and broken loyalties. Potential challenges to the new dynasty issued from a number of quarters. Al-Manṣūr met these challenges with a plan to build the gleaming new center of Islam where the immediate and troubled past would not exist, a new site where a new order might be established from the ground up. He was firmly meticulous in his plans, strategically brutal in eliminating perceived threats, and highly effective in establishing a new type of regime. Part of this construction involved the literature and thought of major groups that had come under the political and (increasingly) linguistic umbrella of Islam. It seems possible that al-Manṣūr chose to do this in order to pacify these groups, to include them overtly in the "riches" of the new city.

The works of earlier peoples were therefore to be an important ingredient in the full rise of Islam. Such was itself a rational and rationalizing vision, and it sought rational materials for its fruition. What was "foreign" in an original sense was to be made "native" in a final sense. The city chosen to fulfill the ʿAbbasid vision was to be Baghdād, a brand new urban creation that would rise, pure and unfettered, out of the Tigris-Euphrates valley, free of any preexisting local traditions and thus able to appropriate and transform the cultural materials from all corners of the new empire. The city that emerged to be Baghdād appears to have been a truly awesome spectacle, if the descriptions offered by the thirteenth-century traveler/author Yāqūt al-Hamāwī in his book *Muʿjam al-Buldān* (Geographical Dictionary) are at all accurate. These descriptions, though late, were drawn from ninth- and tenth-century sources and suggest a metropolis of stunning magnificence, built with geometric precision:

> The city of Baghdad formed two vast semi-circles on the right and left banks of the Tigris, twelve miles in diameter. The numerous suburbs, covered with parks, gardens, villas and beautiful promenades, and plenti-

4. Lapidus 1988 (81). Much of the foregoing discussion is based on information found in this excellent survey, as well as in the following works: Hitti 1989, Hodgson 1974, Kennedy 1981, and Gibb et al. 1960–1994.

fully supplied with rich bazaars, and finely built mosques and baths, stretched for a considerable distance on both sides of the river. In the days of its prosperity the population of Baghdad and its suburbs amounted to over two millions! The palace of the Caliph stood in the midst of a vast park several hours in circumference which beside a menagerie and aviary comprised an enclosure for wild animals reserved for the chase. . . . On this side of the river stood the palaces of the great nobles. Immense streets, none less than forty cubits wide, traversed the city from one end to the other, dividing it into blocks or quarters, each under the control of an overseer or supervisor, who looked after the cleanliness, sanitation and the comfort of the inhabitants. . . .

Every household was plentifully supplied with water at all seasons by the numerous aqueducts which intersected the town; and the streets, gardens and parks were regularly swept and watered, and no refuse was allowed to remain within the walls. An immense square in front of the imperial palace was used for reviews, military inspections, tournaments and races; at night the square and the streets were lighted by lamps. . . . The long wide estrades at the different gates of the city were used by the citizens for gossip and recreation or for watching the flow of travelers and country folk into the capital. The different nationalities in the capital had each a head officer to represent their interests with the government, and to whom the stranger could appeal for counsel or help. . . .

Both sides of the river were for miles fronted by the palaces, kiosks, gardens and parks of the grandees and nobles, marble steps led down to the water's edge, and the scene on the river was animated by thousands of gondolas, decked with little flags, dancing like sunbeams on the water, and carrying the pleasure-seeking Baghdad citizens from one part of the city to the other. Along the wide-stretching quays lay whole fleets at anchor, sea and river craft of all kinds, from the Chinese junk to the old Assyrian raft resting on inflated skins. The mosques of the city were at once vast in size and remarkably beautiful. There were also in Baghdad numerous colleges of learning, hospitals, infirmaries for both sexes, and lunatic asylums. (Davis 1912–1913, 2.365–67)

Little surprise that in such a cosmopolitan nexus, a flowering of intellectual activity would take place. Indeed, the range of such activity that came to center itself in Baghdad between the late eighth and early tenth centuries constitutes one of the great intellectual movements in human history. Beginning with literary (mainly poetic) and folk (e.g., star lore) traditions, sophisticated in certain respects yet also steeped in restricted oral traditions, Arabic-based civilization appropriated the most advanced, technical, and thus challenging textual materials to be found anywhere in the ancient Near East, and did so while simultaneously developing a highly literate, worldly, text-based intellectualism that not only encompassed the Qu'rān but extended deeply into such areas as history, biography, satire, prose-writing in general, grammar and linguistics, geography and travel writing, and all the sciences. Translation thus took place within a larger context of effort and interest that spanned nearly the whole of knowledge, as it then existed.

The progress of translation began, as noted, with works in Syriac, Pahlavi, and, to a lesser degree, Sanskrit—for the most part, with materials that were close at hand and available in languages (especially Syriac and Pahlavi) with which many Arabic-speaking Muslims were familiar. During the ninth century, however, there was a distinct shift in interest toward Greek texts and Greek authors, and both the abundance of works and the selective specificity of their authorship became very impressive (Endress 1982). This seems to reflect a perception on the translators' part that the greater portion of the material with which they were working had originated from such "sages" as Galen, Ptolemy, Euclid, Aristotle, Theon of Alexandria, Porphyry, and Plato, among others. The name and influence of Aristotle, "the Philosopher," became particularly dominant in Islamic philosophy and in many schools of theology, which took up to varying degrees the scholastic methods of exegesis earlier embodied by Syriac patristic literature, also translated (Peters 1968; Rosenthal 1975). Similarly, Ptolemy and the *Almagest* became during the ninth century the acknowledged touchstones of mathematical astronomy, being translated at least five times (twice during the reign of al-Ma'mūn) and continually revised and commented upon (King 1996). In medicine, the correlative "sage" was Galen, whose massive corpus was rendered nearly *in toto,* epitomized, and given its due commentary, by the greatest of all translators during this era, Ḥunayn ibn Isḥāq (A.D. 809–873).

Such assignments to individual intellectual fields are necessary and important, but give little indication of the wider effects many of the translated works acquired. These effects, indeed, extended far beyond the bounds of science. For example, a number of literary figures, not just physicians and astronomers, incorporated the newly translated material into their writing. Among these are to be counted Abū Hanīfa al-Dīnawarī (d. 895), nearly as well known and influential as the famed al-Jāḥiz (usually said to be the first great prose stylist of Islam). Abū Hanīfa was a true polymath, whose works helped shape and advance narrative techniques in several fields of prose. His *Kitāb al-Anwā* (Book of the Rising and Setting of the Stars) was among the most widely read popularizations of its day. Aristotelian scholasticism, meanwhile, in its Syriac versions, was mentioned above with regard to its influence on theology. Beyond this, one sees that Porphyry's epitome of Aristotle in the *Isagoge* was used to introduce Aristotelian concepts of language into early writings on Arabic grammar (Carter 1990). More generally, Aristotle's larger categorization of knowledge into various disciplines was adopted, modified, opposed, or adapted by nearly every domain of study and writing (Peters 1968; Rosenthal 1975). Neoplatonism, an essential part of Porphyry's writings (and of those of many Christian theologians whose works were also translated), was a crucial influence on Sufi mysticism. On a much smaller scale, but equally revealing, are local borrowings from translated works that can be found in the writing of many historians, who sought to combine information from a host of different sources on the past. Indeed, some of the examples in this area reveal how sophisticated Islamic historians were in identifying and employing such sources, for example as shown by the great al-Masʿūdī (A.D. 890–955?), who used Theon of Alexandria's astronomical handbook to Ptolemy's *Almagest* (rendered as a *zij* in Arabic) as a reference work for the dates and names of the "kings of the Greeks" in the *Kitāb at-Tanbīh* (Book of Admonitions) (Dunlop 1971, 108–9).

Historians, in fact, must be counted among the more active translators of the time, particularly from Persian. An instance is offered by the *Taʾrīkh* (History) of al-Yaʿqūbī, an early ninth-century Shīʿite author who seeks to give a historical account of the pre-Islamic and Islamic eras in two books, the second of which contains an extended discussion of the scientific and philosophical works of "the Greeks" and thus offers a rare sampling of what was then known. As noted by one recent scholar, the amount of space granted each Greek author might even be taken as some indication of his reputation at the time: Hippocrates, 22 pages; Ptolemy, 11 pages; Aristotle, 6 pages (Dunlop 1971, 50). This is particularly interesting in view of the far-reaching influence Aristotle was soon to have. In the early and most productive phase of translation, it would seem to have been medical and astronomical works that were favored. More importantly, al-Yaʿqūbī is our sign that a full awareness then existed, and was often expressed, regarding the historical significance of the translation episode—indeed, of translation itself—to Arabic culture.

If positive endorsements of "foreign learning" reveal something of their daily presence, it is in the more negative evaluations and outright rejections that one gets a glimmer of how prevalent, even on the most mundane level, this influence really was. One of the best examples is offered by the prose writer and conservative scholar Ibn Qutayba (A.D.

828–889), who criticizes even al-Jahiz for his presumably tepid critique of "the arguments of the Christians." In point of fact, al-Jahiz was fairly hardened toward the Byzantines but neutral on the subject of the new translations, even slightly satirical of their perceived importance. In one work, he gives a general list of these texts, without comment, as including the more scientific writings of Aristotle (*Organon, De Generatione et Corruptione,* and the *Meteora*), the *Almagest* of Ptolemy, plus various works by Euclid, Galen, Democritus, Hippocrates, and Plato as well (Dunlop 1971, 50). Ibn Qutayba, however, has no patience for such neutrality. To his mind, the "foreign sciences" have poisoned much of Arab intellectual society. In *Kitāb Adab al-Kātib* (Book on Education of a Secretary), he takes to task any appreciation of these sciences, posing them directly against the teachings of Islam, but in a highly suggestive fashion. As described by Dunlop (1971, 51), Ibn Qutayba

> first criticizes the modern *kātib* who is content if he can write a fair hand and turn a poem in praise of a singing girl or in description of a wine glass, and prides himself on knowing enough astrology to cast a horoscope and a little logic. This kind of man . . . finds it tedious to examine the science of the Qur'an and the accounts of the Prophet, and equally the sciences of the Arabs and their language and literature. . . . He is the sworn enemy of all such studies, and turns away from them to a science of which there exist purveyors to him and his like and but few opponents. . . . When the inexperienced and the young and heedless hear [this man] speaking of "generation and corruption" (*al-kawn wa'l-fasād*), "physical ausculation" (*sam' al-kiyān*) . . . "quality" (*al-kayfiyya*) and "quantity" (*al-kamiyya*), "time" (*az-zamān*), "demonstration" (*ad-dalīl*), and "composite propositions" (*al akhbār al-mu'allafa*), they think that what they have heard is wonderful and imagine that under these expressions are all kinds of profit and all kinds of subtle meaning. . . .

It would be very difficult indeed to find a better statement of how the "foreign sciences" fared at this time in Baghdad or Damascus. Ibn Qutayba tells us these sciences are part of the ordinary equipment of the average 'Abbasid government clerk (*katib*), whom he characterizes as an intellectual dandy. His description, however, is deliciously ironic. It reveals to us that such a man was not only worldly and fashionable, but also widely educated and accomplished, an individual who revealed a combination of influences from many lands—skilled in calligraphy and poetry, both dependent on translations from Persian literature; knowledgeable in Indian, Persian, and Greek astrology/astronomy (one should note that casting a horoscope was no small thing); more than a little acquainted with the writings of Porphyry and Aristotle.

Ibn Qutayba's apparent criticism of "foreign" learning, meanwhile, is focused mainly on Greek knowledge and its "theoretical" character. Elsewhere, in a famous passage, he holds up as a positive example the more practical skills the Persians demanded of a civil servant. This learning, he notes, covers irrigation, measurement, basic astronomy, surveying, geometry, and accounting (Lecomte 1965). Inasmuch as this knowledge had been a requisite part of Sassanian administrative training for centuries, it had long before incorporated many Greek elements (e.g., Euclidean geometry), since at least the time of Anūshirwāan. Viewed from this angle, therefore, Ibn Qutayba's commentary appears more of a plea for favoring certain types of "foreign" learning. His ninth-century *kātib,* if he brings together even a few of the "vices" noted above with some of the favored skills just mentioned, was truly an accomplished human being. When we consider that a mere two centuries had elapsed since the time when the average Muslim leader was largely or wholly illiterate and attached above all to the holy mission of conquest, we begin to understand the tremendous changes that had taken place. Indeed, the *kātib* is such an established presence that he can be accused of decadence. Ibn Qutayba also indicates that a good deal of Greek philosophical terminology had by this time been fully appropriated into Arabic. One notes that the terms listed are not transliterations from Greek or Persian, as was often the case during the earlier phase of translation. Through the window provided by Ibn Qutayba's invective, we get a glimpse of just how common and even traditional the "foreign sciences" had become within a sizable sector of Islamic intellectual culture.

The term "foreign," therefore, had become an ironic designation by this time, and this in several ways. True, the sciences of astronomy/astrology, mathematics, medicine, philosophy, and so forth were not allowed entrance into the mosque-centered *madrasa* educational system, which took children from about the age of six and instructed them in the words and meanings of the Qu'rān. The *kātib* Ibn Qutayba describes, or anyone else desiring similar knowledge, acquired his learning through a system of private tutoring, which had sprung up during the eighth century and flourished thereafter. This system existed alongside, and even prior to, that of the *madrasa,* being an offshoot of the growing bookstore culture in the larger cities from the late eighth century on, as well as the "literary salons" supported by various wealthy nobles and, often enough, by the caliph himself (Tibawi 1954; Nakosteen 1964). By any account, despite whatever criticism it may have attracted, secular learning was a central activity in Islamic intellectual culture. Applied to the substance of its teachings, the term "foreign" thus takes on ideological significance. On the other side of the coin, it might be emphasized that such "native sciences" as *ḥadīth*, religious poetry, Qu'rānic jurisprudence, and philology (again based on Qu'rānic study) were all made possible or greatly advanced by non-Arabic influences—biblical styles of exegesis, for instance, Syriac poetry and Persian literature, Greek models of language analysis, all of which contributed fundamentally to the founding of a highly literate, text-based culture without which *kalām* or *ḥadīth* would barely have been possible (Peters 1968; Nakosteen 1964).

Persian literature, to take only one of these examples, had no less far-reaching an effect upon Arabic learning than did Greek, Syrian, and Indian science. If the latter (especially the Greek element) have attracted more attention in scholarly circles, this is perhaps because of the systematic, often centralized policies at work in the translation of scientific works—in this realm, at least, it seems that 'Abbasid caliphs picked up where the great Library at Alexandria had left off. But this has meant, in turn, that Persian literary influences have been too often viewed as more or less "internal" to early Islamic civilization. In fact, within the larger space of translation, what Persian texts may lack in total quantity, they more than make up for in terms of broadness of material, exceeding perhaps that of any other language. We have already noted that some of the earliest astronomical works rendered into Arabic came from Persian sources, and that this included a version of Ptolemy. It has also been mentioned that Greek literary works—fiction, prose, poetry, history, biography, the entire larger humanistic canon—were only very sparsely translated into Arabic, and this includes even the heavens of that most famous of Hellenistic authors, Aratus.[5] Such subjects, instead, were generously covered by translations and adaptations of Persian works, which, by the ninth century, had the effect of introducing, alongside the simpler, more terse forms of expression typical of classical Arabic, a much more courtly style, "polished and elegant . . . rich in elaborate similes and replete with rhymes"[6] and thus able to render a

5. On this subject, see Honigmann 1950. This brief article, based on no small degree of scholarly detective work, can be considered an indication of how unimportant, generally speaking, was this most famous of Hellenistic science popularizations.

6. Hitti 1989 (403). Like some scholars, Hitti sees the 'Abbasid period as one of intellectual decline, mainly due to this idea of a gilded literary style. In reality, however, a diversity of specific styles existed, each adapted to its genre of writing. This can be easily seen in the works of a single author such as al-Jāḥiẓ, whose satires (e.g., the *Kitāb al-Bukhala,* or Book of Misers) vary from being terse, pointed, and incisive accounts of individual actions to more windy and even burlesque disquisitions on the state of society generally.

greater variety of stories, tales, events, and contemplations. Writers like Ibn al-Muqaffa', Abu al-Fārāj, and al-Jāḥiẓ made abundant use of Persian source material as models of style, form, and content. Al-Muqaffa' in particular, through his own translations of Persian histories on Sassanid imperial traditions, helped establish new genres of writing on the ethics and machinations of statecraft, known as "Mirrors for Princes" (Bosworth 1990). Even (or perhaps especially) so well known a work as the *Thousand and One Nights* (*Alf Laylah wa-Layhah*) represents an elaborate adaptation of a much older Persian book, the *Hazar Afsana* (Thousand Tales), which in turn contains a number of tales of Indian origin (Horovitz 1927). The history of this work, in fact, seems an adequate metaphor for Arabic textual culture in its early phases: accumulating contributions from Indian, Greek, Hebrew, Egyptian, Syrian, and other sources, thus growing into a creation at once supple and expansive, able to reconstitute in varied fashion literary materials brought in from nearly every corner of a vast empire.

In the end, therefore, whether applied to literature or to the sciences, "foreign" was among the most domestic of terms used by Arabic-speaking intellectuals. One even finds so great an historian of the early medieval period as ibn-Khaldūn devoting an entire chapter in his classic *Muqaddīmah* (Prolegomena) to the simple fact that "Most of the Learned Men in Islam were non-Arabians."

Sociocultural Aspects: The Idea of Mobility

Islamic court society under the 'Abbasids recognized in the multilingual communities under its purview an extremely valuable resource. Scholars, astronomers/astrologers, physicians, artists, poets, and prose writers constituted a type of wealth, one with the ability to broaden the caliphate's own powers of influence and, more generally, to elevate to new levels of magnificence traditions of patronage that formed a fairly venerable tradition in the Near East. The Sassanid dynasty in Persia, which had long supported human achievements of the mind as a mirror to its own worldly splendor, was perhaps the more immediate predecessor of the 'Abbasid caliphs, who were themselves Persian by origin. The 'Abbasids, however, had a far greater spatial-cultural domain from which to draw their potential intellectual riches than did any previous ruling body. From Alexandria to Antioch, from Nisibis to Jundishapur, rulers such as al-Manṣūr and al-Ma'mūn had it within their power to select and collect scholars and texts from cities that had been for centuries individual nodes of intellectual activity, each with its own history of efflorescence (and possibly decay), its own libraries, educational institutions, and traditions—in short, its own technology of the written word.

Such "wealth" became available, even visible, because of certain practical realities. Though space is lacking here to enumerate these in detail, a few general points might be made. Above all, there existed in early Islamic society conditions favoring the mobility and circulation of textual goods and their bondsmen, scholars. In addition to the umbrella of empire, which allowed for increased flow of men and material, evidenced for example in the movement of commercial caravans and armies, trade goods and any spoils brought back from campaigns (including manuscripts), the ʿAbbasids established or extended certain forms of bureaucratic organization favoring mobility, to significant effect. Examples include the vast system of taxation that was set up to ensure the ready transfer of capital from all provinces to Baghdad. Another was the postal system, initiated in Umayyad times but greatly strengthened and enlarged under Hārūn al-Rashīd, who used it as a basis for espionage as well as for encouraging intercommunication between different portions of the empire. Existing roads were improved, new highways built, way-stations erected, and maps or distance logs produced centered on Baghdād, the new capital. All of which facilitated travel and trade to no small degree.

This appears to have aided, in turn, an important aspect of ʿAbbasid intellectual culture: the *Lehrreise,* or pilgrimage of study, which became "part of the regular preliminaries to a successful career as a member of the learned class" (Dunlop 1971, 99). Such a journey normally included visits to the main intellectual centers of the empire in Egypt (Alexandria), Syria, Iran, Iraq, and might also include areas as far disparate as East Africa, Samarqand on the Chinese border, and Ceylon. Both scholars and students spent much of their early lives embarked on these journeys, moving from one city to another (Netton 1992, 31). The historian al-Masʿūdī , as an example, in a search for varieties of experience and knowledge he could then turn into books, traversed nearly the entire limits of eastern Islam in the early tenth century, from Egypt to Armenia, then to Indonesia and back again. While historians, geographers, and poets (who sought to collect material from various traditions) were perhaps the most ardent travelers and travel writers, students of *ḥadīth* and of the "foreign sciences" were no less prone to follow "the grand tour." Partly, this was due to the fact that the highest learning was offered not by schools, such as had existed in Athens or Alexandria, but by individual masters scattered across the empire. Such men were accorded considerable renown; in the case of those such as al-

Fārābī (A.D. 870–950) and al-Kindī (A.D. 800–870?), known as "the second master" (after Aristotle) and "the teacher," varieties of legend came to surround their fame.[7] In a sense, such masters, in their lifetimes, represented a web of order and promise in the midst of the political instabilities that characterized the ʿAbbasid period. The image of the famed scholar, drenched in the sunrise nobility of his quest for knowledge, at times reduced to various sacrifices (poverty, prison) for the sake of his search, was a common one among the learned in early medieval Islam, and is to be contrasted (though not entirely) with the more lithic dignity of the holy warrior, prized by the early Umayyads.

In the year 765, suffering from a stomach ailment his Baghdād physicians had been unable to cure, the caliph al-Manṣūr sent a message to Jundishapur, summoning the head of the hospital and teaching center there, Jurjīs ibn Bakhtīshūʿ, to come and attend to his illness. Jundishapur lay nearly three hundred miles to the southeast, in northern Khuzistan. Historians regularly note that ibn Bakhtīshūʿ was quickly brought before the caliph, recognized the nature of the problem (indigestion?), and alleviated it by means of a medicine he had made up for the purpose. Not discussed is the fact that this physician was able to make such an extended journey over rugged terrain in a very short amount of time—all the more abbreviated, no doubt, by the urgency of the caliph's summons and the consequences to be expected from extending his sufferings. As the embodiment of the Jundishapur academy, with its centuries of accumulated fame and expertise, ibn Bakhtīshūʿ was transferred across an enormous tract of desert, hill, and canyon territory within a matter of weeks, abruptly resettled in Baghdād, there to become the new court physician and to found a family line of royal doctors that continued to exercise medical authority for more than two centuries.

The episode is revealing, for it shows that the mobility of knowledge, whether personified in single individuals or in manuscripts, was well established in the Near East by the time of the Muslim conquerors and only grew thereafter. The "wisdom" of any particular people, whether of the past or the present, could become the property of another. The only final boundaries to such appropriation within an empire like that of Islam were the preservation of sources and language. Given the "wealth" of scribal traditions and linguistic expertise within Islam's major cities, this was no real barrier at all. Indeed, two of the most im-

7. An interesting example here is the recasting of al-Fārābī, known (among many other things) for a great work on music, in the image of Orpheus. See Netton 1992 (5–6).

portant mobilizing forces for knowledge during this period were the introduction of paper making and the spread of Arabic as the language of learning. The city of Samarqand, center of Chinese paper manufacturing, was taken by Muslim armies in A.D. 704; before the end of the century, Chinese paper-makers had been brought to Baghdād, where a mill was then established (Hitti 1970, 414–15). Use of cheap, locally produced paper, which entirely replaced papyrus and parchment before A.D. 900, allowed for more widespread duplication and distribution of existing manuscripts, thus bringing about an increase in the demand for scribal labor. This helped, in turn, make possible the creation of many large public and private libraries, a thriving book trade, and thus the spread of textual culture generally. No better evidence for this can be found than the famous *Fihrist* (lit. "catalogue" or "index," mainly of technical works) by Muhammad ibn Isḥāq al-Nadīm (d. A.D. 975), without doubt the most important single work in existence on Arabic writing from the seventh through the tenth centuries. The *Fihrist* is largely an annotated list of works and authors in all branches of knowledge, reaching over a thousand entries in number and including translations from Persian, Greek, Coptic, Syriac, Hebrew, and Hindu. It has been often suggested that al-Nadīm's text represents an example of a learned book dealer's catalogue (this was indeed his profession)—an idea that certainly gives one pause regarding the enormous potential diversity and general availability of books in Islamic urban society (Nakosteen 1964).

The triumph of Arabic, meanwhile, took place in complex fashion, on at least two major levels, and was by no means an overnight process. In the vernacular sphere, its victory required a long time. More than a few local cultures within the empire, Syrian among them, rejected for centuries the eclipse of their native tongues, which had enormous literary momentum behind them. However, once the textual dimension of these cultures—i.e., their poetry, history, and holy works—began to be written regularly in Arabic, this hold on the past began to loosen and gradually give way, though not altogether. On the plane of intellectual activity, however, the ascent of Arabic was rapid and sure. Due to consistent patronage of the arts and, eventually, the sciences too under the later Umayyad and the early ʿAbbasid caliphs and their wealthy courts, Arabic became the speech of both religious and secular intellectualism throughout most of the empire by as early as the late eighth century (Carter 1990). This was hugely aided by, and in fact largely stood upon, political and commercial realities, since the language of government, civil authority, taxation, and trade was Arabic, which quickly replaced Greek and Syriac in the Near East.

Linguistically, however, it was the greater effort of translation that enriched and broadened the language in a host of ways, for example to include new forms of syntax and greater grammatical flexibility on the basis of Syriac, Persian, and possibly Greek models and to introduce entire new realms of vocabulary in every major discipline (Rosenthal 1975, 1–14; Endress 1989; Carter 1990). The effect of these changes in making Arabic a *lingua franca* of intellectualism—connected to political forms of power—cannot be overestimated. It is no exaggeration to say that the language underwent an evolutionary growth as swift and far-reaching as that of English between 1400 and 1600. Within two hundred years of the Prophet's death, at the end of al-Ma'mūn's reign in A.D. 833, literary and scientific works were being produced in Arabic by Turks, Egyptians, Syrians, Sabians, Persians, Armenians, and Nestorian Christians alike. Even as political unity began to crumble, with the empire breaking apart from the ninth century on into a range of regional principalities and dynasties, strong ties of social unity were maintained through Arabic, the medium of the Qu'rān, Islamic religious practice, and intellectual pursuit (Hitti 1989). Such a triumph among scholars did not occur smoothly: a distinct rebellion took place beginning in the late eighth century, centered in Persia, proclaiming the superiority of ethnic non-Arabs in matters literary and artistic and drawing its name from the Qu'rān itself (Shu 'ubiyah, "belonging to the non-Arab peoples"). Yet it is indicative of the outcome that the debates, especially over literature, took place *in Arabic* and yielded in their wake some of the best examples of *adab* (roughly, "belles lettres") from that time.

To all this, one should perhaps add the encouragements toward knowledge contained within the Qu'rān itself. There are a number of often-quoted statements along these lines, and while their precise meaning is always open to interpretation, it is exactly this aspect that may have given them the suggestive power they seem to have attained. "Seek knowledge from the cradle to the grave . . . even as far as China," counsels one such *ḥadīth*. Another commands: "The quest for knowledge is obligatory for every Muslim." Still another: "Verily, the men of knowledge are the inheritors of the prophets."[8] And even "Obtain knowledge; its possessor can distinguish right from wrong; it shows the way to Heaven; it befriends us in the desert and in solitude, and when we are friendless, it is our guide to happiness. . . . The scholar's ink is holier than the martyr's blood." What seems interesting about these (and

8. These and other sayings of similar type have been collected in Nasr 1976 (6–8).

other) statements is their emphasis on the active character of learning: acquiring knowledge (*'ilm*) is integral to the holy search for self-betterment and is by no means confined to pedagogic situations (Rosenthal 1970). Nothing within the religious basis of Islam stood openly opposed to intellectual work, whether this involved Qu'rānic commentary or the pursuit of texts written by sages (*hukuma*) of the non-Islamic past. Indeed, in this regard, one might well consider the words of the eleventh century intellectual historian Sa'id al-Andalusi, whose *Kitāb Tabaqāt al-Umam* (Book of the Categories of Nations), widely read among religious and secular scholars alike, provides a survey of contributions to human knowledge by all major peoples known to Islam:

> The *tabaqāt* [group of nations] that cultivated science is comprised of the elites that Allah has chosen from among His creatures. They have focused their attention to achieve the purity of soul that governs the human race and straightens its nature [behavior]. . . . For example, in the fine construction and the perfection of shapes, the bees prove their superiority in building the cells for storing their food. And the careful spiders construct the strands of their homes to harmonize with the circles that they intersect. . . . For the noble reason and the honorable quality that demonstrate human values, [and thus] the nobility of human virtues . . . scientists came into being. (al-Andalusi 1991, 9)

Conditions of mobility favored translation, and translation enhanced the mobilization of knowledge. Once begun in earnest, the exchange, appropriation, and confiscation of "foreign" texts by Arabic-speaking scholars was a self-enforcing process.

The Translators and their Texts

What are some of the basic facts regarding this era of translation? What texts, traditions, and translators were involved, for example, particularly with regard to astronomy? Abundant recent scholarship in this area allows one to answer such questions with an introductory eye. It has been common to characterize the translation movement as taking place in three major phases (Peters 1968), the first beginning in the late eighth century, under the caliphate of al-Manṣūr (A.D. 754–775) and al-Rashīd (A.D. 786–809), and reaching a peak in the first several decades of the ninth century with al-Ma'mūn (A.D. 813–833), the most active patron of this activity among all the 'Abbasid rulers. Indian and Persian works were very important early on, but texts in Syriac soon came to dominate, an increasing number of which represented translations from late copies of Greek scientific texts. During the second phase of effort, reaching from the mid-ninth to the tenth century, there was a pronounced tendency to return to the "original" Greek texts (which had themselves undergone centuries of copying), to translate these first into Syriac and then into Arabic, to revise and correct already existing translations, and to begin the task of standardizing technical nomenclature as a means of consciously ensuring higher-quality versions. A third phase of more dispersed activity, extending roughly from the eleventh to the thirteenth century, involved the editing of selected works, their recombination into new textual forms, and retranslation of a few major texts (or portions thereof). Historically speaking, the first two phases were by far the most important, with the first phase yielding versions that were revised, corrected, and adapted by second-phase translators.

This general picture, emerging from scholarly studies during the past century or more, offers a useful (if somewhat crude) approximation. In reality, however, the situation was considerably more complicated. In some cases, true enough, a tracing of individual texts—the *Almagest,* for example—will broadly support the three-stage interpretation. But this is a notable exception. Far more often, no such neat partitioning of effort is supported by the relevant details. With regard to Euclid's *Elements* or various mathematical texts by Archimedes, for instance, multiple translations were made throughout the period spanning the ninth and tenth centuries, with later versions sometimes exhibiting no improvement over (or knowledge of) earlier ones, no patterns of translation technique evident, and no clear standards being consistently applied (Clagett 1964–1976, vol. 1). Moreover, the nearly exhaustive amount of specific information provided by Endress (1982) strongly militates against any easy chronology for the sciences in general. For the sake of simplicity (and to avoid undue historiographic entanglements), however, and because the *Almagest* forms one of the basic texts of our discussion, the three-phase scheme will be loosely followed here for astronomy, yet with the proviso that this be considered a short-cut until that time when the required scholarly work has been done to provide a more complete characterization.

As to the names of the translators, a fair number have been preserved, at least from the late eighth century on, either by survival of their actual texts (rare, and usually in much later copies) or by citation in much later works, notably the *Fihrist.* The earliest texts recorded appear to have been Greek alchemical writings, translated into Arabic by one Stephen at the request of an early Umayyad prince (Khālid ibn Yazīd) in about A.D. 680 (O'Leary 1949; Rosenthal 1975; Endress

1982). In the area of astronomical translation, the first names of significance are undoubtedly lost; one begins, by default, with those of the late eighth century such as al-Fāzārī, whom we have already discussed (see previous chapter), and Abū Yaḥyā al-Biṭrīq, who translated Ptolemy's *Tetrabiblos*. With the rise to power of the ʿAbbasids, of Persian background, astrology became important and many texts of this sort were rendered into Arabic under the auspices of the second ʿAbbasid caliph, al-Manṣūr. This ruler appears to have followed in the footsteps of the Persian kings of former centuries, who were ardent patrons of astrological and medical practice. Most of the major texts of this type circulating in the Near East in several languages were brought into Arabic by the early reign of al-Maʾmūn. Thereafter, however, the major emphasis on study of the skies passed to astronomy proper, specifically theoretical and observational astronomy, where it was the tradition of the *zij* (with its mixture of Hindu, Persian, and Greek elements) and the translations of Ptolemy that mattered most.

Above all, interest came to focus upon the *Almagest*. No fewer than five individual translations of this work were in existence by the end of the ninth century, with all of these still available three hundred years later when Ibn al-Salah wrote a critique and correction of Ptolemy's star catalogue (King 1996). The oldest version was in Syriac, and it remains unknown who produced it and exactly when. The first rendition in Arabic, produced around A.D. 800, is also anonymous. Al-Nadīm in the *Fihrist* states that Yaḥya ibn Barmak, son of a Persian court official under al-Manṣūr and widely known for his dedication to learning, became interested in Ptolemy's masterwork and had it sent to the renowned Bait al-Ḥikma ("House of Wisdom"), where "they made sure [of its meaning] and persevered in making it accurate, after having summoned the best translators" (Nadīm 1970, 639). In A.D. 827–828, a second Arabic translation was made by al-Ḥajjāj ibn Maṭar (flourished A.D. 786–833) in collaboration with one Sergius, son of Elias, presumably a Nestorian Christian, thus suggesting that a version in Syriac was used as an intermediary. Following this, another Nestorian, Ḥunayn ibn Isḥāq (A.D. 830–910), the most productive and remarkable of all Arabic translators, created a version of the *Almagest*, evidently somewhat reduced in excellence from his more numerous renderings of Aristotle and Galen (Ḥunayn was himself a physician). One can say this, since it appears that Ḥunayn's version was compared unfavorably with that of al-Ḥajjāj and was revised accordingly by the last translator of this period, Thābit ibn Qurra (A.D. 836–901), a non-Muslim from Ḥarrān, who also translated and revised works by Euclid, Galen, and others. One final known ver-

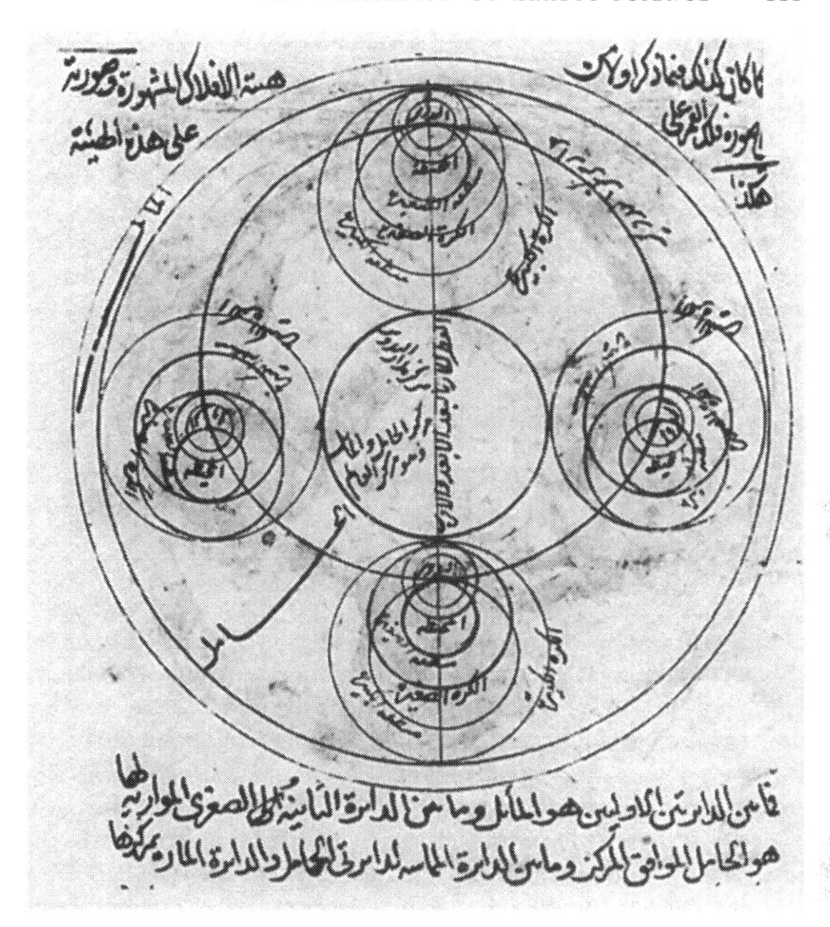

Figure 6. Diagram from a fifteenth-century manuscript illustrating an important Arabic emendation to the Ptolemaic system of planetary motion. This "correction," which improves prediction of planetary positions, was derived by the thirteenth-century Persian thinker Naṣīr al-Dīn al-Ṭūsī and is known as the "Ṭūsī-couple." The diagram shown depicts different stages of the lunar epicycle for half a synodic month. University Library, Cambridge, Add. 3589, fol. 155r.

sion of the *Almagest* produced during the medieval period was not a translation per se but instead a redaction by the astronomer and mathematician Naṣīr al-Dīn al-Ṭūsī (1201–1274), who used Thābit ibn Qurra's corrected translation of Ḥunayn's edition as his source (fig. 6; Kunitzsch 1974). By the time al-Ṭūsī came to it, Ptolemy's text had thus gone through a number of linguistic interpretations. Moreover, it had become the centerpoint to a considerable tradition of exegesis, involving not only commentary but commentaries on commentaries, as well as continual updating and correction of its details (fig. 7; see also fig. 6), clarification of its more difficult sections, shifting of individual sentences, and other internal changes (Kunitzsch 1974).

Regarding these translators and the texts they worked on, there are certain essential points to consider. First, the translations into Arabic took place as an essential ingredient in a broad movement that transformed Islam, generally speaking, into an extremely rich, intensely active textual culture—engaged not just in study and teaching, but in writing specifically, in the production of an ever greater number and variety of manuscripts and the accumulation of these works into sometimes massive libraries (fig. 8).[9] The translations added vastly to this movement; literally thousands of new texts entered Arabic in the eighth and ninth centuries (more than a hundred each from Galen and Aristotle). But they were also dependent upon it, upon the growing demand for intellectual materials, the spread of literacy, the market for books among the aristocracy.

Second, the Arabic versions of scientific works produced during the first several centuries of effort very often represented *second-hand* translations. This means that they were rendered from Syriac or (less often) Pahlavi texts that were themselves translations of earlier works, mainly Greek and Indian. There are complexities to be considered here. One needs to distinguish, for example, the rendering into Arabic of already existing Syriac works (e.g., the fifth-century translations of Sergius Reshaina) from the conscious use of Syriac as an intermediary language, such as was done by Ḥunayn ibn Isḥāq and his students in the ninth century. The two processes are very different, procedurally and epistemologically, particularly with regard to the notion of the "original" text. Removed (indeed, nonexistent) in the first case, the Greek "original" is immediately present in the second, both as an object of erasure and as a source of continual reference. Consciously or otherwise, translators like Ḥunayn recognized, in a historical sense, the ideal status of Syriac, which had been deeply influenced by (and contained many adapted elements of) late classical Greek and which, in turn, had itself been a formative influence on Arabic during its transformation into a written, urbane, and sophisticated language (Carter 1990).

Third, the great majority of active translators were neither Muslims

9. Brief but excellent discussions of Islāmic libraries can be found in Nakosteen 1964, Shalaby 1954, and Thompson 1939. In general, the enormous loss to history from the destruction of the early libraries of Islām—due to varied causes, but, in Baghdād, the center of a flourishing book trade for centuries, a result of systematic burning by the Mongols—would seem to equal or surpass that of the great library at Alexandria.

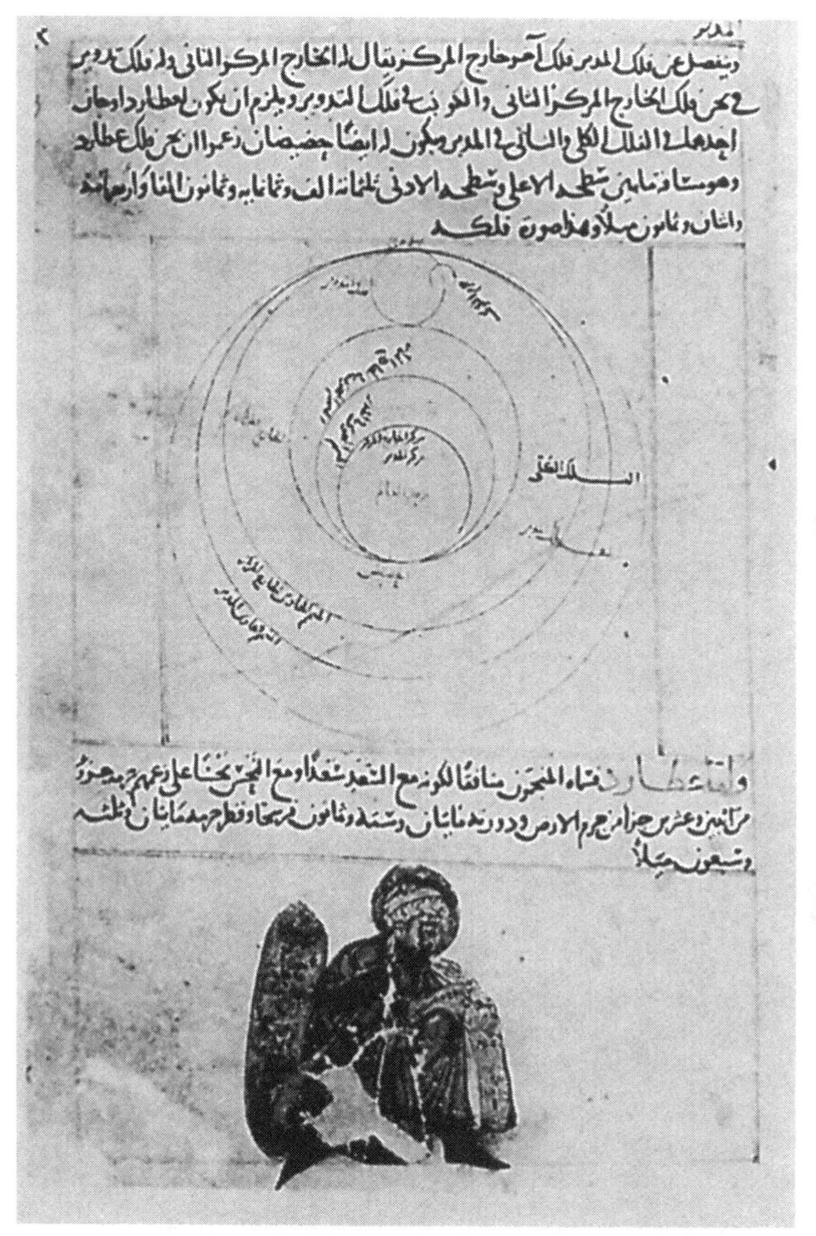

Figure 7. Page from a thirteenth-century Arabic manuscript depicting one of Ptolemy's physical models from his work *Planetary Hypotheses*, here related to the motion of Mercury. The model has been adopted by the author, al-Qazwīnī, to suggest the existence of a series of spheres (which Ptolemy did not delineate), and is included in a book titled *Wonders of Creation*. Bayerische Staatsbibliotek, Munich, MS Arab 464, fol. 13r.

Figure 8. Depiction of a medieval Arabic library, from a thirteenth-century manuscript. Books were stored flat side down, in a large number of niches, possibly for purposes of classification and preservation or because they were often unbound. Bibliothèque Nationale, Paris, MS arab 5847, fol. 5v.

nor Arabs, but instead Nestorian (less often Jacobite) clerics of the Eastern Christian Church, most of whom had been educated in towns and cities where the remnants of classical learning existed or even flourished. It is a crucial point that bears repeating: those translators who brought the natural sciences into Arabic between the seventh and tenth centuries were people for whom Arabic itself was a second language. In cultural-linguistic terms, science, and a considerable portion of prose scholarship and writing in general, were established in this language by nonnative speakers. These translators, furthermore, were anything but passive conduits. They often chose the texts they wanted to work on, stretched or deformed the existing Arabic language to their specific needs, and built the central vocabularies of a dozen fields by choosing nomenclatural equivalents to Syriac or Greek words. Through a mixture of mercenary intentions and the desire to keep alive (indeed to nurture) those intellectual traditions they held most dear, these translators helped influence directly a shift of interest among court patrons toward Greek sources above all others. In equal measure, the success of their efforts greatly increased the need for any student interested in "foreign wisdom" to learn Arabic, while commensurately reducing any demand (or desire) among intellectuals to learn the languages of the source texts, like Greek, Persian, and Hindi. The result was thus a specifically Arabic textual culture. The translators, meanwhile, were employed as personal tutors, physicians, and advisors within ʿAbbasid court society and were sometimes paid handsome sums for their work. According to the *Fihrist,* the Banū-Mūsā brothers, who were both important patrons and active scholars of the new learning, paid monthly sums of five hundred dinars in gold (roughly a man's weight) to Ḥunayn and others for their translation work alone. Offered the power to build and staff libraries, set up observatories, effect changes in the existing calendrical system, and to be included among the "wise men" of the age, consulted and questioned (and sometimes imprisoned) by the caliphs themselves, the translators acquired considerable legitimacy and even renown.

This, then, also calls up the question of motive: was utilitarian interest, as so often stated, the guiding principal for the translators and their patrons? The issue cannot be easily settled and surely deserves reexamination. In some part, it seems to turn upon the degree of independence allowed the translators themselves. Taken as a whole, the evidence suggests that, in the earliest stages, the choosing of works for translation was almost entirely under the conscious control of rulers and the court. Astrology, with its abilities of prediction; alchemy, with its promises of unlimited wealth and control over the secrets of nature; medicine, alleviator of suffering and extender of potency and life: these were areas of knowledge that promised enhanced power and immediate benefit. They were the areas selected for the sporadic translations made under the Umayyad caliphs and the earliest of the ʿAbbasids. With the reign of al-Manṣūr, however, things begin to change. A more systematic interest in the knowledge of other peoples began; translators were given more freedom, as experts, to search out and choose the best examples of "ancient wisdom" they could find. This tendency reached a peak in the early to mid-ninth century under al-Ma'mūn and such wealthy patrons as the brothers Banū-Mūsā, who sent out "missions" of translators, including the great Ḥunayn ibn Isḥāq and Thābit ibn Qurrah, to collect works from the remaining libraries of antiquity. At this stage, judging from the existing reports, some translators had achieved such renown that they were often given almost complete liberty to select their source material. In part because of their Nestorian background, those like Ḥunayn seem to have favored the texts included in the syllabi of the old Greek schools and the famed academy at Jundishapur. While the Arabic translations were "ultimately intended for caliphs, government officials, highly literate theologians, and educated laymen" (Rosenthal 1970, 8), they were also—and this point, I believe, has been overlooked—collected sometimes by the translators themselves, as the writings of Ḥunayn make clear (Bergsträsser 1913). These men must have assembled sizable personal libraries. Certainly their taste in source material was broad indeed, typically ranging across philosophy, mathematics, astronomy, optics, mechanics, and other subjects.

In short, an early proclivity toward immediately useful material soon evolved to include the most abstract, theoretical, and intellectually challenging sources. As already noted, the *Almagest,* the most demanding work of classical Greek science, was translated no less than four times between A.D. 800 and about 890 (Kunitzsch 1974, 1996). If only the most practical portions of this work had been preferred, the *Handy Tables* alone, or the epitome and commentary by Theon of Alexandria, would have easily sufficed. Instead, a greater vision, acquisitive to be sure, was at work. Only if one expands the notion of "practicality" to include an idea such as "enriching Islam with the 'wisdom of the ancients'"—that is, "wisdom" as a combined source of material, intellectual, and spiritual power—can the designation perhaps be retained. In reality, there is evidence that this was the case. The books of the ancients, especially the Greeks, seem often to have been viewed as "riches" in a very concrete way. One of the best examples is offered by al-Nadīm, who reports in the *Fihrist* a story told by Abū Isḥāq ibn Shahrām of his recent mission to the Byzantine emperor Basil II sometime around A.D. 980:

> There is in the Byzantine country a temple of ancient construction. It has a portal larger than any other ever seen with both gates made of iron. In

> ancient times . . . the Greeks exalted this [temple], praying and sacrificing in it. I [Ibn Shahram] asked the emperor of the Byzantines to open it . . . but this was impossible, as it had been locked since the time that the Byzantines had become Christians. I continued, however, to be courteous to him, to correspond with him, and also to entreat him in conversation during my stay at his court. [The emperor finally] agreed to open it and, behold, this building was made of marble and great colored stones, upon which there were many beautiful inscriptions and sculptures. I have never seen or heard of anything equaling its vastness and beauty. In this temple there were numerous camel loads of ancient books. [There must have been] a thousand camel loads. Some of these were worn and some in normal condition. Others were eaten by insects. . . . I saw there gold offering utensils and other rare things. After my exit, the door was locked, causing me to feel embarrassed because of the favor shown me. (Nadīm 1970, 2.585–86)

Books could thus be the stuff of legend no less than—and in the same breath as—gold, jewels, and other "rare things" of untold value.

With regard to astronomy in particular, it should first be said that study and discourse were largely divided between a preexisting "folk astronomy" (an unfortunate term), developed in the deserts of Arabia proper and concerned mainly with time-keeping and weather predictions (especially rain signs), and a more mathematical-observational astronomy derived from translation of Indian, Persian, and Greek works (King 1993, 1996). While there was some overlap between these two species of astronomical thought, they did not merge, at least not on any consistent basis. Star narratives came to be associated, historically speaking, with the Bedouin Arabs and the origins of Islam and were favored by scholars involved in sacred studies such as *fuqaha* (religious law). Mathematical-observational astronomy—by which is meant an astronomy based on sophisticated geometric methods, which accepted the classical Greek version of the constellations, and which soon came to involve recorded observations of celestial motions—was a secular pursuit. The primary source texts here appear to have included: (1) the Indian and Persian works described in chapter 2, notably the *Sindhind* and *zij-Shahriyaran;* (2) Ptolemy's *Handy Tables* and, above all, the *Almagest;* (3) epitomes and commentaries on Ptolemy's astronomical works by Theon of Alexandria (fourth century A.D.), mainly in Syriac translation; (4) adaptations and summaries of Theon, such as those by Severus Sebokht; (5) various treatises by Greek authors in the "small astronomy collection" (Aristarchus of Samos, Menelaus, Theodosius, Autolycus), which served as an introduction to study of the *Almagest* and which was available in a mixture of Syriac and later Greek versions; and (6) a number of lesser works, mainly in Syriac and Pahlavi, that took up different portions of Greek or (to a lesser extent) Indian astronomy in intermediary fashion and that are now lost (King 1997; Kunitzsch 1989; Morelon 1996). This list is probably far from complete. Portions of it are even uncertain, since so much is known only through the citations of later authors, especially the great *Fihrist* of al-Nadīm (late tenth century).

What such a list highlights, however, are the importance of Greek astronomy in particular and the diversity of its specific forms. Most of these forms had traveled considerable distances in time, space, and language before entering the linguistic gates of Arabic. In the early phase of translation, up through the time of al-Ma'mūn, such diversity of source material coupled with the variety of cultural background among the translators themselves, as well as their individual command of Arabic, led to a very heterogeneous, inconsistent, and error-ridden corpus of astronomical works. Discrepancies in terminology, syntax, accuracy, and even comprehensibility among different versions of a single work became obvious. Methods of translation, though generally literal early on, still varied significantly. In many cases, individual passages that seemed too complex or obscure to a particular translator were either summarized, glossed over, or simply left out altogether. As a result, "a strong need grew up toward verification, refinement, and correction of this work" (Kunitzsch 1993, 214), a need that helped give rise to new versions that selectively edited and retranslated earlier texts. Among the most important of these were Thābit ibn Qurrah's edition of the *Almagest* (already mentioned), an updated and greatly improved version of the *Sindhind* zij by al-Kwarīzmī (d. A.D. 840), and the various corrections to Ptolemaic astronomy produced by al-Battānī (A.D. 868–929), all of which were used as standard texts until the time of al-Ṭūsī (thirteenth century).

Techniques of Translation

What were some of the methods used by the translators to produce Arabic versions of Greek, Persian, and Hindu texts? How might these have changed over time? The short answer here is that these methods, and the attitudes embodied in them, were very diverse in the early phase of work, but became more standardized later on. This might seem at first inevitable; the abilities, experience, and interest of the individual translators were at first variable, and, as just noted, resulted in texts and terminologies that required considerable correction and refinement. It was really only under the auspices of al-Ma'mūn and subsequent caliphs, when the translation movement began to become more centralized and the work of a relatively few individuals became dominant, that the trappings of standardization in method and nomenclature resulted. But these factors do not explain the whole of the matter.

The early translators of the eighth and early ninth centuries were not really craft professionals. During their lifetimes, they usually worked on only a few texts chosen for (or imposed upon) them by their patrons. They used only one or perhaps two versions of a particular source work, e.g., one in Greek and one in Syriac, and did not compile any lists of terms, names (e.g., of the constellations), or definitions that might have helped their readers or later translators. In general method, they were prone to a combination of literalism and a desire for readability: this means that they held fast to the source text where it was comprehensible to them and were prone to omissions and simplifications, as well as misreadings, where it was not. This is shown, for example, by al-Ḥajjāj in his version of the *Almagest*, where a tendency to summarize or eliminate the more difficult details of many passages has the effect of erasing entirely their process of mathematical deduction (Kunitzsch 1974, 66).

Single titles or terms in this work often appear in several different forms. For example, the constellation Bootes is translated in certain instances as a direct transcription (*bu'utīs*), in others by identification with an ancient Arabic star name (*al-awwa*), and in still others through an etymological description of the Greek name—"its meaning is 'the one who calls'" (Kunitzsch 1974, 174–75). At the same time, in other astronomical works—and certainly in astrological ones—there occur many direct borrowings not only from Greek, but from Syriac and Persian, or from Greek through transcriptions in these other languages. Coining of new terms in Arabic, or assigning existing words new meanings, indicates a search for relatively concrete, accessible language that might make a particular translation more friendly to its patron (Endress 1982, 1989).

These methods are reflective of certain cultural realities regarding scientific works during the eighth and early ninth centuries. Most important perhaps is the fact that Greek works had not yet attained the supreme status they were later to acquire. This meant a flexible attitude toward the rendering of Greek texts and terms, as well as a liberal adoption of Syriac and Persian words. Persian texts may well have been most numerous at this time, while a majority of Greek works came into Arabic via Syriac. This alone would have guaranteed the strong influence of these languages upon the formation of an early technical vocabulary in Arabic.

By the later ninth century, this situation had changed. Consistent patronage, not only by the rulers and their immediate family but by significant portions of the court aristocracy, military leaders, merchants, and others had legitimized the intellectual, spiritual, and economic value of the translations and helped create a true interpretive textual community of scholars and teachers centered on Greek works above all, whose primary source texts thus became ever more paramount. In sheer quantity and diversity, Greek texts came to be favored far beyond those of other nations (Endress 1982). Their quantity and difficulty kept the later translators and their readers of the ninth and tenth centuries busy. In a sense, the Greek corpus offered a magnificent challenge, even resistance, to scholarly conquest and completion. No less a translator-scholar than Thābit ibn Qurrah confesses as much in a foreword to his edited version of to Ḥunayn's *Almagest:*

> The work was translated from the Greek into the Arabic language by Isḥāq ibn Ḥunayn ibn Isḥāq al-Mutatabbīb for Abū s-Saqr Ismaīl ibn Bulbūl and was corrected by Thābit ibn Qurra from Ḥarrān. Everything that appears in this book, wherever and in whatever place or margin it may occur, whether it constitute commentary, summary, expansion of the text, explanation, simplification, explication for the sake of clearer understanding, correction, allusion, improvement, and revision, derives from the hand of Thābit ibn Qurra al-Ḥarrānī. (Kunitzsch 1974, 68)

One would be hard pressed to find a more revealing confession regarding how open the translators conceived their field of endeavor and their opportunities for textual change to be. Whether Thābit's statement here is one of humility or attempted possession does not matter; his words show that many recognized and acceptable chances existed for an individual translator and editor to impose himself upon whatever material lay at hand, to enter into a textual dialogue with the "ancients," and to thereby make this material a living, contemporary presence in a culture and time far removed from those of its first composition. In a sense, this is how Greek works were displaced: Ptolemy, author of the *Syntaxis Mathematica,* was transformed into "Ptolemy," a community of translator-interpreter-revisers.

This type of transformation is especially characteristic of the second wave of translators (late ninth to tenth centuries). In contrast to their predecessors, these men developed sophisticated methods for assembling an individual source text. They took great care to search out and procure various versions of a single work and to collate the best (most complete and error-free) sections of each into a "superior" source that would then serve as the basis for the translation. Ḥunayn, in particular, developed this method of working and embarked on a number of extended journeys to various cities in order to acquire different versions of individual works (Bergsträsser 1925). The trouble spent on acquiring such material was matched by the care such translators showed toward their linguistic training as well as their renderings into Arabic.[10] By this time, the need for standardized dictionaries and vocabulary lists had become apparent, and such works had been assembled by authors like al-Kindī, al-Kwarīzmī, and al-Fārābī. In addition, the fame of some translators helped make their versions of certain works standard references for later workers, with the effect that their terminology also became relatively fixed. To follow our earlier example, Ḥunayn's rendering of Bootes as *bu'utīs wa-huwa l-baqqar,* "Bootes, the ox-driver," was retained by almost all succeeding versions of the *Almagest* (Kunitzsch 1974). The later tendency was to reject older Arabic titles in favor of transcribed Greek constellation names. This was done mostly for practical reasons: the older Arabic titles and star imagery simply did not match the configurations of classical astronomy and thus presented too many difficulties to be retained. In general, due to the increasing prestige of Greek sources, there was a tendency to retain earlier loanwords (e.g., *baraksis* for *παραλλαξιζ* [parallaxis]), to add Arabic prefixes and suffixes to a loanword core, and to coin or re-coin terms of more abstract content (Endress 1989). The overall pattern reveals a shift toward more direct appropriation of Greek terminology.

It is interesting that, in direct contrast to Roman authors, the Arabic translators have almost nothing to say about their philosophies and methods. While it is sometimes claimed that Ḥunayn wrote extensively of the translator's art, this is not really true. All we have from him, with a few exceptions, are a series of discussions of how he came to translate certain works, judgments on other translators, and some very general statements about the condition of various works and about collating manuscripts. In one instance, he tells us that he sat down with one of his patrons to perform a detailed comparison between a Syriac translation of a work by Galen and a Greek version, and that the two men pro-

10. Biographical information suggests that Ḥunayn, a native speaker of Syriac, spent time in both Alexandria and Constantinople to perfect his Greek and then returned to Baghdād to further refine his Arabic and begin translation work. There are a number of picturesque stories of Ḥunayn's life, character, and intellectual career that discuss this and other aspects. See, for example, the discussions in Hitti 1989 (312–14), Moussa 1980, Meyerhof 1930.

ceeded by having the patron read out loud the Syriac and Ḥunayn noting for him each place that needed correction. This method proved far too cumbersome for the patron, who soon gave up and commissioned an entirely new translation (Bergsträsser 1925). The implication seems to be that oral methods were sometimes used by the translators themselves, particularly when Syriac served as an intermediary language, and required considerable patience. But nothing much more definite than this can be said. There is little among the writings of any of the translators resembling the type of discussion one can find in Cicero, Quintilian, or the younger Pliny, all of whom speak directly of the needed purpose, content, and function of translation, as well as its methods.

Only in one particular case does this seem to occur. An often-cited, unique passage by the historian as-Ṣafadī (A.D. 1296–1363) speaks of "the two methods used by the translators" and outlines specific procedures in a manner that no other author seems to have done. The first of these procedures is the more literal: "the translator studies each individual Greek word and its meaning, chooses an Arabic word of corresponding meaning, and uses it. Then he turns to the next word." For the other method, associated with Ḥunayn ibn Isḥāq and his school, "the translator considers a whole sentence, ascertains its full meaning and then expresses it in Arabic with a sentence identical in meaning, without concern for the correspondence of individual words. This method is superior" (Rosenthal 1975, 17).

How is one to understand these characterizations? Until very recently, they were accepted by scholars as reasonable and accurate and, because wholly unique, were quoted with the regularity of clichés. It is now recognized, however, that they should probably be dismissed almost out of hand (Mattock 1989; Endress 1989). Reasons given for this are straightforward. First, there was no possibility for anything even approaching a one-to-one, word-for-word correspondence between Greek and Arabic; the languages were simply too different. To take only one example, early written Arabic lacked a copula ("to be"), which had to be invented, largely to accommodate Greek technical literature (Versteegh 1977). The evidence, moreover, shows that even the earliest transla-tions are full of adaptations—transposition of words, restructuring of phrases and sentences, omissions, rewritings, and so forth (Endress 1989; Kunitzsch 1974). Moreover, it is simply untenable to argue that "superior" translators—or any translator, for that matter—worked "without concern for the correspondence of individual words." If this were true, no technical vocabulary in Arabic worthy of the name ever would have resulted. As it happens, at least one detailed test of as-

Ṣafadī's portrayal has been done (Mattock 1989). Not surprisingly, the same conclusion was reached. Indeed, in comparing two translations of the same text (Aristotle's *Metaphysics*), this study notes that both versions (one by Ḥunayn) are more free and idiomatic where the original is "comparatively simple," and hug the shore of Greek terminology when the source is difficult. Moreover, Ḥunayn's text, while better in most aspects, is sometimes verbose and overly literary, and seems to err on the side of making the Arabic appear more natural, especially in an idiomatic sense, than in transferring the rhetorical force of the original (Mattock 1989, 81).

A diversity of method, therefore, is typical for individual translators and translations, as one might expect. Such diversity reflects, in no small part, a form of adaptation based on a continual shift in epistemological loyalties (source vs. target/target vs. source languages). This was entirely necessary in order that works as complex and variable in content as those of Greek science might be transferred to an entirely new linguistic context. Doubtless the ability of individual translators to become conscious of this process and to take control over it helps define their level of expertise. Ḥunayn is obviously at a high level in this regard. Indeed, the great deal of effort he seems to have taken to essentially *create* his own sources through repeated collation—to assemble, that is, from a range of now vanished originals a superior recombinant version—shows that he well understood the dependencies at hand. In turn, the passage from as-Safadi just discussed appears as much an attempt to contribute toward Ḥunayn legend as anything else.

Forms of Adaptation and Nativization

Besides the lack of discussion on methods and philosophies, Arabic translation literature shows little evidence of the change from imitation to rivalry so apparent in the case of Roman and, to a lesser extent, Syrian culture. While a movement to reclaim older Arabic star names did occur later on, in the late tenth century, no obvious impulses can be found among translators and their students to diminish or deflect the influence of past "sages." Instead, the most famous of the ancients—Aristotle, Ptolemy, Galen, Euclid, Archimedes—were venerated throughout the translation period, accorded respect and attached to ideas of "wisdom" at the highest levels by Arabic scholars of the "foreign sciences." Indeed, so great was this veneration that it effectively eclipsed appreciation for non-Greek authors, such as Severus Sebokht. Certainly a mixture of legend and lore very often surrounded such veneration. Writing in the tenth century, for example, al-Nadīm cites in the *Fihrist* a famous dream of the caliph al-Ma'mūn, in which Aristotle appears as an almost mystical source of wisdom and guidance who persuades al-Ma'mūn to devote much new effort to the translation of foreign books.[11] Several centuries later, one finds the historian al-Andalus formalizing what had long since become standard understanding: "The Greek philosophers are the highest *tabaqāt* [class] . . . and the most respected among the people of knowledge, because of the true care that they have demonstrated in cultivating all the branches of knowledge, including mathematical sciences, logic, natural philosophy, and theology" (al-Andalusī 1991, 21).

Displacement of Greek authors took the form in Islam of appropriation and nativization. This meant, among other things, the creation of specifically Arabic textual traditions that took the early translations as raw material. First came the great systematizers in Arabic, such as al-Kindī and al-Fārābī, Ḥunayn ibn Isḥāq in medicine, the "second physician" (after Galen), and al-Kwarīzmī in mathematics—authors who brought together a wide range of influences from translated works, rewriting and resynthesizing to produce new interpretive systems. Then there were subsequent writers like Thābit ibn Qurrah, al-Kwarīzmī (again), al-Battānī, Abū al-Wafa, and al-Haytham, who edited, updated, and—most importantly—corrected significant portions of such works as the *Almagest* between the ninth and eleventh centuries, thus producing new texts—in essence, new "Ptolemies"—that formed the

11. As recounted in the *Fihrist*, the dream occurs under the heading "Mention of the reason why books on philosophy and other ancient sciences are so plentiful in this country," and is described as follows:

> al-Mamun saw in a dream the likeness of a man white in color, with a ruddy complexion, broad forehead, joined eyebrows, bald head, bloodshot eyes, and good qualities sitting on his bed. Al-Mamun related, "It was as though I was in front of him, filled with fear of him. Then I said, 'Who are you?' He replied, 'I am Aristotle.' Then I was delighted with him and said, 'Oh sage, may I ask you a question?' He said, 'Ask it.' Then I asked, 'What is good?' He replied, 'What is good in the mind [i.e., according to reason].' I said again, 'Then what is next?' He replied, 'What is good in the law.' I said, 'Then what next?' He replied, 'What is good with the public.' I said, 'Then what more?' He answered, 'More? There is no more.'" According to another quotation: "I [al-Ma'mūn] said, 'Give me something more!' He [Aristotle] replied, 'Whosoever gives you advice about gold, let him be for you like gold; and for you is oneness [of Allāh].'" (Dodge 1970, 2.583–84; see also Rosenthal 1975, 48–49)

Al-Nadīm says that "This dream was one of the most definite reasons for the output of books [i.e., translation activity]" during this period, since shortly thereafter the caliph sent a delegation to select and retrieve "books on the ancient sciences from those preserved in the libraries of Byzantine territory" (Rosenthal 1975, 49). Aside from romantic appeal, this type of claim has little to recommend it. In his support of translation, Al-Ma'mūn was obviously continuing a tradition that had proven helpful, even necessary (as an excellent source of support), since the days of al-Manṣūr.

foundation of an advanced astronomy. At the same time, handbooks to this new Ptolemaic-Arabic astronomy were also written, most notably by al-Fargānī (d. ca. A.D. 850), establishing its contour and availability for nonspecialist readers. Finally, a third type of displacement can be found in a genre of writing that culminated in the work of al-Ṣūfī (d. 986). This author not only corrected the Ptolemaic star positions to his own time, but also presented a list of ancient Arabic star names against their correlative Greek titles and, most strikingly, an entire atlas of beautiful, elegant drawings of the constellations, thus adding an unprecedented pictorial dimension to the textual tradition of astronomy (Ṣūfī 1953). Ptolemy's work was thus frozen in concept as the most elevated type of "classic" and used as a work that inspired continual commentary, updating, and improvement. If observational methods were devised that came to point up difficulties in the *Almagest* or *Handy Tables*, if al-Haytham in particular could write significantly in the eleventh century on "doubts about Ptolemy" with regard to planetary motion and position, such efforts were never intended to pack in clay the fame of Ptolemaic astronomy. Indeed, the style of such fame, and its encouragements toward engagement, were essential to the nativization of this astronomy, and the other sciences as well, within Islam. As once stated by al-Kindī, in his great work on Euclidean optics:

> We wish to complete the mathematics and set forth therein what the ancients have transmitted to us, and increase that which they began and in which there are for us opportunities of attaining all the goods of the soul. . . . We ask the reader to whom this book of ours may come, if he finds anything which we have not spoken of sufficiently, to be patient and not hasten to think ill of us, till he has understood truly all the previous treatises . . . —for this book follows them—and to supply what he thinks we have omitted, according to what the men of his age require. (Dunlop 1971. 228)

There is an impressive sense of history in these words, a sense that knowledge cannot be allowed to stand still. Al-Kindī is speaking for a great deal of Arabic science when he uses the words "complete" and "opportunity": the great minds and texts of the ancients were viewed as a fundament, a starting place from which to advance by steps and fits toward a greater perfection. Moreover, this perfection was not merely intellectual but moral, spiritual, and material—"the goods of the soul." Nativizing the thought of "the ancients" does not mean to Arabic writers like al-Kindī accepting an "inheritance" or "legacy." It has a far more active meaning, like taking in hand a great, unfinished enterprise.

By the ninth century, and certainly afterwards, the favor granted

Greek scientific authors exceeded that shown toward any other people. From the very beginning, however, adaptations were everywhere apparent, on almost every level, and were by no means confined to linguistic phenomena. For example, among the earliest "documents" revealing of Greek astronomical influence is a version of the constellations and ecliptic painted on the dome of a bath in the desert town of Qusaīr Amra (southern Jordan), dated at around A.D. 715. The images are backwards—that is, they appear as they would from above (outside the celestial sphere)—which means that they were copied from a celestial globe (Saxl 1932). The probability that this was done "by Byzantine artists and craftsmen in service to the caliph who brought with them an insufficient understanding of astronomy" (Kunitzsch 1993, 209) is exactly the point. Muslim rulers at this early date were eager and still largely unschooled with regard to the "foreign sciences." A backwards heaven is perhaps the ideal reflection of this earliest phase of appropriation.

On the level of language, we might pick up an earlier topic and return to the names used for rendering the Greek constellations in Arabic. These, in fact, provide an excellent sampling of the general styles of linguistic adaptation found elsewhere. Fortunately, the relevant information has been examined in detail (Kunitzsch 1974) and can be summarized easily. Several different adaptations were used: (1) phonetic transcription, with varying degrees of fidelity (e.g., *qaytus* for Cetus); (2) transcription with added prefixes (*ad-dulfī*, for Delphinus); (3) replacement by common Arabic words (*al-asād*, "lion," for Leo; *al-aqrab*, "scorpion," for Scorpio); (4) substitution by Arabic words that are related to, but do not exactly parallel, the Greek (*ad-dagaga*, "hen" or "fowl," for Cygnus; *al-tīnnīn*, "serpent," for Draco); (5) substitution by older Arabic names associated with the principal star in a constellation (*al-gabbar*, "the powerful," for Orion); (6) simplifications for the sake of memory (*al-faras*, "horse," for Pegasus; alternately *al-faras al-awwal*, "the first horse," for Equulus and *al-faras at-tanī*, "the second horse," for Pegasus); and (7) descriptive phrases that explain a transcribed name according to the Greek myth (e.g., *barsawus*, followed by the phrase "bearer of the demon's head")(Kunitzsch 1974, 169–203). For a given translation, two or more of these adaptations might appear for a single constellation. The versions by al- to Ḥajjāj and especially Ḥunayn appear to employ each technique with a degree of conscious appropriation. Ḥunayn is particularly fond of transcribing the Greek name and adding a tag phrase that might aid the reader's association (e.g., *irgus wa-huwa smu safina* "Argo, the name of a ship"; ibid., 198). Finally, it is interesting to note that, whereas Greek titles in the *Al-*

magest are given in the genitive form (e.g., *Ωριωνοζ αστερισμοζ*, for Orion), these were most often changed to nominative when transcribed in Arabic (*uriyun*), showing that the translators were well aware of the linguistic processes at their command.

On the plane of image, too, the Greeks were adapted to Arabic understanding through a range of revealing misconceptions. These surface most strikingly in historical works of the thirteenth and fourteenth centuries that sought to produce surveys of all knowledge and the contributions of various nations to it. One such source is al-Andalusī, whom we have just quoted as saying "The Greek philosophers are the highest *tabaqāt* [class] . . . and the most respected among the people of knowledge." With the same stroke of the pen, this author contends, in matter-of-fact style, that "most of the Greeks were Sabians who venerated the heavenly bodies and persisted in their worship of idols" (al-Andalusī 1991, 21). The "Sabians" of which al-Andalusī speaks were actually natives of Ḥarrān, a well-known, even legendary center of star worship under Zoroastrian influence from which such thinkers as Thābit ibn Qurrah had come. The idea that the Greeks mostly hailed from a landlocked desert area in eastern Turkey appears odd, given what must have been known about them during the first few centuries of Islamic conquest. Yet, al-Andalusī is hardly an eccentric exception with regard to such errors. Writing roughly a century after him, al-Khaldūn in his great *Muqaddīmah* (Introduction to History) claims that "The intellectual sciences are said to have come to the Greeks from the Persians, when Alexander killed Darius. . . . At that time, he appropriated the books and sciences of the Persians" (Rosenthal 1970, 373). If one goes back to al-Nadīm and the *Fihrist* (tenth century), meanwhile, a host of smaller misconceptions regarding individual Greek authors can be found, especially where the author is passing on traditional lore. Thus Ptolemy "was the first person to make . . . astronomical instruments, measurements, and observations" (Nadīm 1970, 639); Hipparchus (who lived four centuries earlier) was his teacher; and his books included nearly a dozen works on astrology, war, imprisonment, disease, fortune, and liquids. These works, in reality, represented either spurious attributions or possibly, in some cases (astrology in particular), sections of his actual writings that had been enlarged upon by other writers.

Such "tales" show that by means of these expressions of esteem, Islamic intellectuals essentially adapted Greek writers to traditions familiar in the Arabic world. Assigning "the Greek miracle" in astronomy to the star worshippers of Ḥarrān, for example, brought the essential achievement to the doorstep of Islam itself. Conceiving Alexander's conquest of Persia as the origin of this achievement provided a historical model for the Muslim caliphs, above all the ʿAbbasids, who were themselves of Persian descent. Finally, attributing books to an author of such prestige as Ptolemy, especially books that had found considerable use among Arabic readers and writers, marked a mode of legitimation, an accrual of high status to work by "lower" authors—a process often seen in classical and medieval literature and no less within Islamic intellectual society.

Other forms of adaptation should be noted. Greek learning was also nativized in terms of how it was categorized and classified, involving certain reinterpretations along Islamic grounds. Influential scholars like al-Fārābī and al-Kindī, for example, viewed such learning as fundamentally divided into two main streams or "schools." The first of these was the Hermetic-Pythagorean school, metaphysical in nature and relying upon symbolic interpretations of nature, commonly through mathematics. This "was regarded as the continuation, in Greek civilization, of the wisdom of the ancient prophets, especially Solomon and Idris; it was therefore considered to be based on divine rather than human knowledge" (Nasr 1987, 33). The second, more "foreign" school was rationalistic, encyclopedic, and included the likes of Aristotle, Ptolemy, and Galen. The existing texts of these authors were viewed as representing "the best effort the human mind could make to arrive at the truth, an effort by necessity limited by the finite nature of human reason" (ibid., 33). The first school was therefore linked to a study of deeper spiritual realities; the second was a manifestation of those realities put to earthly use. Both were required to procure "the goods of the soul," in al-Kindī's words, or to delimit what can and cannot be known by human beings, as stressed by al-Fārābī in his *Kitāb Iḥṣā al-Ulūm* (Book on the Classification of the Sciences) and *Kitāb al-Saʿāda* (Book of the Path to Happiness)(Netton 1992).

Knowledge as a concept in Islamic intellectual society was extremely diverse, multiform, and invested, and, as such, was the subject of much discussion (Rosenthal 1970). We have spoken of *kalām,* a sort of umbrella term used to denote all branches of scholastic theology, including *ḥadīth* (traditions of the words and actions of the Prophet). *Adab,* referring mainly to belles-lettres, gained new meaning with the continual increase in translated Persian literary, historical, and other works during the ninth and tenth centuries, coming to indicate general "refinement" or to designate a "cultured person." Under the Umayyads and in the earliest ʿAbbasid period, meanwhile, the words *ʿilm* (knowledge in the broadest sense) and *maʿrifa* (gnosis; knowledge of the divine) were largely interchangeable.[12] *ʿIlm* was used often in the Quʾrān to denote unspecified knowledge of the world, religious law, the place of human beings within the divine cosmos. Following the major phases of translation in the ninth and early tenth centuries, however, these words began to break apart: *maʿrifa* became associated with mysticism, with *bāṭin* (hidden, esoteric knowledge) due to its use by the Sufis, while *ʿilm* was taken up as a central term of the new "foreign sciences" even as it was retained by followers of *kalām*. The word *maʿqulāt,* meanwhile, was used to speak specifically of philosophical subjects, those accessible to rational inquiry and reflection, separable from *maʿrifa,* and most often approached through books. Finally (but this is hardly an exhaustive list) there is the term *ḥakīm,* which originally meant "wisdom," the knowledge of sages. Following establishment of the Bait al Ḥikma in Baghdād, however, with its staff of physician-librarians, and whose founding was related to the import of doctors from Jundishapur to serve the caliph, the word *ḥakīm* came to also mean "medical practitioner" (Nasr 1987).

Such terms were formalized, in part, by various influential books that sought to classify and categorize all knowledge, such as those produced by al-Kwarīzmī, al-Fārābī, and later, al-Khaldūn, or that are found in various encyclopedias compiled by scholars of the tenth century. These works consistently divide the "foreign sciences" from those of "religious law"; even so, a term like *ʿilm* remained too slippery, too broadly connected with the Quʾrān itself, to be confined to a single series of meanings. In some part, perhaps, its uptake by the translators and the followers of "ancient wisdom" represents an effort to link such knowledge with established tradition, to legitimize it on Islamic grounds in other words. But during the time of the early ʿAbbasids, when court patronage reached its height, this was hardly necessary. It seems to have become more important later on, when it was associated increasingly with scientific knowledge and "mere" book learning. Yet *ʿilm* never lost its multifaceted qualities, and in this, it expresses something of the diverse place scientific understanding itself attained within Islam.

Linguistic Changes in Arabic as a Result of Translation

Across the vast territory of peoples, cities, lands, and cultures that composed Islam's empire in the first several centuries following Muḥam-

12. A detailed and illuminating discussion of these terms is offered by Rosenthal (1970, 164–68).

mad's death, the only unifying force outside of religious custom—and upon which such custom depended—was the Arabic language, the mother tongue of the Muslim faith. By the ninth and tenth centuries, however, this was no longer the language of the Prophet and the Qu'rān. It had become something much larger, more worldly, flexible, teeming with new terminologies, densely textual, written with use of a new diacritical system, grammatically and syntactically evolved. A large part of this evolution came as a direct result of the translation episode we have been discussing. Indeed, the words used to denote "translation" itself are an excellent expression of this. These terms included *targama,* a loanword from Syriac that originally meant "biography," suggesting that a sizable portion of such literature in Arabic was a result of linguistic transfer. The word *tafsīr,* on the other hand, at first referred to exegesis of the Qu'rān, therefore to "holy explication," but was adopted by al-Ḥajjāj, for example, in his versions of both Euclid's *Elements* and Ptolemy's *Almagest,* as an equivalent for "translation" (Kunitzsch 1974, 17–18), thus implying the central role elevated interpretation played in such textual activity. A third term, no less revealing and more often employed, was *naql,* which literally meant to "convey" or "transmit" (much like *transfero* in Latin), but which came to also mean "copy," "transcribe," and, more generally, "carry on a tradition" (Endress 1989).

From a broader linguistic point of view, meanwhile, the episode of translation took place at a crucial juncture in the transformation for speakers of Arabic from a mainly oral society to a more fully literate one. This it did in two main ways. First, the sheer quantity and utility of the written material translated from Sanskrit, Persian, Syriac, and Greek helped shift an earlier privilege granted the spoken word as the essence of revelation to an increased focus on textual embodiment, written interpretation, and exegesis. Muslim schools continued to emphasize the study of language, especially that of the Qu'rān. But the influence of the new learning had profound impacts on spoken and written Arabic, helping evolve the language away from older, "classical" patterns.

Nativizing the corpus of Greek and Syriac texts inspired Arabic to develop in certain new directions. As already noted, orthographic improvements, including dots to differentiate consonants, marks to distinguish short and low vowels, and other diacritical additions to designate long vowels and double consonants, were adopted directly from the Syriac script (Carter 1990, 119). While this began fairly early, in the eighth century, and helped eliminate ambiguities in Qu'rānic spelling, the new orthography became firmly established for all writers of Arabic only in the early ʿAbbasid period, when it was taken up by the translators and scholars of *falsafa* (Greek philosophy) and other writers of mainly secular literature.

On the other hand, use of the Arabic language as a medium for interpreting and expanding the knowledge of other peoples led to inevitable changes in its detailed character. To render much of the object-centered language of Greek science, for example, translators had to introduce the copula ("to be" forms), previously absent. Certain verb tenses, such as the conditional, were used much more frequently. To produce the needed vocabularies, meanwhile, translators also made creative use of prefix and suffix forms for nouns and adjectives, adding much new flexibility. A large number of Greek and Persian words were borrowed and altered. An example is the title of Ptolemy's *Almagest,* known in Arabic as *Kitāb al-majastī* ("the greatest of books") and apparently derived from a Middle-Persian calque (*mgstyk*) of the Greek word *megiste* "greatest" (Kunitzsch 1996). New sentence constructions were needed and invented for the description of phenomena, for abstract analysis, and for mathematical discussion. In the capable hands of these translators, then, Arabic evolved creatively well beyond the mainly oral literary traditions of poetry and prayer into new types of syntax more useful in deductive logic (Versteegh 1977). These changes, moreover, soon had their effects elsewhere.

By the mid-tenth century, spoken and written Arabic on topics such as medicine, engineering, and astronomy had adopted most or all of the new expressive styles. From here, they expanded into commerce, philosophy, and literature. Existing literary forms were modified to take advantage of, and further develop, these new types of expression. Without doubt the most influential was the *risāla* (essay, epistle), a "letter to an enquirer" form that had been earlier introduced from Persia (Allen 1998). In the hands of its most famous practitioner, al-Jāḥiẓ (d. 869), this became a kind of fertile alloy, showing fulsome mixture between poetic and technical styles. Al-Kindī, a contemporary of al-Jāḥiẓ, was the first to adopt the *risāla* as an important medium for scientific discussion, altering its meaning to include "treatise" or "exposition." The enormous fame and many imitators of al-Jāḥiẓ and al-Kindī helped further the changes Arabic had undergone in the hands of the translators. The *risāla* form was quickly taken up in the later eighth century and used by many of the most famous translators and those who used their work: al-Fārābī's *Epistle on the Intellect* and *Epistle Concerning the Sciences,* both highly influential texts, might be cited as examples (see Ashtiany et al. 1990).

Overall, then, Islam ended up nativizing Greek knowledge and linguistics both. The larger effect of translation was to help rapidly broaden what was already an extremely flexible mode of communication, closer to Greek in its multilayered capabilities than to Latin. But it did this not as a simple matter of "influence" from Hellenistic thought. The changes noted came from within Islam; they were not in any way inevitable and they reflect not the power of "Greece" but the creativity of Muslim thinkers, who used Greek thought for their own purposes. The distinction, as we have implied, is crucial. Responsibility for "the flowering of Arabic science" belongs to Islam.

Al-Ṣūfī: A Unique Nativizing Influence

By the latter part of the tenth century, Islamic astronomers felt they had accurately comprehended and absorbed the greater part of Greek knowledge. When they gazed at the skies, it was no longer the brilliance of "the ancients" they saw reflected; it was now their own world of observation, calculation, and textual expression. Indeed, the consciousness of this was reflected in an interesting new movement among some intellectuals to go back and reclaim Islamic star names from the earliest age of Arab culture. Islam, now under the sway of somewhat more conservative political and religious influences, sought to look inward and find "legacies" of its own:

> Arabic philologists and lexicographers, in their efforts to collect the genuine ancient Arabic terminologies and vocabulary, composed special books in which they collected all the star names they could find in those old traditions. And it was the astronomer al-Ṣūfī who then made an attempt, in his book on the constellations composed in A.D. 964 [the so-called Uranometria, in Latin translation, or "Description of the Fixed Stars"], to identify the respective celestial objects according to the scientific Ptolemaic tradition. (Kunitzsch 1989, 4.264)

Al-Ṣūfī thus put himself in the role of mediator between two textual/literary traditions, two "pasts": one Greek, the other Muslim—with the former now being, ironically, the more established and therefore nativized.

Al-Ṣūfī's book, *Kitāb Ṣuwar al-Kawākib,* is both a linguistic and a pictorial atlas and represents one of the pinnacles of the whole of Islamic astronomy (Ṣūfī 1953). For each of the forty-eight constellations, it offers a critique and partial correction of the Ptolemaic star positions (though for many of the smaller of the total 1,025 stars, no changes were made); a list of native Arabic star names and their correlative

Greek titles; and, perhaps most important of all, beautiful and elegant drawings of each figure. Two drawings, mirror images of each other, were made for every figure, to represent it as it would appear "in the sky," i.e., from below, and on a celestial globe, from above. A copy of the *Ṣuwar al-Kawākib,* dated at A.D. 1005, is among the oldest illustrated Arabic manuscripts known to exist and thus occupies an important place in the history of Islamic art (Gray 1994). Al-Ṣūfī's images are an unparalleled example of the nativization of Greek astronomy, with each of the classical constellation figures given a detailed depiction in the visual terms of tenth-century Baghdad society. Perseus appears in profile as a heroic (mythical?) Bedouin aristocrat, dressed in flowing silk, cuticles perfectly trimmed, bearing in one hand a slim ornamental sword and in the other, the head not of Medusa but of some bearded male giant. The beauty and precision of this image no doubt helped ensure its use as a model—not only in Islam, but, no less significantly, in Europe as well (fig. 9).

Al-Ṣūfī's work represents nothing if not an incomparable merging of sources and media, revealing in such blend the high degree to which Greek learning had been absorbed and transformed into something else. To assemble his catalogue of Greek and Arabic star names, the author drew from many sorts of texts: astronomical, mathematical, literary, historical, artistic. By so doing, he helped establish all these as essential ingredients to further scientific work. As a grand document, standing at or near the high point of Arabic astronomy, al-Ṣūfī's work reflects the many connections of this knowledge within Islamic culture generally. In its wake, an increasing number of Arabic names for individual stars came to be used throughout Islam. Indeed, so exact were the star positions of the *Uranometria,* and so aesthetically appealing its images, that it became a standardizing source for the next four hundred years of Arabic science and also in medieval and Renaissance Europe as well, up to the seventeenth century in fact. It is the ultimate source of nearly every Arabic star name now gracing the sky of contemporary astronomy (Kunitzsch 1989).

Conclusions

Al-Ṣūfī's work, like that of Thābit ibn Qurrah's *Almagest,* al-Bīrūnī's Book of *al-Maʿūdī* (a canonical encyclopedia of Greek/Islamic astronomy), or ibn Sīnā's *Kitāb al-Shīfa* (Book of Healing), all show that Greek, Persian, and Indian learning underwent a transformation by becoming part of Islamic culture. This transformation was an essential el-

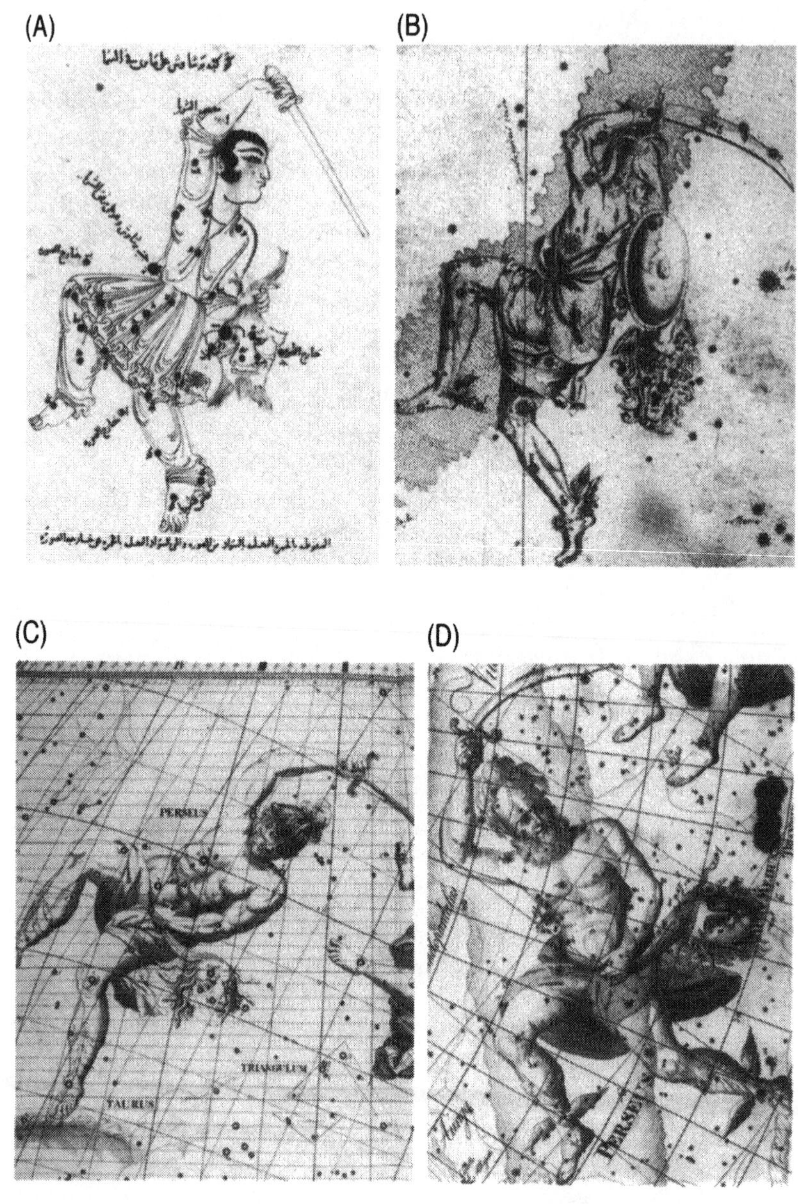

Figure 9. Juxtaposition of images representing the constellation Perseus, revealing the long-term influence of Arabic writings in Europe. Image A is from a version of al-Ṣūfī's eleventh-century *Book of the Fixed Stars* (Bodleian Library, MS Marsh 144, p. 111). This version of the hero is clearly the model for Image B (British Library Maps C.10.a.17, f.11), which appeared in Johann Bayer's *Uranometria* (1603), the first great star atlas published in Europe. Bayer's atlas became a standard reference and this fact is reflected in image C (British Library Maps C.10.c.10, plate 16), from the more elaborate and accurate version of the heavens by John Flamsteed (1729). Finally, image D (British Library) comes from the *Uranometria* by Johann Bode, published in 1801, one of the last of the great celestial atlases done in artistic style.

ement to the greater flowering of this culture and worked its effects upon all regions of intellectual activity, even those of traditional religious practice. It deeply affected not only all areas of science—indeed, it could rightly be said to have founded most of them—but also such areas as history, belles-lettres, grammar, lexicography, and commentary in general; and it made lasting changes in the Arabic language itself. The ultimate cultural power of the translation episode lay not merely in the wealth of learning it introduced, but in how it helped compel a larger transition toward the written word in Muslim society. Yet, again, inasmuch as this took place, it did so because individual translators, patrons, authors, and others in this society did not simply "absorb" the "ancient wisdom" now made available; they adapted it profoundly in turn, continually, as befitted their historical moment. No text of science stood still during these centuries. No work was so esteemed that it failed to draw unending commentary and alteration, at times in overflowing measure (fig. 10). By the time medieval Europe began to take an avid interest in the intellectual resources of Islamic culture, it was no longer possible—or advisable—to divide (for example) "Greek" and "Arabic" contents from one another. Greek knowledge did not come into Europe "trailed by clouds of Moslem and Jewish commentaries," as is sometimes unfortunately described (Cantor 1993, 359). What the Latins, in their turn, translated and transformed to their need was a vast amalgam—if we speak only of astronomy, for instance, an Arabic-Greek-Persian-Indian alloy whose internal boundaries had long since blurred or else melted entirely. In addition to the types of textual changes we have noted, this astronomy came with advances in instrumentation, the use of art, a new mathematics. Such advances were anything but parasitic or secondary. They were, instead, entirely integrated into the "wisdom," the *ḥakīm*, of what Islam had created, and what European translators found so appealing. The growing reaction within Islam itself against the "followers of foreign philosophy" could not alter this.

Such reaction culminated in the great eleventh-century debate between the Asharite theologian al-Ghazzālī and the "rationalist" Aristotelian ibn Rushd (Averroes). Al-Ghazzālī and his followers won this debate; "foreign" philosophy was thereafter sent down to a lower order of importance, made to support Divine Principle, and greatly reduced in its ability to attract high-level patronage. Most historians have noted this debate as marking the end of the "golden age" of Arabic science. This is something of an exaggeration; the work of the great al-Ṭūsī did not appear until the twelfth century. Thereafter, it is true, the level of scientific study and writing declined by almost any nonreligious standard.

Figure 10. Opening page to a highly glossed Arabic epitome of an Arabic commentary on the famous *Book of Healing* by Ibn Ṣīnā (Avicenna). Here, several writings are present. The larger script (lower left) describes the opening chapter and occurs with an interlinear Persian translation. This is completely enveloped by comments and annotations written by several hands, in several different orientations. The work is an excellent example of how individual manuscripts did not represent fixed or final objects, but could be altered physically over time. Such alterations were common in scribal culture, both in medieval Islāmic society and European society. Wellcome Historical Medical Library, MS Or 10a, fol. 1v.

But the important point to make is that Islam had done pretty much what Hellenistic Greece had done: absorbed influences from a number of cultures and solidified them into a new whole. Arabic science, perhaps, rose and fell just as Greek science had, if for different reasons. The fact is, however, that the greater corpus of *scientia* now existed in the Arabic language and that this is where it had come to belong.

4 Era of Translation into Latin

Transformations of the Medieval World

Introduction

Between the time al-Ṣūfī was writing, in the late tenth century, and the period when al-Ṭūsī performed his sophisticated improvements on Ptolemy, two hundred and fifty years later, the vast majority of Arabic science passed into the Latin language and thus entered Europe. This next great era of translation, like its Arabic predecessor, had its reasons, its heroes, its motives, and its evolving effects. Related work began quietly, mainly in southern France, where a few books on the use of the astrolabe and other subjects were translated as a result of contact with Mozarabs from Spain, immediately to the south. The earliest signs were, perhaps, evident in the sporadic interest in Arabic learning indicated, for example, by the travels and writings of Gerbert of Aurillac (945–1003), whose influence was centered in Lotharingia (Welborn 1931), and by contacts between the monastery at Lorraine and the city of Toledo, a great center of learning in Arab Spain (Thompson 1929).

This activity, which included some translated materials from the Arabic, can be viewed as necessary preparation for the Latin translation movement that began a century later. As recently indicated, Gerbert's influence reached as far as England (Burnett 1998), though apparently only in localized fashion. Certainly no waves of textual transfer began until the decades after 1060 or so, at which point a veritable burst of relevant energy commenced. This energy was itself part of a much grander series of expansions within European culture that medieval historians have been fond of calling (not without reason, but with certain ideological leanings) the "Renaissance of the 12th century"[1] (see Haskins 1927).

1. The term was established by Haskins (1927), whose work with regard to the era of Latin translation has had enormous influence and remains a crucial source for study of this period. A number of his conclusions have since been modified, extended, or discarded, as one might expect, but the overall thrust of his interpretation—that the twelfth-century era of translation represents a critical turning point in the history of Europe—

Background: A Time of Social Change

Impelling this rapid development were forces both large and local in scale, operating on every level of daily life, which cannot be summarized easily.[2] Major upheaval was taking place in medieval society. New technologies, based on water power, shipbuilding, gear-machinery, and other areas, were being put in place. Introduction of the cam at the end of the tenth century, allowing for the use of water-driven hammers at mills, led to the increasing mechanization of many industries: tanning, textiles, beer making, mining, iron making. The compass made its appearance near the end of the twelfth century, introduced from the East. Economic change, including the liquification of capital and the establishing of international trade with Islam, were in progress. There was a strong expansion in agriculture and forestry, which stimulated the building of new roadways and led to a reinvigoration of town life. There were, too, the Crusades and the taking of Spain, with its great centers of Muslim cultural life: above all, Granada, Cordova, and Toledo.

Europe at this time, like Islam several centuries before, was engaged in more fully urbanizing itself, in building a worldly and a more fully textual society. Goods, currencies, students, teachers, knowledge—these were all given a new force of mobility and exchange (Waddell 1934; Ridder-Symoens 1992b). Many things began to migrate rapidly into the cities from the countryside and from abroad—objects, families, books, languages—there to achieve great concentration and metamorphosis. In these new centers, artisans and craftsmen gained new freedom and productivity, being liberated from previous feudal servitude. This period also saw the beginning, as a regular phenomenon, of long-distance pilgrimages to holy places, and the accompanying writing down and transfer of stories and experiences. Finally, too, the late medieval era (tenth to thirteenth centuries) was when the great transformation from oral to written culture was finalized, and in its immediate wake, the first great wave of the "vernacular revolution" and the spread of silent reading as a private act (Ong 1977; Saenger 1982; Stock 1983). That such phenomena should have coincided with the introduction of foreign learning, just as had happened in Muslim society, is no mere fortuity.

---

remains a fixture among medievalists and intellectual historians and, indeed, has been profoundly deepened by a range of new work. For a review of Haskins's ideas and their modifications, see, especially, Al-Daffa and Stroyls 1984 and d'Alverny 1982.

2. In what follows, I have drawn on a number of sources, including: Le Goff 1997, Cipolla 1994, Benson, Constable, and Lanham 1982, Long 1985, Stiefel 1985, Stock 1978, 1983, Weimar 1981, Southern 1962, White 1962, Gimpel 1983, and Gies and Gies 1994.

The mixture of stability and flux that pervaded Europe at this time was dependent on an appreciation, at all levels, of the powers of knowledge. On the one hand, this involved manual skills: artisans and their nascent guilds proved the value of trained craft, of a type that required teaching, practice, and standardization of terms. On the other hand, there was a shift toward literacy and its daily uses in many areas of business and politics, e.g., accounts-keeping and inventories, navigational logs, diplomatic and legal records, architectural plans, royal biographies, patronage lists. According to Henri Pirenne, "By the middle of the 12th century, the municipal councils were busy founding schools for the children of the burghers, which were the first lay schools since the end of antiquity" (1952, 231). Education and learning, in short, were no longer rarities or jeweled luxuries outside the monasteries. Textuality, in various forms, was rapidly taken up and placed—as a mode of documentary consciousness—at the very core of the new urbanizing Europe.[3] This took time and required new institutions, such as the university, which became a center of manuscript use, and the lay scriptorium, a primary supplier of such texts. Literacy and teaching, moreover, were eventually helped by the introduction, as in Islam before, of papermaking and related writing implements, which were brought in from Egypt, Damascus, and Baghdād. Again, this did not happen overnight: the first paper mill on the continent proper was established by 1151 in Jativa, a town in recently conquered Valencian Spain. Yet the quality of the paper was generally poor, and resulting documents perishable, leading at least one monarch of the next century (Frederick II, in 1221) to declare that any such document was without legal validity (McMurtrie 1943). Yet, in less than a century, paper was cheap and readily available and was being used to record more "transitory" documents, such as sermons, speeches, poems, and letters, than ever before, as well as helping to stimulate the keeping of more diaries, copybooks, notebooks, and the like (Febvre and Martin 1958; Eisenstein 1979; Martin 1988).

The twelfth century as a whole saw the beginnings of what many have called a "book revolution." Manuscript (not "book") production now also expanded well beyond the walls of the monastery. It was contracted to lay scribes and copyists after about 1150, many of whom came to work for the new urban schools. Only slightly later, after 1200, it moved into the universities as well. As a form of labor, copying had been rendered much easier and faster by improvements in the technology of writing, including new methods of line ruling, better writing im-

3. Some interesting details related to this process are discussed in Bäuml 1980, Clanchy 1979, and Ong 1977.

plements and ink, and most important of all, the practice of fully separating individual words and paragraphs from one another in a text (Diringer 1982). This last advance, established in the eleventh century, greatly aided reading as well, making the production and consumption of books a far less ponderous, more accessible activity than ever before. There is no greater indication of the new fluidity granted to knowledge than the growing circulation in manuscripts that ensued by the twelfth century, which in many ways made possible the idea of a "university"—a guild of teachers and learners of the written word. "Twelfth-century intellectuals did not feel bound to any particular school or curriculum; they freely chose their discipline and teacher, and the first universities arose in Paris and Bologna from the influx of students from the four corners of Europe" (Ridder-Symoens 1992b, 282).

Writing in the middle part of the century, Bernard of Chartres, one of the great teachers of the age, expressed in two brief passages what must have been essential sentiments among the learned. The first of these passages is, to say the least, well known: as quoted by Bernard's student and colleague, John of Salisbury (in his *Metalogicon* III, 4), it states that the men of this era can see further than the ancients, not by being taller or possessed of sharper vision, but by standing as dwarfs upon the shoulders of giants. This very phrase, of course, with its self-conscious pride, was taken up by an even greater thinker (Newton) during a later period of innovation, the Scientific Revolution itself. Yet the second passage from Bernard may be the more significant. Delivered in poetic form, it says, "With humble spirit, eager learning, and peaceful life; in silence and poverty, to explore the most distant lands; many now endeavor to unlock through study what has long been unknown" (Classen 1981, 25–26).

The Scene of Activity: Workers and Wanderers

All of this therefore serves as background to the great episode of translation that began in the late eleventh century and continued for approximately a century and a half thereafter. The main phase of activity lasted from about 1100 to 1220 and was centered in Spain (esp. Toledo), Italy (esp. Pisa, Salerno), and Sicily. Works in Arabic were translated first, mainly in Toledo, a city of great libraries, by such indefatigable men as Adelard of Bath, Gerard of Cremona, John of Seville, Hugh of Santalla, and Dominicus Gundissalinus. Between these individuals, during the years 1140 to 1180, a truly enormous portion of the Arabic-Grecian synthesis of scientific thought was rendered into Latin, including Ptolemy, Thābit ibn Qurrah, Euclid, al-Kwarīzmī, Archimedes, al-Kindī, Galen, al-Rhāzī (Rhazes), Geminus of Rhodes, al-

Fārābī, Aristotle, and al-Haytham, to name only the most well known. The latter part of the twelfth century and the early thirteenth century saw an increase in translations from the Greek in Italy and Sicily, stimulated in part by diffusion of the earlier work from Arabic (especially that by Gerard of Cremona) and by several intellectual delegations to procure manuscripts from Byzantium. A fair number of the names to be considered here are unknown; Burgundio of Pisa, Michael Scot and William of Moerbecke are notable exceptions. Generally speaking, translations from the Arabic were focused upon the sciences (including astrology and alchemy), philosophy (Aristotle in particular), and related commentary, while those from the Greek especially came to favor the Aristotelian corpus (D'Alverny 1982). Science and philosophy were not, however, the only subjects involved; religious works were also sought out and translated. An important project along these lines, for instance, was that of Peter the Venerable, Abbot of Cluny, who enlisted Robert of Ketton and Hermann of Carinthia to render into Latin the Qu'rān itself, along with a variety of other Islamic writings, in order to provide Christendom with the written knowledge it required to better and more precisely argue its case against Muslim thought (Jolivet 1988).

Profound similarities and differences mark this episode relative to its predecessor in early Islam. Primary among similarities are the scale of work involved, the choice of subject matter (sciences and philosophy), the heroic labors of individual translators, the travels these men undertook for the sake of acquiring requisite knowledge and manuscript material, and, in the realm of attitude, the basic sense that something of great "wealth" from another people and time was being discovered and appropriated. To this brief list, we might also add certain effects the resulting translations imposed, for example the vast enrichments of textual culture, new stimulations to authorship (to scholarly writing in general), the creation of entirely new vocabularies and thus influences on the Latin language, and the support for new educational institutions (in this case, the universities). All these things had also come to bear in Islam, though with different weight and contour.

But there were also important contrasts in the varieties of translation experience in Europe. First, this was work performed by individuals, operating under their own directives and ambitions, without any organized support. No such thing as the Bait al Ḥikma, the famed House of Wisdom established under al-Ma'mūn, whose auspices supported those such as Ḥunayn ibn Isḥāq, ever existed in Europe on the same scale. Nor did the Latin translators profit from anything resembling the widespread forms of patronage that existed in Islamic society. Since the nineteenth century, medieval scholars have often proposed that such an example was present in a "school of Toledo," operating under the patronage of Archbishop Raymond (flourished 1125–1152), to whom certain works were apparently dedicated. Yet, as discussed by D'Alverny (1982, 444–46), close examination of the relevant documents shows that such attributions have been based largely on errors and mistaken assumptions, and, in any case, have failed to note that the major period of translation in Toledo occurred after Raymond's episcopate had ended. The only clear example of royal patronage comes from Sicily (a rare and brilliant site of state patronage for the new learning), where Frederick II engaged the services of Michael Scot, known for his translations and knowledge of astrology (Haskins 1927). For the rest, some of the translators had ecclesiastical careers; some were civil servants or physicians; some made a living through their teaching or sale of their writings or through the occasional patron. Many appear to have lived with some degree of economic insecurity (see Haskins 1927, 1929).

The Latin translators were not simply individuals in their economic self-sufficiency and liberty from direct authority. They were wanderers, homeless intellectuals of a sort, who had embarked upon their careers out of private interest, ambition, and greed for knowledge. As pilgrim scholars and autodidacts, moreover, they were self-conscious heroes. Many of their writings begin with open expressions of contempt or rhetorical dismay for the "penury of the Latins" in all things of the mind. This had sometimes been done by Islamic writers too (al-Fārābī, for instance); but among the Latin translators it became a routine invocation, a type of spell for clearing ground. The sense of history here is palpably different from that, say, of al-Kindī, who speaks not of the past but of those future scholars who may improve upon and perfect his own work. In Europe there was a more overt need to declare the new, to propound discontinuity from the existing order, even to disparage it.

The late eleventh and twelfth centuries mark a time of enormous textual effort generally, when writing became essential to the exercise of power in society (Stock 1983; Classen 1981; Burnett 1998). Set down and fixed, language was rendered mobile: in a dozen different ways—from manuals on goldsmithing and astrology to new forms of literature such as diary keeping and the romance epic—the written word now fully entered the vast new circulation of *things*. Translation, then, was a necessary and inevitable part of all this. With the opening and expanding of international trade, it was of immediate practical value. No powerful merchant of the 1100s was without one or more interpreters (commerce with Islam demanded passage through several languages—

Arabic, Coptic, old-Italian, Latin, Spanish, French). Border regions and cities, such as major ports along the Mediterranean, Sicily above all, became places of profound cultural porosity. With the conquest of Spain, moreover, the libraries of western Islam—much smaller and more meager in their holdings than those in Egypt, Syria, and Iraq, but spared systematic destruction by the Mongols—became fully available and suggested the textual riches that might lie to the east. Ambitious young students, perhaps lured by the promise of fame or the excitement of new intellectual riches, left their homes in the north to visit famous centers of Arabic learning in the south, seeking to mine the new knowledge (D'Alverny 1982). By the mid-twelfth century, a trade in texts had begun, making translators the new merchants of the word. As in Islam, their efforts were intended to mobilize the textual wealth of one culture for the consumption and, later, naturalization of another.

The Scene of Ambition: Of Texts and Translators

Indeed, where men and women went, so went words, knowledge, and writing. And with this great new complexity of exchange came expansions of opportunity, yet, at the same time, routine failures of certainty. For many among the new learned class, this meant a restlessness with the given order and a desire for new intellectual materials (Lemay 1962; Classen 1981; Ferruolo 1984; Stiefel 1985). "It is worthwhile," wrote Adelard of Bath (d. circa 1146), a pioneer of the new translation movement and one of the great travelers of the late Middle Ages, "to visit learned men of different nations, and to remember whatever you find is most excellent in each case. For what the schools of Gaul do not know, those beyond the Alps reveal; what you do not learn among the Latins, well-informed Greece will teach you" (Thorndike 1923, 2.20). Though he speaks here of Greece, Adelard and those translators who came immediately before and after him more often praised the "learning and reason" of their "Arab teachers" as a source and model. Peter the Venerable, in his influential *Liber contra sectam sive haeresim Saracenorum* (Book against the Saracen Heresy), spoke of the Arabs as "skilled and learned men" who have huge libraries at their disposal filled with books on the liberal arts and the study of nature, far beyond anything existing in Europe (Jolivet 1988; Kritzeck 1964). A half-century earlier, at the dawn of the new translation movement, Constantine the African (mid- to late eleventh century) established a pattern by journeying widely in the East and arguing for a great transfer that would uncover the "more famous things which the Arabic language contains, the hidden secrets of philosophy." Only a decade or two later, Petrus Alfonsi, a converted Jew from Al-Andalus whose writings on Arabic learning were instrumental in helping further the translation movement, noted that "some of those men who investigate wisdom . . . prepare to traverse distant provinces and exile themselves in remote regions" (Tolan 1993, 175).

This idea of "hidden" or "distant" knowledge, allied with a great and untold "wealth" of intellectual material or a "rare, sweet, and delicious" abundance, is no less prevalent among the Latin translators than it was among their Arabic counterparts. A particularly excellent example is offered by Dominicus Gundissalinus, one of the most prominent of the twelfth-century translators in Toledo, who wrote in his *De anima* (On the Soul):

> I have carefully collected all the rational propositions about the soul that I have found in the works of the philosophers. Thus . . . a work hitherto unknown to Latin readers, since it was hidden in Greek and Arabic libraries, has now, by the grace of God and at the cost of immense labor, been made available to the Latin world so that the faithful . . . may know what to think about it, no longer through faith alone but also through reason. (Jolivet 1988, 142)

Such a passage is enormously suggestive of the translator's own historical self-image. Yet also crucial here is something embedded in the last few phrases: the advent of a new, secular rationalism that can be applied both to the natural world and to faith itself. This new approach to contemplating and investigating the universe of material and spiritual realities—and its expression in the arts, as a distinct naturalism—is a decided innovation of the late medieval period, deeply dependent on the new intellectual substance provided by the episode of translation.[4]

By the early to mid-twelfth century, a profound dissatisfaction had set in with the old, monastic centers of study. Bernard of Chartres moved to revise the traditional curriculum to include many new secular works of pagan Latin literature, but also works dealing with natural philosophy, with nature in general, which had been allowed little or no place in the *artes liberales* until that time, dominated as they were by the influence of Martianus Capella and Isidore of Seville (John of Salisbury 1955; Colish 1997). Among the most vehement in expressing his feelings of dissatisfaction with the reigning system of teaching was Daniel of Morley, an intellectual pilgrim who helped diffuse the new translations in northern Europe, particularly into England. Daniel

4. A great deal has been written on this topic, which essentially lies outside the present discussion. For several relevant surveys, see White 1948, Lindberg 1978, the essays contained in Weimar 1981, and those in Benson, Constable, and Lanham 1982, particularly by Ladner, Benton, d'Alverny, Martin, Dronke, Kitzinger, and Sauerländer.

wrote of his "disgust" with his teachers in Paris, "brutes" he called them (*bestiales*), who sat reading from books too heavy to move (*importabiles*), and whose speech was "childish" when it attempted to be learned (see Thorndike 1923, 2.172–73). "When I discovered things were like this," he wrote, "I did not want to get infected by a similar petrification" (Burnett 1997, 61). The connection between dead-letter study and immobility is clear. True knowledge was movable; its ability to be traded and transferred was the sign of its greater value. Adelard is no less clear on this: "What is an authority to be called, but a halter? As the brute beasts, indeed, are led anywhere . . . and have no idea by what they are led or why . . . so the authority of writers leads not a few of you into danger, tied and bound by brutish credulity" (Daniels 1975, 266). Only slightly less radical in tone were the comments of Petrus Alfonsi, who states how he has "discovered that almost all Latins are devoid of knowledge of this art of astronomy" (Tolan 1993, 174), or William of Conches, who dismisses all scholars that rely upon divine explanations of natural phenomena. John of Seville, Plato of Tivoli, and the great Gerard of Cremona, all of whom reportedly spoke of uncovering "the secrets of Arab learning" and overcoming "the poverty of the Latins," also argued the case of "reason."

It seems clear, therefore, that the Latin translators were entirely conscious of what their work might mean, both in historical and political terms. They comprehended the need for it and understood it as a way to mark a distinct and unprecedented place in history. In reading the various prefaces and commentaries of these writers, one hears time and again the complaint regarding the failed or shallow learning within Latin culture.[5] To those, meanwhile, who objected to Arab learning for its heretical aspects—for example, its placing God in the role of architect or craftsman, an idea that later came to have enormous consequences for the history of science—William of Conches expressed open scorn: "Because they are themselves ignorant of nature's forces and wish to have all men as companions in their ignorance, they are unwilling for anyone to investigate them, but prefer that we believe like peasants" (Lindberg 1992, 200).

The translators and those who sought immediate use of their work saw themselves not as laborers or mere scholars, but as crucial media through which the Greco-Arabic past would pass into the Latin present and future. The very scale of the translators' work, their large, at times remarkable output of texts, is ample testimony to their ambition and

5. A number of these prefaces are quoted (mainly in Latin) by Haskins (1927).

sense of adding to posterity. No less than Ḥunayn and his school, a number of these men, such as Gerard of Cremona and John of Seville, brought dozens upon dozens of highly complex works into Latin in a range of disciplines. Inspired by the Arabic example in another way, they also wrote their own commentaries upon these works for the sake of popularization (and also for money). Translation may have been their central activity, their life's work, yet it brought them at best a meager and occasional living. This suggests all the more their sense of mission. The translation movement took place through the cracks as it were—as a profession without an institution. Where Ḥunayn had been paid his own weight in gold, and by the caliph himself, set at the head of a school of translators, equipped with a workshop of assistants, paper, and utensils, and sent on distant missions to collect important texts, his correlative in the twelfth century, the prolific Gerard of Cremona, worked with a few collaborators, supported only by the sale of his work and contributions from his ill-trained students (Lemay 1978).

Literalness and Authority

What types of translations were produced during this episode? What philosophy or philosophies of language were embodied in them? What happened to the knowledge at hand when it was brought into Latin? To answer such questions, if only in summary fashion, we need to look at one other important difference between the Latin translators and their Arabic predecessors. Where the latter were active scholars, often contributing specialists in the field of their translations and excellent (though variable) linguists besides, the Latin translators frequently were not. In fact, they were "scholars" in the sense of being learned generalists, intellectual wanderers among the disciplines whose Arabic and Greek had in many cases been learned later in life and though advanced, remained imperfect. There were notable exceptions, to be sure: Adelard of Bath, whose Arabic was excellent enough for him to include marginal "jingles" and various mnemonic devices in this language on his Latin translations; or John of Seville, whose command of Arabic appears equally excellent and who focused almost entirely on astrological works; William of Moerbeke, who knew Greek well enough to retranslate nearly the entire Aristotelian corpus. But the rule holds in the main: these were magnificent amateurs who acquired the greater part of their learning as they went along.

This is reflected, to a degree, in their methods. From Adelard and Gerard in the twelfth century to William of Moerbeke in the thirteenth, translations often tended to err on the side of literalness, with frequent Arabicisms and Greekisms in terminology and syntax and with explanatory paraphrase reserved for only a few special terms (Opelt 1959; Murdoch 1968; Lindberg 1978; Busard 1977). In the case of certain translators, such as Gerard of Cremona, the "slavish fidelity" seen in such texts as the *Almagest* seems partly a result of the difficulty of the material, partly due to the use of one or more intermediaries who translated the source work orally into an intermediary vernacular (Kunitzsch 1974). Gerard, in fact, if the few small indications about his life are true, appears to have worked on this one text for decades, suggesting he used it as a primary means to learn Arabic and that he lingered over nearly every passage like a true exegete (Lemay 1978). The worst cases of literal renderings were practiced upon Greek sources, where even individual particles and specific word order were obeyed (Murdoch 1968). Such deformations in Latin syntax and grammar usually did not lead to creative innovation; the result was too idiosyncratic and often of dubious legibility. A result like this represented a kind of mediatory document, located somewhere between the source and target languages.

Yet the case for literalness, as guiding method, has been also considerably overstated. Some authors, such as William of Conches and Hugh of Santalla, composed works of great eloquence and creative vocabulary. Others, like the highly influential Constantinus Africanus, Adelard of Bath, and Dominicus Gundissalinus produced translations that were closer to adaptations, rewriting, omitting, or expanding portions of the source work in a manner that suggests no particular method or systematic approach (McVaugh 1973; Clagett 1953, 1972; Jolivet 1988; Burnett 1985, 1998). In Adelard's case, three strikingly different versions of a single, extremely important work—Euclid's *Elements*—were produced (Clagett 1953, 1972; Busard 1977). One of these is a close, literal rendering of an Arabic original, complete with Arabicisms in terminology and syntax; another is a combination showing close renderings of certain passages and wholesale rewriting of others (esp. proofs); a third version is more in the manner of a commentary, but quotes version 2 in a number of places. Of these three renderings, the second had the most influence on succeeding late medieval readers.

One well-regarded summary of the twelfth-century Latin translator's art has characterized the result as follows:

> There was little variety in the vocabulary employed by the literalists; it was safer to render a given Greek or Arabic phrase in a fixed and determined manner. Transliteration was frequent when the translator could

> find no exact Latin equivalent, and this had the benefit of introducing Arabic and Greek technical terms into the Western vocabulary. (Lindberg 1978, 78)

This is both accurate and mistaken, as we will see below in the case of Gerard of Cremona. Certainly the uptake of many Arabic and Greek terms occurred in this way, just as it had in Islam. Such was a pattern of no small consequence: words such as azimuth, nadir, and zenith entered Latin directly from Arabic (specifically in early translations on the astrolabe), along with dozens of others that did not survive. Greek words (or cognates) were even more numerous, in part because of transliteration by Arabic translators, but also for reasons of favoritism from the thirteenth century onward. Yet it must be said, too, that the changing, sometimes unsystematic manner in which most Latin translators worked, as well as their variable grasp of Arabic or Greek, guaranteed that individual terms or phrases would find a range of linguistic faces (or none at all) in any single work. While certain translations and translators do clearly show a high consciousness devoted to nomenclature, to consistency, and to completeness, this was not the rule: Gerard of Cremona spent much time and effort putting back original passages, standardizing terms, and rewriting passages in works earlier rendered by such scholars as Dominicus Gundissalinus and John of Seville (Lemay 1978). This does not disavow the goals of fidelity entirely, only their consistent achievement.

The Latin translators were not strict idealists. They had few of the resources, textual and otherwise, that their Arabic forebears possessed. Both literalness and adaptation were practical choices for them, backed by a long, if complex, tradition of scholarly argument and linguistic philosophy. In the end, their methods and techniques were as varied as their backgrounds, specific circumstances, linguistic competence, and textual preferences. As the case of Adelard shows, the work of individual translators could reveal nearly as much diversity in approach as existed among the translators as a whole. It is a grave error, indeed, to impose any single characterization on a corpus of work so multifarious in origin and aspect as these translations.

Medieval Theories of Translation: A Brief Review

Traditions centering on the idea of *translatio* began with a double legacy left to the European Middle Ages by patristic writings on translation, especially those by St. Jerome (Copeland 1991). Much returns to his "Letter to Pammachius," composed sometime between A.D. 405

and 410: "I admit and confess most freely that I have not translated word for word in my translations of Greek texts, but sense for sense, except in the case of scriptures in which even the order of the words is a mystery. Cicero has been my teacher in this" (Lefevere 1992a, 47). This condenses a good deal of what constitutes an extended discussion on translation philosophy.[6] Jerome advises the utmost textual fidelity for sacred works, whose divine qualities come from the very order of the words as well as their meaning. For secular writings, however, such fidelity is a mistake; it "conceals the sense, even as an overgrown field chokes the seeds." Jerome even conceives the metaphor of the translator taking meaning as an unwilling captive into his own language.

Yet, as recently noted (Copeland 1991, 45–55), such a philosophy of literary conquest was later turned entirely on its head. Writers like Boethius and John Scotus Eriugena, whose influence on later medieval translators became far greater, argued the case for the *fidus interpres* in secular as well as biblical writings, with an overarching regard for accuracy in all things (Schwartz 1944). The reasons for this inversion are not clear, but seem suggested by the following quotation from Boethius' translation of Porphyry's *Isagoge,* a crucial summary of Aristotelian thought:

> I fear that I have incurred the blame of the "faithful translator" as I have rendered it word for word, plainly and equally. And here is the reason for this procedure: that in these writings in which knowledge of the matter is sought, it is necessary to provide not the charm of a sparkling style, but the uncorrupted truth. Wherefore I count myself very successful if, with philosophical texts rendered into the Latin language through sound and irreproachable translations, there be no further need of Greek texts. (Copeland 1990, 52)

As was the case with Roman translators like Cicero and Quintilian, Boethius considers the deeper work of the *fidus interpres* to be historical and cultural. The best translators have the power to reconstitute and replace the source text entirely—not merely to take it prisoner in a new language but to lock it away in a golden cage of forgetting. Yet the Romans here are also themselves done away with or ignored by the literalism thought best able to achieve this reconstitution. The crucial point here seems to be the image of knowledge, as a sort of sacred substance, and its juxtaposition against the more superficial (i.e., corrupting) finer-

6. Much of the subtlety of this discussion, as worked out mainly in biblical and literary texts, can be found in the various scholarly essays presented in Ellis 1989 and Beer 1989, 1997.

ies of style. Those who devote themselves to translating philosophy (which includes science) must reach for "uncorrupted truth"; anything less in motive and act is itself impure, debased. Words, to this view, were the very essence of learning, the primal bearers of the material universe in the realm of human experience. Language was not merely the organ of truth but the substance of its possibility, and therefore of any and all ability to possess it. This was a view fully institutionalized in the trivium (grammar, dialectic, rhetoric), the cardinal realm of medieval learning. And, as such, it was transposed to study of the *quadrivium* as well. Writing continued to be guided by "the interconnections between words, thoughts, and things" (Stock 1983, 243). If it was true that certain authors, such as St. Hilary (d. A.D. 367), had earlier spoken of the failure of the written word (*sermo*) to adequately convey its object (*res*), this did not prevent writing itself from being the conveyor of such ideas, nor the means through which they might be entered into the canon of timeless thoughts.

This hallowed belief in the authority of language was embodied in an equally ingrained system of textual veneration. Such was the *auctoritates* system, in which authors deemed as important were viewed in patristic terms, as bearers of timeless wisdom and value. This was medieval Europe's equivalent to Islam's idea of the "sage," whose knowledge was of a semi-divine character. In the medieval schools and, later, the universities, as historians have often pointed out, canonical authors from various periods of antiquity, the early Christian period, the Carolingian era, and so on were studied side by side with little or no regard to chronology (e.g., Rashdall 1936; Delhaye 1947; Curtius 1953). Such an approach was not merely maintained but more deeply ingrained by the twelfth-century curriculum revisions of Bernard of Chartres and others, when a number of new secular works from Roman literature and natural philosophy were introduced (Speer 1995). Commentators and writers of *glose* of the twelfth and thirteenth centuries, such as William of Conches, wrote works on classical, early medieval, and contemporary authors without any necessary effort or interest expressed in specialization. Among the famed commentaries produced by the abbey of St. Victor in Paris, it was apparently common to include all manner of classical references (Cicero, Seneca, Virgil, Juvenal, etc.), as well as those from Jewish sources, in discussions of specific biblical passages (Curtius 1953).

Such styles of reverence also make it hardly surprising that a literal approach to translation was often adopted by most of the translators. To be literal, even perhaps to the point of absurdity, meant respect and reverence with regard to source material and thus authorial power. To render texts mobile meant to decant them as they were, without spilling a drop of their greater contribution toward the causes of knowledge, authority, and fame—including that of the translators themselves, many of whom sought for themselves entrance into the realm of *auctoritates.*

Yet, as noted, *fidus interpres* is by no means the final word on medieval translation theory and craft. The other half of Jerome's method—that most often proclaimed by Cicero and other Roman translators—also became a constant intellectual guide for translators, at least in certain circumstances. The common notion that literalism ruled in all things, especially religious ones, turns out to be quite false (Kelly 1997; Dembowski 1997; Shore 1995). Indeed, one finds in scientific texts a wide spectrum of method, from abject attempts at fidelity to ardent paraphrase. More, such variety of technique frequently occurred within single texts, just as we saw in the case of the Arabic translators. The impossibility of absolute *fidus interpres* between such dissimilar languages as Greek, Arabic, Latin, Catalan, and Old English ensured in every case that significant changes would occur, whether by design or (more commonly) by necessity.

Another frequent presumption, that medieval translators did not think very much about their craft but instead worked mainly as craftsmen, through imitation and a certain desire for the safety presumably guaranteed by literalism, is also difficult to maintain. Even during the twelfth century, the translation of a major work by an authoritative source, be it holy or secular, was a rare event of labor, a self-conscious undertaking. There was no such thing as a guild of translators; there were no large-scale demands for translated works. The men who performed this labor did so as individuals, with specific ambitions—whether to sell the work in question for a decent sum, to perform a requested service, to act as a transmitter of "secret" wisdom, or to enter the ranks of *auctoritates.*

An elegant discussion on early Anglo-Saxon renderings of religious texts into Old English, for example, shows a very different style of thinking about the subject from Boethius, and a very high level of consciousness of the problems involved (Amos 1920). The translation work of Alfred in the late ninth century and Aelfric in the tenth shows a direct influence stemming from the Romans, as a likely result of the Carolingian *renovatio* of classical learning. The emphasis is here "now word for word, now sense for sense," with a specific intent to popularize the original Latin text and to enrich the vernacular tongue as well. Interestingly enough, Boethius is one of the authors chosen for this purpose (specifically the proem to his *Consolation of Philosophy*). It is in the preface to Alfred's version of St. Augustine's soliloquies, however, that we find what can only be one of the most extended and wonderfully suggestive descriptions of the translation process in the whole of medieval literature:

> I gathered for myself cudgels, and stud-shafts, and horizontal shafts, and helves for each of the tools that I could work with, and bow-timbers and bolt-timbers for every work that I could perform, the comeliest trees, as many as I could carry. Neither came I with a burden home, for it did not please me to bring all the wood back, even if I could bear it. In each tree I saw something that I needed at home; therefore I advise each one who can, and has many wains, that he direct his steps to the same wood where I cut the stud-shafts. Let him fetch more for himself, and load his wains with fair beams, that he may wind many a neat wall, and erect many a rare house, and build a fair town, and therein may dwell merrily and softly both winter and summer, as I have not yet done.[7]

The image of translator-as-builder seems a truly excellent one. Besides the magnificent literary qualities of this passage, it very neatly describes not only the physical difficulty of the labor and the cultural construction (and importance) of the result, but the process involved as one of mixed peasant toil and aesthetic pleasure, based on selection and adaptation as much as on a simple carrying over. The translator is a builder of textual towns and cities; his yeoman's work involves the cutting down and physical reshaping of a material that has grown up, over time, in another setting. He chooses, omits, assembles, and leaves to others what improvement or rebuilding might be necessary. Against Boethius' mason, working brick-by-brick, Alfred poses the journeyman carpenter.

Taken together, the words by Boethius and by Alfred read like general manifestoes for the late medieval translators of Arabic and Greek works. If many sought faithful transfer of Ptolemy, Euclid, Archimedes, al-Kwarīzmī, ibn-Haytham, and so forth, under the belief in a direct correspondence, even inseparability, between knowing and speaking, others felt no such literal reverence for their sources, but instead saw them as materials to be reformed in some sense for better use and comprehension. Taken as a whole, translated works of this time period indicate a wide range of specific methods; no generalities apply across the board. Diversity came from differences in conditions and locations, use or nonuse of intermediaries, demands made by patrons or by the in-

7. I have quoted this passage, rendered into modern English, from Amos 1920 (3–4), who in turn borrows it from Hargrove 1904 (2–3). The interested reader is referred to both of these works for further discussion along these lines.

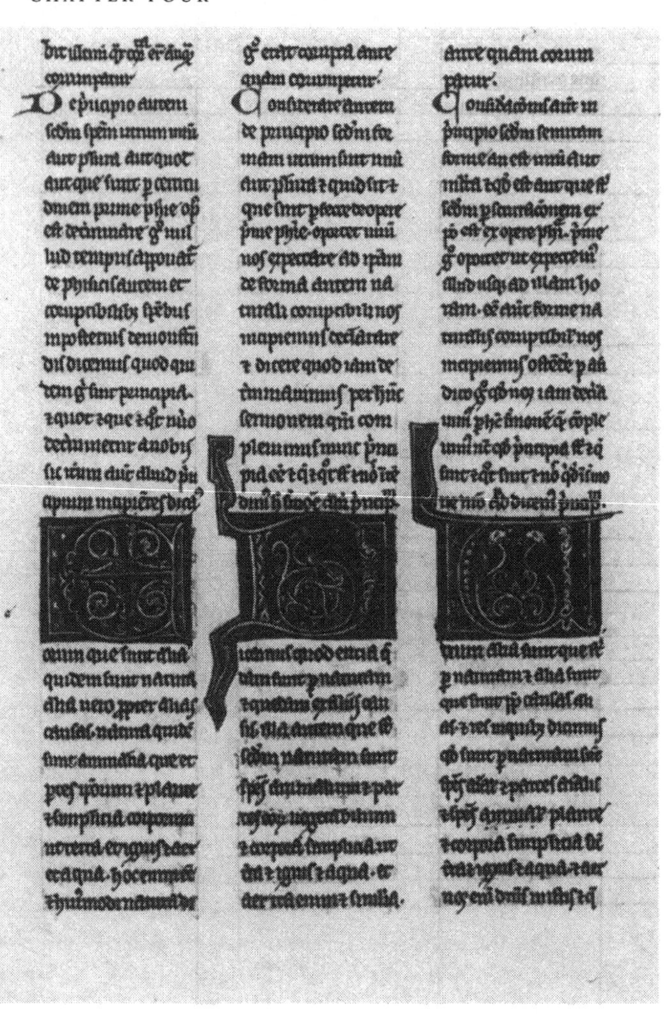

Figure 11. Manuscript page showing three separate Latin translations of Aristotle's *Physics*. The version on the far left was made from the Greek, the other two from the Arabic, with the far right translation by Gerard of Cremona. The existence of such a work, though rare, suggests that evaluation of different translations was common enough to produce a text for study of this phenomenon. Bibliothèque Nationale, Paris, lat 16141, fol. 25r.

tended use of a work, levels of difficulty and quality among source works, and also the individual intelligence, training, and sensibility of each translator. There is evidence to suggest that translations were often compared, sometimes even for study or training (fig. 11). Whatever the case, *fidus interpres* no more ruled the landscape of technique than its opposite.

Gerard of Cremona: Attitudes and Methods

Some of the translators were born in Spain (Mozarabs) and converted to Christianity, seeing in the fall of Toledo (A.D. 1085) an opportunity to make themselves useful to the new conquerors (Millás Vallicrosa 1963). Jewish scholars such as Abraham bar Hiyya or Ibn Ezra established careers translating Arabic works into Hebrew. This they did for the burgeoning Jewish intellectual communities in southern France and Italy. Most of the Latin translators, however, came from the north or east, from England, France, and Italy, drawn (it seems) by the news of fabulous intellectual resources. In these areas, the image of Islam had long held an aura of material riches generally, vast palaces filled with wealth, harems, the glitter of unchecked indulgence (Southern 1962). In the twelfth century, however, this stereotype was turned on its head in the realm of learning and textual knowledge—into overwhelming aspiration for a magnificent plenitude of mind.

Gerard of Cremona, for example, according to what paltry biographical information exists, left Italy for Toledo around 1144, specifically to find a copy of Ptolemy's *Almagest,* which was "nowhere to be found among the Latins," to learn Arabic, and to render this work into Latin (Lemay 1978). It was this text, with its enormous scope and obvious potential importance—of whose existence many scholars had become aware, in part through earlier translated works on the astrolabe—that seems to have fired his original ambition. It by no means exhausted such ambition, however. According to a eulogy written sometime after his death (A.D. 1187) and attached to one of his translations of Galen, during four decades of activity, he introduced into Latin no less than seventy-one major works in the fields of medicine, astronomy, mathematics, optics, logic, geometry, and natural philosophy, as well as astrology, alchemy, and geomancy (Sudhoff 1914; Grant 1974; Lemay 1978; d'Alverny 1982). In addition to the *Almagest,* many of these works were Arabic versions of what had once been classical Greek texts by Aristotle, Euclid, Archimedes, and Galen. But more than half of the works associated with Gerard were original writings by Arabic authors: al-Kwarīzmī, ibn Yusūf, al-Farghānī, Thābit ibn Qurrah, al-Kindī, and Ibn Sīnā, including the latter's *Book of Healing,* as well as numerous books of the occult sciences. The enormous amount of effort represented by this work, which included acquiring and preparing all the various individual texts, may have required Gerard to use assistants or collaborators of one sort or another.

One point should be emphasized regarding Gerard's work: the tremendous contribution he made to the Western intellectual tradition (for this must be acknowledged) was based on bringing into medieval Latin a large portion of the Arabic-Indian-Persian-Greek synthesis in science, mainly as it had been forged in the Arabic language. In the words of Richard Lemay,

> Gerard's transmission of this heritage to the West had an exclusively Arab garb that directly reflected scholarship in the Arab world at the time of the Latin's contact. Furthermore, Gerard's influence perpetuated this interplay throughout most of the Middle Ages. Even the Greek interests [stimulated] by William of Moerbeke's translations in the late thirteenth century remained minor in comparison with the weight still carried by the Arab tradition. . . . Latin science and philosophy of the twelfth–fourteenth centuries was totally dependent on the Arabic tradition. This is in sharp contrast with the situation of scholarship in the Renaissance, which [focused on] the original Greek texts. (1978, 176)

Gerard seems to have purposely tackled many of the most potentially significant and difficult texts in existence: Euclid's *Elements,* the *Almagest,* al-Kwarīzmī's *Algebra,* Aristotle's works on natural philosophy, the *Optics* of al-Haytham, the medical works of al-Rāzī. There may well have been some type of plan here; Gerard may have been advised by his Mozarab collaborator(s) or some other authority, likely a Toledo Arabic master, about which texts and authors to select.

Gerard was the Ḥunayn of his day—with certain differences, of course. What he produced, with the help of assistants and students, was an entire university of textual material, nearly all of it crucial to the development of Western science thereafter. His methods of collation, too, were close to those of Ḥunayn. He searched for different versions of a particular text in order to assemble a manuscript he felt was most "complete": his versions of Aristotle's *De caelo* and Ptolemy's *Almagest* indicate that he translated passages from two or perhaps more versions (Opelt 1959; Kunitzsch 1974). Yet, if it is possible to speak cogently on the basis of meager evidence, his guiding vision was somewhat different from that of the Arabic translators. He seems to have viewed his effort less as a matter of archeology, of retrieving "the wisdom of the ancients," than one of actual "discovery." Certainly he seems to have been less than humble about his historical role and accomplishments. According to the report by Daniel of Morely, he described himself as being born under a royal sign: "'I who speak am king' . . . and when his students ask, with true irony, Where then does the 'king' reign? He responds: 'In the mind, for I serve no living man'" (Lemay 1978, 174).

Despite such statements, Gerard did not work alone. Daniel of Morely also tells us that Gerard enlisted the aid of a Mozarab named

Galib, who helped with the *Almagest* translation by reading the text out loud, probably into Spanish, which Gerard then rendered into Latin (Sudhoff 1914). This type of method makes sense in light of the great quantity and variety of Gerard's translations. Also, "he had learned Arabic, but was probably not proficient enough to fulfill such an ambitious program by himself; it is likely that he enlisted not only Galib but other interpreters to hasten the work" (d'Alverny 1982, 453). Even more than his Nestorian counterpart, Gerard was interested in knowledge linked to certain images of power and use, knowledge that was new, challenging, capable of adding unanticipated textual wealth to the culture of his origin. Where Ḥunayn was transmitting a corpus that had long been part of the scholastic tradition of his own sect, the Nestorians, and that in some part must have been familiar to him, Gerard was embarked on a mission to appropriate what was truly unknown, or known mainly through hints and hearsay, to his own world. His work of "discovery" was made somewhat easier by the simple fact that Arab scholars had collected, organized, and taught scientific material in a way that roughly corresponded to the medieval *quadrivium*—mathematics, astronomy, geometry, and music. The most significant texts were those that students had to master and that therefore offered a sort of ready-made curriculum for Gerard to mine. "It would seem," Lemay (1978, 188) writes, "that Gerard of Cremona sought out such collections, particularly in mathematics, astronomy, and medicine, in order to translate them as a corpus in each branch of the *quadrivium*. . . . [The] larger number of his translations in those fields frequently agrees with the order of those Arab scientific collections."

Indeed, this may understate the matter significantly. Gerard, it now appears, was following a specific guide that laid out in detail each text that should be included within the ideal curriculum of higher learning, particularly in natural philosophy. This guide was none other than al-Fārābī's *On the Classification of the Sciences* (*Kitāb Iḥṣa' al-ʿUlūm*). "If one looks at the translations of Gerard of Cremona it seems that he was methodically working his way through Alfarabi's list of titles, to make sure that the Latins would have a complete set of authoritative texts" (Burnett 1997, 71). In reality, Gerard went beyond these limits as well, venturing into astrology, magic, the Arabic commentary literature, and more. But the notion that, in much of his work, he was following a distinct program seems irrefutable. Indeed, it appears evident that after his death in 1187, at least one important translator, Alfred of Shareshill, stepped in to help complete the project (Burnett 1997). Thus, there is some evidence that the Latin translators tended to prefer those versions of a particular text that had been developed for pedagogic use. The sources they used for their translations, in short, were most often those that Arab scholars had assembled, corrected, edited, and sometimes reorganized or even partly rewritten for students. Certainly a great many of the new texts were very soon taken up by masters at various schools, copied, studied, and commented upon by scholars in such intellectual centers as Oxford, Paris, Chartres, and Bologna. Within half a century, by the early 1200s, the new textual "wealth" had been rapidly (if variably) taken up as a foundation for the new universities. Indeed, as a new institution on the landscape of Western intellectual culture, the university appears to have been a principal mode by which the new knowledge from Islam was absorbed.

TRANSLATION TECHNIQUES: THE *DE CAELO* AND *ALMAGEST*

What of Gerard's methods, as expressed in his translations? What sort of translation techniques did he use? It is fortunate, indeed, that some effort has been given to this topic.[8] As it happens, the "slavish literalism" so often attributed to Gerard turns out to be true in certain instances but wholly insufficient and a bit smothering as a blanket characterization. Two astronomical texts that show this are his versions of Aristotle's *De caelo* and Ptolemy's *Almagest.* The former he rendered from a text by Yaḥyā ibn al-Biṭrīq (late eighth century), who had possibly used a sixth-century Syriac version (attributed, mistakenly perhaps,

8. Detailed treatments on Gerard's astronomical works can be found in Opelt 1959 and Kunitzsch 1974. Opelt concentrates on Gerard's translation of Aristotle's *De caelo.* The source text he used appears, on the basis of various comparisons, to have been a late copy of a translation by Abū Yahyā ibn al-Bīṭṭrīq (late eighth century A.D.) of the earlier Syrian version by Sergius of Reshaina. This work by Sergius, meanwhile, included large portions of Aristotle's book (as it then existed) but not, apparently, the entire text. What was transmitted to Europe, through Gerard's hands, was thus no longer a work specifically by Aristotle, but something else, a hybrid. This is made all the more clear by the various errors and mistranslations it contains. An example cited by Opelt 1959 (137) is the name of the Greek author Hesiod, which became in Arabic "cosmitros," due to a misreading by an Arabic copyist, and was then taken up by Gerard himself and transliterated in his Latin version.

Despite the valuable information it contains, Opelt's study suffers significantly, as it too often compares Gerard's specific renderings with their *Greek* equivalents. This seems misguided, since Gerard's source was obviously in Arabic: the Greek "original" (which, of course, was nothing of the sort, even when it was introduced centuries later) did not exist for Europe at this time and has no historical relevance to Gerard's specific work and influence.

Another text dealing in detail with Gerard's translation methods that does not suffer from this flaw is Minio-Paluello 1961. I have not discussed this study here as it examines Aristotle's works on logic and not astronomical work(s). Many of Minio-Paluello's major points, however, agree with those made by Opelt. Finally, Lemay (1978, 181–82) also gives a brief look at some related issues regarding Gerard's translation of philosophical terms. It would seem that a good deal more scholarly work might be done in this area.

to Sergius of Reshaina) as his own source. As we saw in the previous chapter, Syriac translators of this time period had themselves made certain alterations from their Greek sources; thus, what came into Gerard's hands was almost certainly a hybrid document, at least twice removed from whatever "original" might have existed in fourth- or fifth-century Greek. For the *Almagest,* meanwhile, Gerard used the versions by al-Ḥajjāj and Ḥunayn. According to Kunitzsch (1974, 104), the al-Ḥajjāj rendering was relied upon mainly up until a certain point in the text, after which the version by Ḥunayn became the dominant source. This is attributed to the possibility that the long-term effort he devoted to this work eventually caused Gerard to perceive certain superiorities in the Ḥunayn version. Yet the evidence makes it is just as likely that the translator weighed both and found them of equal merit for different portions of the whole.

The *De caelo*

In any case, with regard to *De caelo,* Opelt (1959) shows beyond any doubt that Gerard's Latin version is characterized by a highly complex mixture of close fidelity with selective paraphrase, fluctuations in terminology, mistranslated words, omissions, theological additions, and corruptions of Arabic versions of Greek names. Moreover, Gerard also included in his version various glosses by Arab commentators that (Opelt tells us) he "mistakenly" took for part of the "original." This may not be true; the translator may well have recognized these brief explanatory passages for what they were and simply decided to include them. Either way, however, such glosses very much *were* part of the Arabic source, and not of its Greek forerunner. They are the very essence of the "Arabic-Greek synthesis," an expression of the state of Arab scholarship as transmitted to Europe at this time.

As to some of the other characteristics noted by Opelt, a few specific examples might suffice. With regard to terminology, there appear to be changes and shifts that do not carry much meaning, e.g., the use of both *potentia* and *virtus* for what was a single Greek term (δυναμιζ). Some terms that are simple in both the Greek and the Arabic are given a much more extensive Latin equivalent, such as δηλον = *wâsih* ("proven"), which became *ostensio illius et verificatio.* An interesting corruption exists in a passage where Aristotle discusses the theory that elephants emerge as a species at both ends of the Earth, i.e., in India and at the Pillars of Hercules: Gerard here appears to transcribe an error in (or misreading of?) the Arabic, which renders the Greek ελεφαντεξ into the incomprehensible *alcobati* in Latin. While most Greek proper names came into Latin correctly, or nearly so, a few exhibit interesting changes: the natural philosophers Anaxagoras and Anaximenes, for instance, are, respectively, curiously replaced with *Pythagoras/Pythagorici* and *Acsimes,* the latter of which appears derived from a careless transcription of a corrupted Arabic rendering in a late copy of al-Biṭrīq's version. Certain abstract terms (e.g., *αχμη* and *μιγμα,* "acme" and "migma"), such as those of the pre-Socratics and those for which Latin equivalents were particularly difficult, were simply omitted by Gerard, though they existed in his Arabic source. In this respect, the tripartite division of asymmetrical movement applied to the planets—acceleration (*εριτασι*), zenith (*αχμη*), and deceleration, or retardation (*ανεσιζ*), retained in the Arabic—was reduced in the Latin version to a contrasting couplet of terms of far simpler and more ordinary sense: *vehementia-mollificatio.*

The greater portion of Gerard's text does, in fact, follow the Arabic very closely. Examples of close rendering are everywhere apparent, in terminology and syntax both. The Arabic phrase *'ilm-un-nugūm* ("science of the stars"), for example, which al-Biṭrīq used for *αστρολογια* ("astrology"), became in Gerard's Latin *ars scientiae astrorum.* More generally, Gerard uses Latin prepositions (*in, ex, sub, super*) to imitate Arabic versions of Greek constructions. An example here is his employment of *ex* consistently in the sense of a partitive, to form sometimes large and complex phrases after the model of the Arabic *min,* intended to translate genitive forms in Greek (Opelt 1959, 150–51). An example of the type of exact parallel Gerard often strove for is shown by the following phrase: *omnis res ex rebus generatis* ("all things are generated from things"), for the Arabic *shai' min-al-aṣyā*. Such parallel repetition of *res/rebus* for *shai/aṣyā,* rendered with a fidelity seen in many other phrases in the text, argues strongly that the translator was fully conscious, even philosophically (certainly linguistically) committed to, the basic tenets of the *fidus interpres* in such specific instances.

The essential point here is that this philosophy was practiced upon Arabic texts, not Greek ones, and thus conveyed to Europe something specifically reflective of Arabic-speaking culture. This is especially and concretely evident in two types of substantive change al-Biṭrīq introduced to the Aristotelian text and Gerard passed on. The first type of change takes the form of additions and glosses made to the introductory and concluding portions of each of the book's sections. Interestingly, many of these added passages express Aristotle's conclusions in much stronger and more certain language than the Greek even suggests, using such formulaic phrases as *Et dico iterum* ("I say once more . . ."), *ostensum est et declaratum* ("as it is stated and shown"), and the still more frequent *manifestum et certum* ("as revealed and proven"). At the same time, there are augmentations of a decidedly monotheistic character: invocations of God the creator (*Deum et eius potestatem creandi*), glosses that suggest correspondence be drawn between certain Pythagorean ideas and hell, praises raised to the Almighty (*deo, cuius fama est sublimis*), and more. Whether some or all of these changes were introduced by al-Biṭrīq, his later copyists, or the Syriac version al-Biṭrīq may have used as a partial source cannot be easily discerned. What *is* clear is that Gerard very often had ready-made Latin equivalents to such phrases at his disposal, due to their abundance in Christian literature. A certain degree of adaptation therefore took place through this aspect of his loyalty to the *fidus interpres.*

Gerard's *Almagest*

With respect to the *Almagest,* meanwhile, other interesting changes can be seen. For his translation of this work, Gerard used both the Arabic versions by al-Ḥajjāj and Ḥunayn, as revised by Thābit ibn Qurrah. Though an earlier translation had actually been produced in Sicily from the Greek, on the occasion of a manuscript being brought from Constantinople (d'Alverny 1982, 434), the result appears to have had little influence compared with Gerard's, which "remained the standard version of the *Almagest* in Europe until Renaissance times" (Kunitzsch 1993, 56). From the detailed analysis provided by Kunitzsch (1974), comparison between the Latin of Gerard's translation and the Arabic of his sources gives one the first impression of an almost abject literalism. As with *De caelo,* the use of Arabic prepositional constructions as a syntactic calque and the transliteration or word-for-word rendering of technical terms and phrases are all evident. Mistakes or corruptions in the Arabic were therefore transferred directly to this most influential Latin text. Examples here include certain terms used in star descriptions, such as *pupilla deferens* (lit. "eclipse, or carrying away, of the orphan girl") applied to the constellation Lyre (instead of Virgo, for instance), or the use of *crinis* and *coma* (both meaning "hair") in descriptions of Scorpio (Kunitzsch 1974, 107). Similarly, a number of the actual names for the constellations, stars, and earlier astronomers were rather clumsily transcribed from Arabic transcriptions of the original Greek titles: *Cheichius* for Kepheus, *anchoris* for Antares, *abrachis* for Hipparchus, *mileus* for Menelaus (pp. 108, 160–62).

In many aspects of the translation, however, Gerard's general fidelity served the cause of astronomical terminology and science very well. The very important terms for "degree" and "minute," for example, which in the original Greek were mainly written *μοιρα* ("part") and *λεπτον* ("small, thin") and were accurately translated as *guz* and *daq'qa* in Arabic, with the addition of *dara'a* (for "degree") in the star catalogue, became under Gerard's pen *pars* and *gradus* (for "degree") and *minutum.* Again, these were terms Gerard did not coin himself but drew from a long tradition of use in medieval science, as adapted from classical writings.[9] The word for "constellation," meanwhile, which had acquired a number of different equivalents in antiquity (*signifer, sidera, stella,* etc.), was now finally standardized by Gerard as *stellatio,* in a loyal rendering of the Arabic verbal substantive *kawkaba* (p. 171). In general, a large part of mathematical astronomical terminology was now purged of any confusing astrological variants or associations and fixed for the future.

It is in the area of names, however, that the contributions of Gerard's *Almagest* seem particularly interesting in view of translation. Here, as noted, some curious corruptions resulted: Gerard does not appear to have had any interest in (or knowledge of) Aratus or Hyginus and apparently either was unfamiliar with Pliny's discussion of the constellations or else ignored it. Indeed, there are enough "errors" and changes made in this area to convince one that the translator did not use any sort of then-standard reference on the constellations but instead went about his work by direct transcription and, possibly, with the occasional aid of memory (enough of the constellations are correct, in terms of common Latin usage at the time, to convince one of the latter). What makes this significant is that Gerard's transcriptions had the effect of helping introduce a number of Arabic star names into European astronomical discourse, where they remain to this day. An excellent example is the translator's direct version of al-Ḥajjāj's title for Perseus: *cuius nomen in latino est perseus et est deferens caput algol* ("Its name in Latin is Perseus, the bearer of the demon's head"), from which the star name Algol ("demon" or "giant" in Arabic) derives. Another example is the star Alphecca in the constellation Corona Borealis, whose title was translated literally from the Arabic as *corona septentrionalis et est alfecca,* with the last indicated as the ancient Arabic name. A total of sixteen other names were introduced in a similar manner—though, as we shall see, their survival to the present was not quite as simple as might be assumed.

Retention of Arabic: A Few Other Examples

The decision to retain Arabic words in Latin translation was a common one among the twelfth-century translators. Often enough, this was en-

9. For example, Pliny, in Book 2 of the *Naturalis historia,* uses *pars* frequently in a similar context with regard to planetary motion.

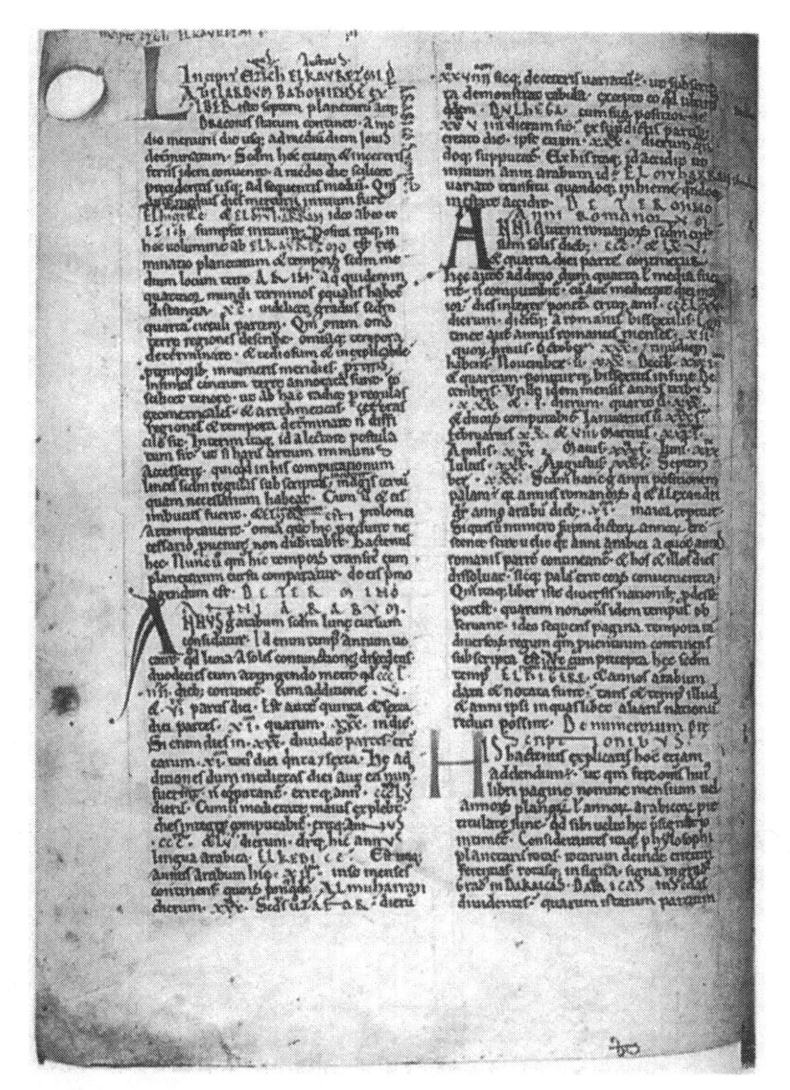

Figure 12. Opening page to Adelard of Bath's translation of the *zij* of al-Kwarīzmī, a sophisticated update and correction based on the *Handy Tables* of Ptolemy. The manuscript retains a large number of astronomical terms in Arabic. These, as well as numerals, are written in red ink. Bodleian Library, Auct. F.1.9, fol. 99v.

tirely conscious and illustrates a mixture of linguistic sophistication, elevated motive, and historical naivete.

Adelard of Bath provides an excellent case in point. As a proclaimed creator of new books for a new learning (esp. in England), Adelard remained very much aware of how he was using language. In his Latin version of the tables of al-Kwarīzmī concerning planetary position—a *zij* of great historical importance in Islamic astronomy—he retains many significant terms in Arabic but apparently (based on close copies of the same era) wrote these in red ink, to distinguish them from the rest of the text (fig. 12). Indeed, use of different colored inks was a common method of separating and highlighting different forms of information in late medieval manuscript culture, particularly when diagrams and tables were employed (e.g., in *computi*). Adelard has adapted this to a linguistic purpose, possibly as an aid to memory. Indeed, in other works he translated or wrote himself, he includes marginal poems to help the Latin reader remember Arabic terms or, more commonly, numbers.

Doubtless one of the most interesting examples of retained Arabic can be found in Adelard's translation of a Book on Magic (*Liber prestigiorum*) written by none other than Thābit ibn Qurrah. One portion of the work contains advice for helping a wife rejuvenate her husband's faded love. For this purpose, the wife must tie an image of a female beauty to a talisman of the husband, while a spell is chanted:

> O fount of honor, joy and light of the world! Mix together the loves of these two people, o spirits, using your knowledge of mixing, and being helped towards this end by the greatest power and the might of *al-malik al-quddus wa al-hayah ad-daima* [the king, the holy, and eternal life], and by the power of him who moves the heavenly circles, giving to them *nur wa iyan* [light and illumination] over this world. (Burnett 1997, 41)

Crucial invocations are kept in their original linguistic form—a sign of Adelard's sensitivity to magic as a language-bound medium (what *is* a spell, if not words?). The issue brings forward a herd of other questions regarding the fate suffered by words of power when utterly changed or abandoned by translation. At the very least, Adelard seems aware that the retained Arabic carries with it a quality of exoticism and the unknown, therefore a certain quality of promised efficacy.

Another interesting example is offered by Daniel of Morely's *Philosophia* (Burnett 1997). This book is not a translation but instead an amalgam of ideas, theories, explanations, and information the author has assembled, possibly from a mixture of translated and untranslated works of Arabic science. Daniel is conversant with Ptolemaic astronomy, probably as revised and updated by Arabic thinkers, for he discusses the courses of the planets in terms of epicycles and eccentrics. He speaks, for example, of Saturn's forward, stopped, and backward motion using the terms *processio, statio,* and *retrogradatio,* but the crucial word "epicycle" is rendered as *elthedwir*—a somewhat Anglicized version of the Arabic *al-tadwīr* (Burnett 1997, see his plate 4). No doubt this reflects Daniel's unfamiliarity with Greek and a lack of contact with astronomical translations from that language. Yet, at the same time, it suggests a motive to consciously adapt important terms to the target language, as an aid to reception.

The Image of Islam

A few general comments are in order regarding the translators' views of Islam. It was under the system of *auctoritates* that the image of this culture as a "safekeeper" of the Greek "legacy" began. As already noted, the new translators commonly spoke of the "hidden," "secret," or "lost" learning they were to reap from the Arabs. It was this "treasure"—a term that betrays the sense of a materialized wealth whose substance remained constant despite any transfer from one set of hands to another—that the translators sought, as Adelard put it, from their "Arab masters." So far as I can tell, only Daniel of Morely wrote of "borrowing" knowledge from the Arabs.

The "Arab teachers" were seen, at first and with few exceptions, as part of the "moderns"—al-Fārābī, al-Haytham, al-Kindī, al-Kwarīzmī, al-Rāzī, and Thābit ibn Qurrah were all considered true *auctoritates,* teachers of "reason," exemplars of a way of thought medieval Europe desperately required. Yet what this meant, in effect, was that Arabic natural philosophy remained, in some sense, exempted from Islamic culture. The Latin translators were appreciative of this culture in certain ways, but were largely uninterested in it, except when commissioned to write works about its religious principles.[10] Their work, as they conceived it, was to generate and provide texts, to transfer a body of knowledge in written form, and to thereby fill a great gap in Latin learning. It was not to engage in cultural diplomacy, of whatever type. Often enough, they were likely to pass on older traditions of unsympathy for Islam, such as the astrological belief that since the Arabs consider Friday sacred, and therefore lived under the influence of Mars and Venus, they were prone to warlike, avaricious, and erotic qualities.

Such provincialism was to be expected from a Europe that had been inward-looking for centuries. Its deeper influence, however, was to give the new translators a double-sided view of Arabic learning. On the one hand, the Arabs were vastly superior in their intellectual resources and in their appreciation of them. Their best authors were the equal, or

10. One of the main examples here would be the works by Hermann of Carinthia and Robert of Ketten commissioned by Peter the Venerable, which included the Qu'rān (d'Alverny 1982, 429).

nearly the equal, of the best of the Greeks. Their libraries were vast "chests of riches"; their science unparalleled; their philosophers, "the wisest in the world." And yet, somewhere within all this appreciation was the sense that these achievements were the work of individuals and owed little to Islam as a society or a civilization. As one enters the thirteenth century, moreover, there seems to arrive the suggestion, if only implicit, that in order for this knowledge to release its full potential, it must be taken forward by "the Latins."

Driven by fervid dissatisfactions and the sense of their own "discoveries," the translators could not perhaps help but indulge the feeling that they were bringing the wisdom of the ancients out of a brilliant darkness. Though Muslim culture, in Adelard's words, covered "half the known world," and, in its vivid cosmopolitanism, combined the peoples "of many nations," somehow the learning available to this great collective still qualified as "arcane" and "hidden." Of course, this really reflected the first side of the equation: what had remained "secret" or "lost" was not such learning, but Europe itself—separated as it was from centers where the civilization of this learning grew and flourished. But this is not how the matter was expressed. As Richard Southern writes:

> For the greater part of the Middle Ages and over most of its area, the West formed a society primarily agrarian, feudal, and monastic, at a time when the strength of Islam lay in its great cities, wealthy courts, and long lines of communication. To Western ideals essentially celibate, sacerdotal, and hierarchical, Islam opposed the outlook of a laity frankly indulgent and sensual, in principle egalitarian, enjoying a remarkable freedom of speculation. (1962, 7)

Such freedoms, in short, were simply too spectacular and overwhelming, too begging of Christian judgment, to be given their cultural due.

It has been correctly observed that "the proponents of the new science were never accused of crypto-Islam" (Daniels 1975, 270). This, no doubt, was beneficial—even necessary—for the success of the translations, since the Christian view of Islam in the larger society was anything but rosy. Despite some leavening influences, the period of the eleventh and twelfth centuries saw the launch and defeat of several Crusades, as well as the ascent to power of the Mongols, an event that served to aggressively re-pose "the problem of Islam" in both religious and military terms. Proclamations by the Church and by its representative scholars sometimes demanded study of the Arabic language—but not in the name of learning or welcome. They were made in the hope of training effective missionaries, "soldiers of God," Crusaders by another name.

The translators' provincialism therefore helped them in the end.

Technical knowledge in Arabic was not seen by Christian society as linked in any essential way to Muslim religion and culture. This was another side to the idea of a "lost" knowledge. For in order to be "found," this knowledge had to be stripped of its history and of contents that would make it the inalienable property of another people.

The Late Middle Ages: New Texts, Old Traditions

As for the new astronomy imported from Islam, it did not really begin to solidify into a new synthesis for nearly a century. Too much material, of too complex a nature, had been introduced all at once. Time, use, and reworking were required for its nativization to Latin culture, just as it had been in Muslim society. The basic process involved had several aspects, some of which no doubt were manifested simultaneously: (1) the new translations, or portions from them, were selectively adopted by individual masters and incorporated into their teachings; (2) these masters, and possibly some of their students, added explanatory glosses and wrote commentaries upon the translated texts; (3) new works, possibly textbooks or other condensations, were produced for university use; (4) commentaries upon these new works were then written, moving the contemporary state of textual use still further from the first translations. This process is essentially an ancient one, reaching back to the Hellenistic period. In late medieval Europe, however, it was renewed with tremendous vigor; the vast quantities of new textual knowledge demanded it.

Such quantity, to be absorbed, also required new social formations, focused on the acts of learning and writing. These formations arose mainly in the universities, with their hunger for texts, but only after a new type of intellectual culture had begun to develop, in which standard sourcebooks of a more or less secular nature were widespread, and where an expanding market for new commentary had grown up. In astronomy, following the early period of translation and glossing, the new material was absorbed primarily in the form of three main works, which show that the subject had become, at least by 1250, divided into introductory and advanced levels. Two of these works, or groups of works, could be called beginning textbooks: they included the *Tractatus de Sphaera* by Johannes of Sacrobosco (first published 1220) and the various texts that went under the title of *Theorica Planetarum*, sometimes attributed to Gerard of Cremona or John of Seville, but more often said to be anonymous (Grant 1992; Pedersen 1978), written by an unknown Latin master who sought to remedy the scant discussion of planetary motion in the *de Sphaera* (fig. 13; see Grant 1996,

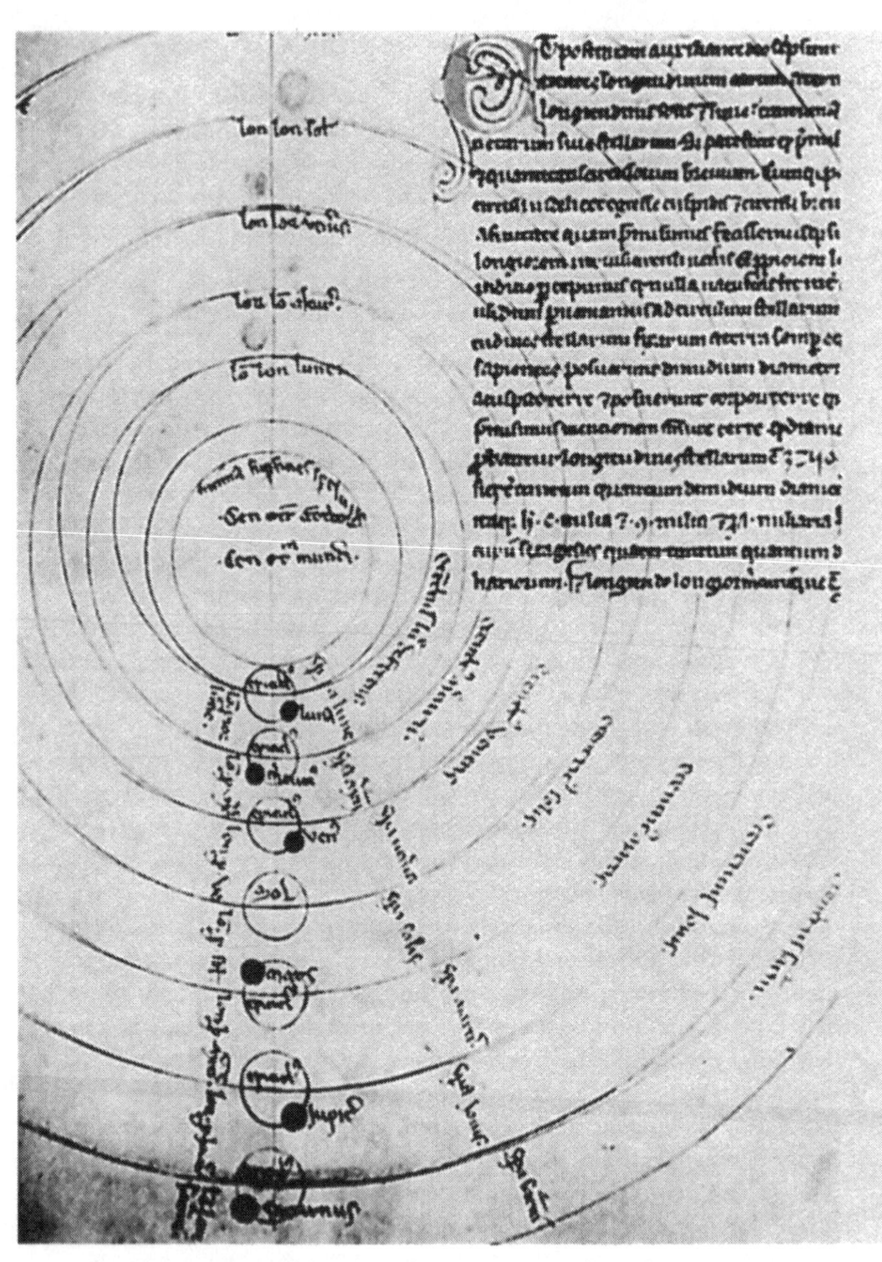

Figure 13. Diagram illustrating planetary motion, from the anonymous thirteenth-century textbook *Theorica planetarum*, an important (if overly simplified and, in places, misleading) introduction to Ptolemaic astronomy. Interestingly, in this image the lines of the planetary spheres run under the Latin text, indicating that the drawing was done first. This suggests that two different stages of copying were involved, one stage for the figures alone, another for the written text. University Library, Cambridge. Kk.I.I, fol. 219v.

45). The third text, meanwhile, was the *Almagest* itself, which was widely available in Gerard's translation by about 1230. This was either taught in small portions, such as the descriptive sections of Book I, or was reserved for the most sophisticated scholars of astronomy.

Sacrobosco's *De Sphaera* and the *Theorica* are among the most influential textbooks ever written in European science. Both were introductions to, and overviews of, Ptolemaic astronomy. They were mainly derived from existing digests of the *Almagest* and commentaries on it by Islamic scholars such as al-Farghānī, Thābit ibn Qurrah, and al-Bīrūnī. The *Sphaera* is said to have been taught "in all the schools of Europe" for as many as three full centuries and could be found in courses at Paris and Oxford as late as the seventeenth century (Pedersen 1978). The *Theorica* was no less "destined to play a remarkable role in the history of medieval astronomy." This work seems to have

> definitely settled the question of the vocabulary of theoretical astronomy. Earlier expositions had reflected the uncertainties of the first translators struggling to create suitable Latin equivalents for technical terms in Arabic or Greek, often with curious arabisms as the result. By contrast, only two or three Arabic terms were retained (and Latinized) in the *Theorica,* which also established a successful one-to-one correspondence between words and concepts, thus doing away with the ambiguities of earlier texts. (Pedersen 1978, 318–19)

By this time, in other words, a new stage in nativizing the Arabic-Greek synthesis in astronomy had begun. Indeed, throughout the field of natural philosophy there had occurred a shift, away from Arabic source works and toward those in Greek. A central part in this new phase was linguistic and involved de-Arabicizing the relevant knowledge. This is wholly in keeping with what we have said above about the image of Islam and the concept of a decultured Arabic science. Of course, it might seem that any such large-scale incorporation of foreign knowledge would have to involve this kind of replacement. But recall: both the Romans and the Arabs, in their vastly different uptake of Greek science, the one popular and the other highly technical, had adopted many terms from the Greek itself, in partial recognition of the true foreignness and newness of this knowledge and the insufficiencies for it in their own languages.

This occurred only at the very beginning for the era of translation into medieval Latin. Soon, many Arabic terms began to disappear. The *Theorica* and Sacrobosco's *Sphaera* include no Arabic words at all; instead, they succeeded in making the Ptolemaic system of the universe standard in Europe by entirely Latinizing it. In modern terms, these brief works were a combination of textbook and popularization—they remade a restricted and largely inaccessible knowledge into something widely available. In historical terms, they carried forward the age-old tradition of writing handbooks on difficult subjects, begun in Hellenistic times and continued in Rome and Islam both and transmitted to medieval Europe by way of the early encyclopedists, such as Martianus Capella. As we have seen, these handbooks form the bedrock of Western education as a whole—the entire liberal arts tradition is a product of handbook literature. As a literary form, they are therefore of inestimable cultural importance, indeed power.

It is in these terms that the new textbooks in astronomy should be seen. Taken up by the universities of the thirteenth century, they helped fix Latin astronomical terminology, including the names for the planets and a great number of the constellations. The constellations, in particular, were now shorn of any Arabic titles and completely Latinized. Those star names and the few technical terms in Arabic we use today are the product of a later time. Words such as azimuth, nadir, and zenith, originally adapted by the Arabs to discussion of the astrolabe, and introduced into the English vernacular by the poet Chaucer in the fourteenth century for his own text on the subject,[11] seem to be the sparse remains of a much greater Arabic content that circulated among the early translations and that hung on only because of their direct relation to a crucial piece of astronomical technology.

An Untold Story: The Purloined "Legacy"

The "Old" and "New" Translations

In the early to middle thirteenth century, the curriculum of natural science in the medieval university consisted of a number of works by Aristotle (the *Physics, De caelo,* the *Meteora, De generatione et corruptione, Metaphysics*), several by Arabic authors (al-Fārābī, Averroes), a few by Latin authors, such as Nicholas of Damascus (his *De plantis*), and one or two others. Most, or all, of the works by Aristotle were used in Arabic-Latin translation (Burnett 1997; North 1992). By the last quarter of the century, this syllabus had changed considerably. Not only had such works as the *Sphaera* and *Theorica* been added, but translated books of the "corpus vetustius" (older works) had been entirely replaced by the same works rendered from the Greek, the so-called "cor-

11. See his *Treatise on the Astrolabe,* a textbook in its own right, written by Chaucer for his son's instruction.

pus recentius" (newer works). The Aristotle of Adelard, Gerard of Cremona, and Hermann of Carinthia had been replaced by that of William of Moerbeke, whose versions were in wide circulation by about 1275–1280 and set a new standard. Legend has it that William of Moerbeke undertook this work at the specific request of Thomas Aquinas, who may have found the existing translations inadequate (see Thorndike 1923, 2.599–600). Be that as it may, the *translatio nova* became the acknowledged standard for use in the schools and by scholars, and thus effectively acted to erase much of what had gone before. This sort of replacement was by no means generalized to all scientific texts by this time. A number of Arabic authors remained highly influential well into the next century and some even beyond. But a tradition had been initiated; over the next two hundred years, these authors grew ever fewer.

The progressive deleting of the Arabs from their own "legacy" is largely an untold story of medieval European history. While there is space here to note only a few generalities regarding this change, it should be acknowledged as a possible area for future study. Scholars of the last fifty years who have written about this period of translation have sometimes noted important realities along these lines. Crombie (1957, 35), for example, emphasized that "in the 14th century translation from Arabic practically ceased when Mesopotamia and Persia were overrun by the Mongols." This phenomenon is widely accepted; but it hardly describes the whole of the matter and leaves out complexities having much to do with textual culture within medieval Europe itself.

Roger Bacon and the Learning of Foreign Languages

In the thirteenth century, as we have noted, there were efforts on the part of certain scholars such as William of Conches and Roger Bacon, as well as the papacy and the new evangelical Dominican and Franciscan orders, to promote the study of Arabic and Hebrew, as well as Greek, in the new universities. While this was usually done in the hope of inspiring new missionary zeal, such was not the case with Bacon, who recommended Arabic for reasons of learning. Nearly all earlier translations from Muslim sources are faulty, says Bacon; to be truly learned, one needs to have access to the "original" in its own language. In 1269, Bacon published a work entitled *De linguarum cognitio* (On the Learning of Languages), in which he notes that the Bible had been translated from Greek and Hebrew, while "philosophy" (including science) had entered Latin from the Arabic, as well as from these other languages. He then goes on to quote St. Jerome on the impossibility of truly faithful translations in secular works, and to say:

> Whoever knows a discipline, such as logic or any other, well and tries to translate it into his mother tongue will discover that mother tongue lacking in both substance and words. Therefore no reader of Latin will be able to understand the wisdom contained in philosophy and in the Holy Scriptures as well as he should, unless he also knows the languages they have been translated from. (Lefevere 1992b, 49–50).

However eccentric, Bacon's is truly a lone voice in this regard and was soon buried. Briefly resurrected in 1320 by the Council of Vienna (for missionary reasons), who advised the learning of Hebrew, Greek, Arabic, and even Syriac in the universities of Paris, Oxford, and Salamanca, the idea died again thereafter. By 1400, questions surrounding the learning of Arabic were gone. The main effort was to ensconce Greek as the language of literature, philosophy, and science (Ruegg 1992; Freibergs 1989).

Arabic was never adopted, except on the most temporary basis, as a topic of study within the universities. Such adoption, of course, would have made a great deal of practical sense, given the vast amount of material in this language—even today, more of Aristotle exists in Arabic than in Greek or Latin. But the "tongue of the Saracens" was apparently seen as being too difficult, too foreign, and in the end too *unnecessary* to become an object of study among schoolmen. Indeed, except for a few proclamations such as Bacon's, the topic seems to have stimulated little or no discussion among scholars.

Two reasons for this might be invoked. The first has to do with the very success of the translation movement: the existence of so many new texts, including multiple versions of even single works, argued against any need for the learning of Arabic. Second, those familiar with these works, particularly in the sciences, felt that the Arabs owed the basis, and often the substance, of their learning to the Greeks, whom Arabic authors had been fairly faithful in citing. Aristotle, Galen, Euclid, Archimedes, Ptolemy—these, it seemed to the new Latin scholars, were the "true" Fathers of this valued knowledge. The avidity for learning Greek shown by William of Moerbeke (a Dominican from Flanders), famed translator of Aristotle's and Archimedes' works, or Robert Grosseteste (Bishop of Lincoln), pioneering late medieval writer on optics, mechanics, and other sciences, arose from a desire to experience the new learning "first-hand," just as Bacon advised. Ironically—but revealingly—it was never matched by any such interest in knowing Arabic in order to make similar contact with the algebra of al-Kwarīzmī, the optics of al-Haytham, the medicine of ibn Sīnā, the astrology of Abū Ma'shar, or the commentaries on Aristotle by ibn Rushd (Averroes), all of which were original works, unknown to Greek science, and of profound importance to medieval study.

Indeed, Bacon himself never went very far (if any distance at all) toward learning Arabic, and if we examine his writings, the *Opus Majus* for instance, perhaps the most comprehensive work on late medieval science before the fifteenth century, we find a marked tendency to cite Greek authorities far more—nearly ten to one—than Arabic sources. Even in his discussion on mathematics, Bacon calls upon the likes of Aristotle, Euclid, and Ptolemy repeatedly (even Socrates makes an appearance), but on Averroes and Abū Ma'shar (Albumazar in Latin) once or twice only and al-Kwarīzmī barely at all. Of course, with the rise of Aristotelianism in university culture during the thirteenth and fourteenth centuries, along with its official condemnation in Paris in 1277, much of the Arab contribution to philosophy was forgotten on the one hand and reduced to the towering heights of Averroes, Albumazar, and Avicenna on the other, two of whom stood in the cooling shade of "The Philosopher." Such authority provided a constant, even central attractor for late medieval authors: Aquinas, Albertus Magnus, Duns Scotus, William of Ockham, Nicholas Oresme, Albert of Saxony, John Buridan, and a dozen other writers on matters both philosophic and scientific recycled "Aristotle" as the core of a vast new textual society, as a type of demanding presence for commentary.

Vernacular Translations and Aftermath

It is also important to take note of relevant effects in the area of vernacular translations from the Latin, most significantly into French. Such rendering of scientific works had begun as early as about 1270, when Mahieu le Vilain produced the first French version of Aristotle's *Meteorologica* (Edgren 1945). This version was based on William of Moerbeke's Latin translation from the Greek, which had been completed only ten years earlier and was receiving, along with his other versions of Aristotle, significant attention in scholarly circles. As emphasized by Shore (1989, 302–3), Mahieu's rendition is hardly a literal one: it virtually brims with commentary adopted from Alexander of Aphrodisias (third century A.D.)—again, as existing in a recent version by William of Moerbeke, thus ignoring earlier versions from the Arabic—and also contains a number of explanatory glosses by Mahieu himself. The work, we are told, was performed for Mahieu's patron, the Count of Eu, who must have found existing versions unusable or else distrusted their accuracy.

Such was the situation, too, for Nicolas Oresme roughly a half century later, when he translated Aristotle's *De caelo* into French. Oresme produced *Le Livre du Ciel et du Monde* for Charles V, perhaps the most important patron of vernacular translations in the late medieval period (Shore 1989), from a Latin version, begun by Robert of Lincoln ca. 1250 and completed by William of Moerbeke (Dembowski 1997). Oresme's rendition was loose and expanded, full of explanations, glosses, and added metaphors, yet also terminologically ingenious, creating many new technical words by simply dropping or slightly changing Latin endings (Menut and Denomy 1968). More significantly, a number of these terms, both in *Le Livre du Ciel* and other translations of Aristotle, as well as in Oresme's own *Le Livre d'Ethiques d'Aristote,* were neologisms derived from Greek: *architectonique, demotique, epyekeye, oligarchie,* among many others (Dembowski 1997). It seems likely that the translator acquired these directly from Moerbeke's versions; certainly there are no equivalent Arabisms. Oresme also includes within his own comments frequent mention of the Greek language in a manner that reveals an effort to appear especially erudite (given the audience the translation was destined for), e.g., "in Greek, it is said that . . ." or "Aristotle here makes a distinction that exists in the Greek language . . ." (Menut and Denomy 1968, 156). On the other hand, while it appears that he used the commentary by Ibn Rushd (Averroes) as a reservoir for his own glosses, Oresme offers little acknowledgment of such a source.

Mahieu, Oresme, and other vernacular translators of the thirteenth and fourteenth centuries thus appear to have followed a particular assumption about translation value. This assumption at first seems quite simple: because Aristotle had been Greek, the best Latin versions of his work (i.e., the bases for vernacular versions) would have to be rendered from the Greek language. Greek, in short, was the native speech of this *auctor* and thus a timeless guarantor of a more "original" version. An important shift has thus taken place: unlike their predecessors of the twelfth century, these translators were not interested, above all, in knowledge per se, as "unknown" or "hidden" wealth to be "discovered." They were beset by a new consciousness of *sources,* therefore of "origins." Such, eventually, was what the *auctoritates* system demanded for Greek thinkers.

Finally, in this connection, it should be stressed that the vernacular translations were produced directly as a service for specific patrons. That they contained much more abundant glosses, added word lists and definitions, as well as a range of adaptations and other changes, expresses this fact. They were to be used by particular, powerful individuals for purposes of self-education and had to be easily comprehensible. To a degree, therefore, these are cases of social conditions determining the form of a "translation," with the translator expanding and amending an "original" in specific ways that would make it eminently usable to the commissioning party. This, too, helped ensure that the most recent, widely praised Latin "originals" would be used as source texts.

Reasons and Rationales

By the late 1300s, any overt recognition of the vast Arabic contributions to Latin science had been largely filed down to a small handful of names and texts. Within the curriculum of late medieval universities, one finds al-Fārābī's *Classification of the Sciences* and portions of ibn Sīnā's great *Book of Healing,* but in astronomy and mathematics, the two disciplines in which the Arabs had especially excelled, there was John of Sacrobosco's *De Sphaera,* as well as his *Computus* and *Algorismus,* the *Theorica Planetarum,* introductory works by Boethius, the *Elements* of Euclid, an anonymous *Almagestum abbreviatum,* and not much else (North 1992). The *Almagest* itself, meanwhile, could now be found as a standard "classic" of the scholar-aristocrat's bookshelf, accorded all the aesthetic finery late manuscript culture could bestow (fig. 14). The historical trend toward favoring the Greeks stands clear and undeniable. Where Adelard of Bath had proclaimed the need to "learn from the Arab masters by the guidance of reason," Leonardo Bruni, writing three centuries later in the mid-1400s, spoke for his own time in saying that for nearly a millenium "no one . . . has been in possession of Greek and yet we agree that all knowledge comes from that source" (Ross and McLaughlin 1949, 619). Such comparisons are enough to suggest the degree of deletion that had taken place.

No systematic, racist cultural policy was at work. Nothing can lessen the reality that during the twelfth and early thirteenth centuries, many Arab authors were revered, at times even above the Greeks. Latin natural philosophy between 1100 and about 1275 was largely a subset of the Arabic tradition, and not a few scholars, even those as influential as Albertus Magnus, acknowledged as much. Some, such as Robert Grosseteste in his *Summa philosophiae* (mid-thirteenth century), offered detailed overviews of the "famous gifts" of Islamic philosophers and scientists. Others, like Bartholomeus Anglicus, writing about the same time in his much used *De proprietatibus rerum* (On the Properties of Things), gave frequent and conscientious mention of Arabic source material. For a period, therefore, a wide range of non-European authors were adopted into the *auctoritates* system.

But as the next century progressed, as material in Greek continued to be transferred from Sicily and Byzantium and to be disseminated, and as the early Renaissance got underway in Italy, the change became

Figure 14. Illuminated opening page from a fourteenth-century manuscript of Ptolemy's *Almagest*, showing complex artistry of Gothic marginalia, including various animal figures on two legs gazing skyward and carrying astronomical instruments. Even the most complex of scientific texts were, by this time, absorbed into manuscript culture at all levels, receiving the same care given literary or religious works. The high degree of illumination given this manuscript indicates that it was not intended for use as a textbook or source, but more likely to adorn some scholar's library. British Library, Burney MS 275, fol. 390v.

undeniable, and the results, perhaps predictable. Arabic writers were increasingly viewed, individually and as a mass, as derivative, dependent upon their "Greek masters." This view, of course, had some truth to it for the earliest stages of Arabic science. But medieval Europeans, bound by their own intellectual traditions, were also propelled by newfound ambitions related to textual influence and power. These ambitions, central to the evolution of late medieval and Renaissance intellectual culture, led to a tendency to see through the originality and depth of Muslim contributions.

As already suggested, such ambitions were linked to the *auctoritates* system, which gave enormous privilege to sacred first sources. Within the rules of this system, the sheer textual presence of the Greeks, heightened by constant reference to them in the writings of the Arabs (though often only rhetorically), constituted evidence for a primal originality. "Aristotle" was therefore divorced from the reality of "his" transmission as a collection of texts in the Arabic language. The story of this transmission was unimportant in the sense that it would only serve to prove all the more the timeless value of "Aristotle," as textual Father. Such tautological thinking was a direct result of the styles of reverence given to *auctoritates*. Roger Bacon, in his complaint against the faith in authority, reveals how deep such thinking could go: as described by Thorndike (1923, 636), "there is a certain irony in the fact that [Bacon's] argument in favor of independent thinking . . . consists chiefly of a series of citations."

Writers of the thirteenth and fourteenth centuries continued to draw directly upon the translated Muslim commentaries at their disposal. These were necessary works to consult, greatly aiding the work of interpretation in many areas. But the Europeans tended to cite them less and less in comparison to Greek works and authors, who became by default the patriarchs of scientific learning. In Norman Daniel's pithy phrasing, "Arabic thought was a major influence, but it was now an influence rather than a source" (1975, 275).

An important influence on the relative erasure of Islam had to do with the standards of late medieval authorship. Following the episode of translation, the value of writing increased enormously. A far more competitive textual culture grew up, mainly centered in the universities. In this climate, it became common for authors to minimize reference to their contemporaries in order that their own contribution should appear all the more original, desirable, in a direct line of descent from the truest origins of "wisdom." Beginning in the late twelfth century and increasing rapidly thereafter, there occurred a weedy growth of new commentaries, separating each new generation of readers from older interpreters. Taken as a whole, Latin authors worked, directly or otherwise, to supplant the older Arabic commentaries with their own. Moreover, a tendency toward something like specialization arose alongside the great encyclopedic tradition. Each subject of the *trivium* and the *quadrivium*, that is, now boasted its own expanding library of texts and was more inviting of extended local commentary. Sacrobosco's *De Sphaera* and the *Theorica Planetarum* replaced a whole series of commentaries on Ptolemaic astronomy by those such as al-Farghānī, Thābit ibn Qurrah, and al-Bīrūnī. In particular, the *Elementa* of al-Farghānī (Latin: Alfraganus), which had served as the crucial epitome of the *Almagest* and one of the most widely used texts in university astronomy, was almost entirely eclipsed by Sacrobosco's far simpler work (North 1992). Finally, since medieval textual culture was manuscript culture, its texts were variable, unfixed, without formal standards of accuracy, and therefore uncertain in many ways. This led to many errors, especially in technical works. Passages too difficult or unfamiliar to a particular scribe, names that seemed foreign or unpronounceable, might well be changed or deleted scribe. Many misattributions may well have occurred in this way, along with Arabic names being changed to Greek.[12]

In the end, the things late medieval Europe found of greatest interest and use—astronomy, mathematics, mechanics, cosmology, philosophy in particular—were eventually attributed to the Greeks. Broadly speaking, only in astrology, alchemy, and, to a significant degree, medicine, did the contribution of the Arabs retain the highest level of recognition, as fully original thinkers, down to the High Renaissance. A brief, if significant, love affair with Arabic learning in seventeenth-century England did little to alter this state of affairs (Toomer 1996; Feingold 1996), and, indeed, even carried forward the tradition of comprehending Islamic science as a brilliant polish on Greek ore (Sabra 1987). In most areas, an entire cosmos of Arabic sources was recast, century by century, until it had been reduced to a few visible orbits around the

12. As Edward Grant has recently written,

> Reliance on manuscripts meant that versions of the same treatise in Paris and Oxford might well differ, perhaps in significant ways. . . . Errors abounded in scientific texts . . . knowledge was as likely to disappear as to be acquired. An enormous effort was required simply to preserve the status quo or occasionally to restore texts that had been inherited from Greco-Arabic sources. Although we may be unable to measure the detrimental effects on medieval science . . . we may rightly judge that they were enormous. It is not a mere coincidence that the Scientific Revolution began shortly after the introduction of printing in Europe. (Grant 1992, 367–68)

gleaming suns of Aristotle, Euclid, Ptolemy, Plato, Galen, and Archimedes.

Epilogue: The Fate of Arabic Star Names

In absorbing and nativizing the knowledge of the East, Europe chose to shed this knowledge of certain contents. The impulse to displace the weight and authority of earlier authorship, which the Romans had practiced so diligently upon the Greeks, the medieval Latins imposed upon many Arabic authors, while leaving the Greeks intact and elevated. Europe chose to gradually define the relevant knowledge as its own by constructing a line of heroic descent. Between the thirteenth and fifteenth centuries, "Greece" was naturalized as the projection of Europe's own intellectual tastes and appetites—"Greece," that is, as a library of sacred textual fathers, a patristic literature of scientific knowledge. And so, with the development of a university culture throughout Europe, with the growth of a new collegiate of scholars engaged in both furthering and adapting older textual traditions, the image of Islam as a critical nucleus of civilized learning faded. In its place, there rose the glowing sun of Greece, a "golden" nexus that warmed and nurtured all others. Ironically, this favoritism stood in direct contrast to the long-standing acknowledgment among Muslim writers of their own debt to Greek texts. Indeed, this acknowledgment, with its own varied styles of reverence, may well have played some part in convincing Latin commentators over time to turn their favored attention to Greek "sources" and to offer nodding appreciation to Arabic "preservation" of these works. If the details of this process—which is no doubt far more complex than I have sketched it here—remain to be explicated, the overall result has long been evident to the interested eye. Even today, there are few or no Arabic texts in the "great books" canon of Western science. One might well read Galen's *On the Natural Faculties* or Aristotle's *On the Heavens* but rarely al-Ṣūfī's work on the fixed stars, Ḥunayn's *Questiones,* al-Kwarīzmī's *Algebra,* or even al-Fārābī's *Classification of the Sciences.*

The process at issue here has its own history, which can be briefly outlined in terms of the language of Latin astronomy, especially names. According to recent scholarship, the canon of titles applied to the heavens, particularly the stars, was early on influenced by Gerard of Cremona's translation of the *Almagest,* which, as noted above, contained a fair number of Arabic names (Kunitzsch 1974, 1983, 1989). Within the next 150 years, however, by the mid-fourteenth century, these names had been deleted, being no longer in evidence on the most influential star charts of that period. The catalogue of stars in Gerard's version of the *Almagest* had been replaced by translations (from the Castilian) of the so-called Toledan Tables, compiled under Alfonso X of Spain (thirteenth century), with star positions updated to A.D. 1252. The accuracy of this chart, as well as its considerable expansion into the Alfonsine Tables, a type of grand *zij* written and assembled in Paris during the 1320s, helped make it the standard reference for European astronomers thereafter, even down to the time of Tycho Brahe (Kunitzsch 1986b; Poulle 1988, North 1995). Both the Toledan Tables and the Alfonsine Tables were populated by Greco-Latin names, with no Arabic elements.

Yet the chapter does not end here. As it happens, a number of Arabic names—as many as twenty-nine, in fact—were reintroduced into the corpus of star titles during the Renaissance, sometime in the fifteenth century. Where did these names come from? All of them, it appears, were added by Viennese astronomers intent on going back to "original sources" (Kunitzsch 1983). These astronomers were members of a famous school begun under Henry of Langenstein (1325–1397) and continued by John of Gmunden (1380–1442), Georg Peurbach (1423–1461), and the famous Johannes Mueller, or Regiomontanus (1436–1476), builder of the first true observatory in Europe. Peurbach and Regiomontanus were astronomers and humanists at the same time. Dissatisfied with what they considered to be the static, low-level university science of their day, they came upon "the idea of reforming astronomy by reverting to its classical sources" (Pedersen 1978, 330).

This meant a new effort of textual adaptation, including translation, that marks the next major step in the nativization of the Arabic-Greek-Hindu-Persian synthesis. On one side, it involved a new rendering of the *Almagest* from the Greek, as well as writing a guide to it (called *Epitome of the Almagest*), and also a new and expanded introductory text, the *Theoricae novae planetarum* (Swerdlow 1996). This work was based on recognition of errors and insufficiencies in the existing community of texts surrounding mathematical astronomy and then taught in the universities. Peurbach and Regiomontanus, in particular, set about intentionally producing a new series of works that would replace all earlier textbooks and set new standards for university learning in astronomy. By all accounts, they succeeded, almost without qualification. With regard to the *Epitome of the Almagest,* for example, it can be said that the book

> is far more than an abridgement of the *Almagest,* for it recasts the work in the form of propositions in which geometrical theorems are proved, models, instruments and methods of observation described, and procedures for deriving parameters from observation explained, all with a pre-

> cision and clarity that for the first time gave Europeans a reliable guide to the full range of Ptolemy's mathematical techniques in spherical astronomy, eclipses and planetary theory. Additional materials were drawn from Arabic works [that criticized and updated Ptolemy's measurements and methods]. The *Epitome* is the great advanced treatise on astronomy of the Renaissance of letters . . . and was studied into the early seventeenth century as a preparation for the *Almagest* and even in place of [it]. Copernicus used it extensively in his work and . . . it may have provided a crucial step on the way to his own principal innovation in astronomy. (Swerdlow 1996, 190)

On the other hand, it involved compiling a new standard chart of stars and constellations by searching through earlier star catalogues, many of them translations, some dating back to the 1200s. And among these was the great work by al-Ṣūfī. This had been used as a basis for the Toledan Tables, nearly all versions of which had deleted the star chart compiled by al-Ṣūfī, with its listing of ancient and more recent Arabic star names, along with the Greek, in favor of a totally Castilianized rendering. Only in very rare cases had the star chart been preserved; it appears that the Vienna school of astronomers discovered just such a chart, adapted into a version of the Alfonsine Tables that had therefore uniquely preserved some of al-Ṣūfī's Arabic names (Kunitzsch 1983). Such a text appeared more authentic to the Vienna astronomers, with their sensibility of a "return to origins." It suited their conscious search, in other words, for material that had survived the centuries of change wrought by adaptation to late medieval European uses. Not long after the work of Regiomontanus, Peter Apian, a well-known Renaissance mapmaker, included a significant number of Arabic star names and constellation figures drawn after the manner of al-Ṣūfī in writings and maps he produced during the 1530s and 1540s—indeed, it appears that Apian actually possessed a copy of al-Ṣūfī's book and used it as a reference (Kunitzsch 1987).

Here, then, the Renaissance impulse to "revive" the learning of the past meant, in effect, a demand to remove the effects of appropriation and nativization stemming from Latin translation—a demand that exceeded the high-walled privilege by then granted Greece and Rome. Only by going back to the era of the earliest translators, with their often awkward urgency for precision, did fifteenth-century Viennese astronomers grant European science its "other" origins.[13]

13. Two hundred years later a similar revival, of similar scale, occurred upon the surface of the Moon. A handful of once-eminent Arabic astronomers, al-Sūfī among them, were granted a place on the lunar map of Giambattista Riccioli, whose version of 1651

Conclusion

The Italian Renaissance differed profoundly from "the twelfth-century renaissance" in its greater focus on literary and humanistic sources and also its interest in classical Latin texts. A crucial question that remains to be resolved is whether, or rather to what degree, the humanistic turn in translation may have emerged out of the earlier era of transmission. It would seem that science, philosophy, and Arabic came first; these were the choices made for Europe by its translators in the initial period of reaching outward to the larger textual world. Once these sources had been rendered, and were undergoing digestion, there was a shift toward favoring Greek "originals," and it is exactly at this point that the earliest work on literary classical Latin and Greek sources begins. Lovato Lovati (1241–1309), one of the first of the so-called Italian humanists, was already writing classically inspired poetry and unearthing ancient manuscripts by Seneca, Martial, and possibly Horace even while William of Moerbeke was performing his epochal translations of Aristotle and Archimedes from the Greek. Indeed, a "new fascination with ancient Rome" can be pushed back even further, and found in scattered twelfth-century interest in classical Latin poetry and the use of Roman-inspired motifs in cathedral sculpture (Haskins 1927; Martin 1982; Bloch 1982).

One should recall, too, that a significant amount of scientific translation in the twelfth and thirteenth centuries had taken place not merely in Spain and lands to the north, but in Italy and Sicily as well. Indeed, it is important in this regard to take note of the names of some of the most famous translators at this time: Gerard of Cremona, Burgundio of Pisa, Plato of Tivoli, James of Venice, Moses of Bergamo. The Norman court of Sicily and southern Italy was a famous center of translation, particularly of Aristotle, Euclid, Proclus, and Ptolemy, while the Angevin government of Naples under Charles I (1268–1285) and Charles II (1285–1309) produced "a steady stream of medical translations," mainly from Greek texts (Lindberg 1978, 74). North and south, Italy was teeming with scientific translation activity both before and during the earliest phases of the classical revival in literary material. The

was to prevail as the technical standard for international astronomy as a whole (see Montgomery 1996). Riccioli went beyond this too. In addition to these names, and the much larger number of Greek "ancients," he also included many "moderns"—Galileo, Kepler, Copernicus, Regiomontanus, and a host of others. Like the work of the Viennese astronomers, this was a kind of renativization of the larger astronomical tradition to a later time, in this case the seventeenth century. In a move both bold and brilliant, Riccioli populated the Moon with its own history as a subject of astronomical thought.

overall historical pattern is at the very least suggestive of a fluid overlap or evolution between "renaissances."

Indeed, viewed from the perspective of translation, and from the physical realities of writing and study, the twelfth to sixteenth centuries appear as a single extended epoch. No doubt this is too large an era for modern historians to consider all at once, in any unified sense. Yet it clearly was a time of particular linguistic focus, during which Europe created, recreated, and transformed a dozen textual traditions by taking into its fold the achievements of the Arabs, Jews, Greeks, Romans, and—secondhand—the Indians, Persians, and Syriac-speaking Nestorians and Monophysites as well. Through the powers of translation—and the adaptations and transformations that came in its wake—the scholars, students, universities, court societies, and book publishers of late medieval and Renaissance Europe forged a series of intellectual communities, centered on specific groups of texts, whose ultimate source materials were hugely diverse, indeed unmatched in this regard. In this, the European translators had the great benefit and example of their Arabic predecessors, who had brought together an almost equal diversity of traditions and who, in turn, had found such ecumenicalism partly to hand in Persian, Hindu, and Syriac textual cultures. It was exactly this vast accumulation, with its nativized aspects in Renaissance Europe, that made the Scientific Revolution possible. The mathematics of Euclid, Ptolemy, and al-Kwarīzmī, including the sine function and use of the zero; the observatory, as an embodiment of the eye-as-discoverer, established by Regiomontanus on the basis of Arab models; the role of precision in measurement, by which observations and theories could be continuously refined, even overthrown; and, finally (though a complete list here would stretch into a dozen or more items), the attitude of scientific dissatisfaction by which the work of the past was there to be improved upon and challenged, not merely revered—such were among the elements of "inheritance" from Arabic intellectual culture that provided the foundation for the thought of Copernicus, Tycho Brahe, Kepler, and Galileo.

The history of astronomy in the West is as much a history of translation as it is a tale of observation, correction, and discovery. This means, as we have seen, that it is also a history of displacement. Such has been the price and the wage of accumulation and "inheritance." The Romans sought to lighten their burden of imitation by translating Greek works in adaptive fashion, and by leaving out the great majority of Greek science from this effort. Early medieval authors rewrote and readapted the results of this work, compiling their own handbooks and commentaries, their own versions of Plinian astronomy for example, thereby replacing Rome, even as they came to praise it. In the East, where astronomical knowledge was transmitted through a mixture of scribal and pedagogic cultures, writers in Syriac moved from free to literal translations of Greek texts, seeking immediate replacements, even as Greek among the populace declined. In both Hindu and Persian literature, meanwhile, direct translations of entire texts may have been few; it appears, instead, that Greek astronomy was rendered partially and paraphrased by authors such as Āryabhaṭa to form new systems of detailed calculation and tabular observation. Finally, with regard to both Arabic and late medieval Latin cultures, the process of displacement (or displacements) was eminently complex—involving, on one side, multiple translations of individual works and their pedagogic disappearance into various handbook introductions and epitomes, and on the other side, their digestion into textual communities that included many new works by authors whose fame stood nearly side by side with that of the "ancients." Finally, in the case of Europe, there was actual erasure of the full debt to Islam.

Late medieval Europe chose "Greece" in a manner, and with needs, quite different from those that had propelled similar choices by Arabic, Syriac, and Roman writers centuries before. Yet, as I have tried to make clear, it was not at all the same "Greece." For each of these cultures and eras, a different version of "Greek science," greatly mixed with other contents of passage, was adopted and renativized. It is interesting to note that in the case of Rome, Islam, Europe, and possibly Persia and India as well, the uptake of this science, as a textual nexus, took place at a point when each culture was poised to advance into a new stage of intellectual life, one driven by the written word above all. In most instances—but particularly that of late medieval Europe—highly literal renderings were done first, reflecting a clumsy yet ardent allegiance to writing. Such clumsiness had many results: its attempted "frozen" qualities helped introduce new words, rank corruptions, new syntactic formations, fertile deformations, even new grammatical constructions into the receiver language. Not all of these introductions survived; in fact, most did not. But such clumsiness shows itself to have been very much a central part of the nativizing process, a revelation of the eager inexpertise and sense of discovery at hand. Stripped of the simplicities of its "Golden Age" halo, "Greek thought" becomes a highly complex and evolving collection of textual matter, with great transformational, urbanizing powers. Again, these powers derived not from an unalterable core of final knowledge, but from an expanding ecology of writings and readings with contributions from a number of cultures and languages—from the many changes that transformed a limited harvest of original sources into a truly vast, cosmopolitan bounty of literature, thought, and pedagogy.

The "Greece" that Europe chose was therefore a vibrant urbanity of great vintage. And much of this urbanity reflected the fact that it was not Greek (or rather, Hellenistic) any longer. One of the most irrefutable signs, meanwhile, that deep changes had occurred in Hellenistic knowledge—changes, again, with massive consequences for Europe—is the simple reality that, in Arab hands most of all, "Greek science" had undergone a major shift away from styles of rhetoric bound up with spoken language toward those focused upon reading and writing, the quiet of study. Aristotle's *De caelo* or Cicero's translation of the *Timaeus* drew on models of oral eloquence: antiquity knew, or desired, no boundary between standards of spoken and written skill. From Plato to Quintilian, one can find endless discussion regarding the goals and powers of speech, but almost none on reading, and not many on writing. It is only with Judaism, early Christianity, and Islam, exegetic religions of "the book," that ideas of interpretation aimed solely at the written, unspoken word gain full precedence. This is a very important fact, not to be underestimated. What Europe adopted from Arabic authors was a corpus of work aimed at the silence of the eyes.

In its journey to the present, the knowledge of astronomy has never really stood still. This is true with regard to "advance," certainly, but much more so in terms of the actual physical presence of this knowledge in books. To the degree that such knowledge has long been embodied in texts, it was constantly in motion, susceptible to change, up until the moment it descended upon the dark of the printed page. Printing ended the uncertainties associated with manuscript culture, or at least rendered them temporary. It is no accident or mere coincidence that Regiomontanus sought to establish a "new astronomy" by simultaneously retranslating, epitomizing, correcting, and printing Ptolemy's works. Print was the final step in the appropriation process, the nativizing of "foreign wisdom" from the Arabs. Regiomontanus tried to combine under his own hand all of the major phases in this process, to redo history in a sense. It is more than fitting that it was partly under his influence that Arab star names reentered the heavens over Europe.

PART II

Science in the Non-Western World

LEVELS OF ADAPTATION

5 Record of Recent Matters

Translation and the Origins of Modern Japanese Science

Introduction

According to Goethe, the creation of a new word in any language constitutes a type of birth: the powers of speech and writing are expanded by a new semantic presence. This may be where the sciences have been most fertile over time. Scientific terms make up as much as half the total vocabulary of most linguistic communities in possession of modern technical knowledge, a reality that cannot simply be shrugged off by complaints of arcane usage. Such terms make the expression of vast areas of knowledge possible; without them, the epistemological substance and command of any particular people are smaller and more frail.

From the point of view of intellectual culture, scientific knowledge has appeared most difficult, threatening, and hermetic to those peoples for whom it was quite literally foreign—peoples who had little or no language for it. The development of such language has obviously been essential to the uptake of modern science throughout the world. Such development has never been a simple matter, even in the case of colonial regions. If China was largely under the sway of British imperialism and technology in the late nineteenth century, this came long after the introduction of Renaissance science by Jesuit missionaries, by Dutch and Portuguese traders, by the import of books from France in the eighteenth century, and so on. One crucial aspect of the spread of modern scientific knowledge in the last two centuries, therefore, would be exactly this nationalizing of linguistic influences. All of the major technological powers in Asia today, for example—China, India, and Japan, to name the most obvious—have individual histories of this type, showing when and how Western science was taken up and, over time, nativized into a cultural-linguistic tradition specific to each setting.

Without doubt, one of the best languages in which to pursue such an inquiry is Japanese, and this for a number of reasons. Modern scientific discourse is as advanced in Japan as anywhere in the world, yet it began relatively late and took place almost entirely under foreign influences. Up until the late nineteenth century, Japanese science combined ideas and terms from both traditional Chinese and Western natural philosophy, in a shifting proportion that moved away from the former and toward the latter, as embodied in the writings and translations of Dutch authors. The tale of this historical change in loyalty is written in the nomenclature of certain fields today, which in this sense contain memorials to their own evolution. The most rapid period of scientific modernization, beginning in the late 1860s, took place under a conscious governmental plan, whose basic idea was to model different fields after their most successful counterparts in the West. During this period, Holland, which had served as a guidepost for more than two centuries, was abandoned in favor of countries such as Germany, England, France, and the United States. Yet the process of adoption and adaptation was neither smooth nor fully under control. Certain disciplines, such as chemistry, developed conflicting attachments to more than one country and therefore came to be split in terms of the languages they followed (German and English in this case). The change in such attachments over time took place according to historical and political realities, especially during the twentieth century.

The Japanese Language: A Historical Overview

Early Developments

The linguistic effects of all this require a larger context, that of the Japanese language itself. More than Western languages, more even than most Asian languages, written Japanese reveals its history on its skin. This is because it employs not just one symbolic system but three, each with a wholly separate aesthetic form reflective of its origin. Moreover, the most difficult of these three systems, that of Chinese characters, can be used in a range of different ways, e.g., phonetically, semantically, or in combination. Over time, the evolution of these complexities has allowed for a truly remarkable variety in approaches to writing and reading Japanese. To a large extent, such variety is itself the direct outcome of a vast and unbroken effort of translation—that involved in adapting the writing system of one language (Chinese) to another, wholly unrelated and profoundly different tongue. What follows is a very brief and selective overview of this process and its effects.[1]

1. The discussion that follows draws mainly on the following sources: Miller 1967, Habein 1984, and Coulmas 1989.

The three writing systems of Japanese include two syllabaries, *katakana* and *hiragana,* and one ideogram system, *kanji.* The latter seems to have been imported from China sometime in the fifth or early sixth century A.D., based on the earliest reliable documents, which date from the early seventh century and indicate that Chinese writing was well established and already undergoing adaptation by that time. Contact between the two countries had encouraged conversion to Buddhism among the Japanese court, with a corresponding interest in Buddhist texts and in Chinese philosophy. Prior to this, Japan appears to have had a purely oral culture. By the seventh century, an interesting division had surfaced in Japan between Chinese writing per se, known as *Jun-Kanbun*—lit. "genuine Han dynasty writing" (since this took place during the Han period in China)—and a deformed style of writing and reading called *Hentai-Kanbun,* or "abnormal Han writing." *Hentai-Kanbun* was a complicated type of adaptation: texts in this form were written with various marks (most of them being small characters designating "up," "down," etc.) telling the reader how to rearrange the given order of *kanji* in each line or phrase so that a Japanese reading became possible.

The ability to do Japanese readings of Chinese characters existed because of two other, more basic adaptations, which remain very much in evidence today. The first was to use the *kanji* ideograms semantically, purely for their meaning, and to give them Japanese sounds; this is called *kun* reading. The second method was to do the very opposite, to employ the characters as phonemes—that is, for their sound value only (translated, of course, into Japanese phonetics)—and to write Japanese words with no regard to the original meaning of the symbols themselves; this is known as *on* reading and was required to express elements of Japanese grammar that had no correlative or parallel in Chinese. The sound-based, *on* system of using Chinese characters is traditionally known as *Manyō-gana,* after the famous collection of poems, the Manyō-shu ("Gathering of Ten Thousand Leaves") compiled around A.D. 759. Both the *on* and *kun* forms also involved the direct translation of Chinese writing into Japanese speech. Indeed, it is common for linguists and scholars to speak of *kun* reading during the early and medieval periods as "translation readings" (Habein 1984, 22).

The relevant complexities do not end here, however. Japanese students, scribes, clerics, and officials very often traveled to China to study (this may well have been a standard practice) and were thus subject to linguistic changes that occurred over time in Chinese society. The Japanese, for example, were fond of bringing back and introducing Chinese loan-words and new uses of older ideograms. Moreover, the

Chinese sound system associated with many characters changed as a result of political shifts in power (Miller 1967, 101–11). The oldest identifiable pronunciations in ancient Japanese, in fact, relate back to the so-called Three Kingdoms period (A.D. 220–265) and probably represent holdovers; much more of *Manyō-gana* comes from sixth-century northern China, seat of the Han dynasty. This was succeeded in the following century by the dialect of Ch'ang-an and Lo-yang, the great urban centers of the Tang dynasty (A.D. 618–907), a sound system that was itself superseded by the more southern pronunciations of the Sung (A.D. 907–1271) and Yuan (1271–1368) dynasties.

In China, there tended to be fairly clear recognition of these political-linguistic turns, as well of as the need to obey them. But no such necessity existed for Japanese scribes and clerics. Attempts to standardize pronunciation in Japan did not eliminate earlier sound readings, partly because of their association with Buddhist texts and ceremonies. The result, needless to say, was accumulative: individual characters came to acquire a range of readings that revealed (to the discerning eye, of course) the history of shifting allegiance to Chinese culture.[2] Even by the late medieval period, therefore, *kanji* in Japanese hands had become fingerprints for a complex process involving the continual translation of Chinese writing into Japanese forms.

Indeed, the details of this process are far more complex and convoluted than even suggested here. Suffice it to say that there developed a substantial difficulty with regard to the correct reading of any particular *kanji,* a difficulty that increased with time. This complexity has never been eradicated; in many cases, it has not even been diminished. There are numerous characters today that have several *on* (sound) and *kun* (semantic) readings, stemming from different historical periods. Individual words that combine two or more characters may, in fact, employ a mixture of *on* and *kun* readings. Context is often everything in determining sound *and* meaning. There are no correlatives to these sorts of problems—or their historical revelations and legibility—in alphabet-based European languages.

Further Adaptations

Alongside these forms of nativization, Chinese itself became and remained, for more than a millenium, the language of the literary and scholarly intelligentsia. Government edicts, chronicles, diplomatic documents, and other institutional forms of writing also employed Chinese

2. A number of specific examples can be found in Miller 1967 (111) and in Coulmas 1989 (126).

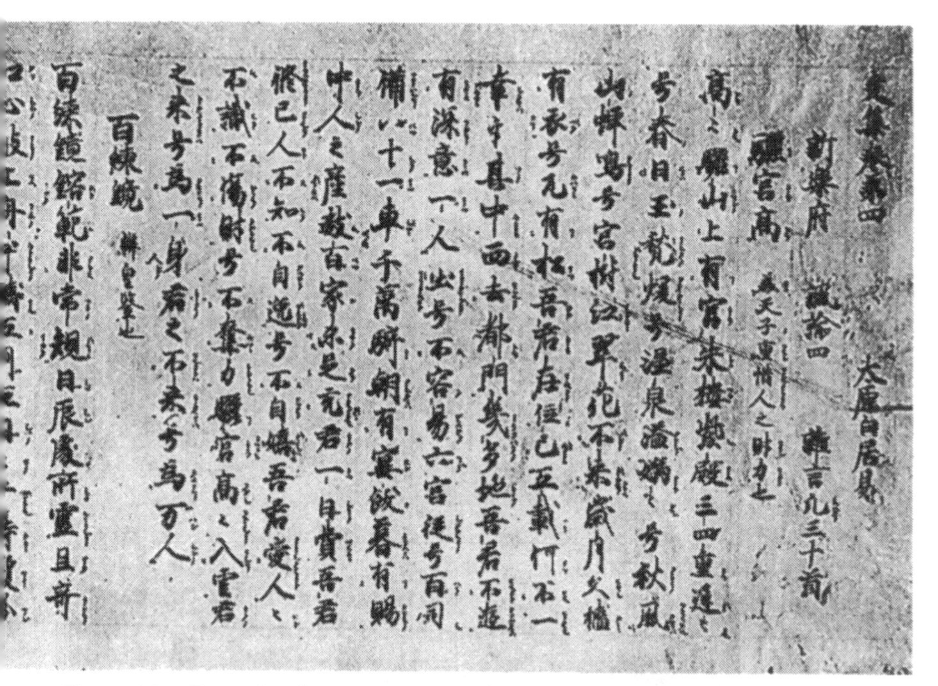

Figure 15. Example of manuscript page from a twelfth-century text in Chinese that includes Japanese reading aids, written in smaller symbols to the left and right of the ideograms. These aids indicate Japanese pronunciations, word order for proper Japanese translation, and compound character combinations. Reprinted, by permission, from Miller 1967, plate 6.

on a regular basis. This continuing presence of Chinese proved of great importance to the further development of written Japanese. Clerics especially, when deciphering religious works out loud in Chinese, commonly slipped into Japanese readings. By the early ninth century, they had developed a separate system of diacritical marks that they placed alongside the *kanji* to aid in this process: such marks clearly show that Chinese texts were being read in Japanese, i.e., that a form of translation-reading was occurring (fig. 15). The marks themselves often took the form of simplified characters or parts of characters, and came to be known as *kana* (lit. "temporary" or "interim name"). By the eleventh century, they had developed into their own system for writing Japanese sounds, eventually called *katakana* (*kata* meaning, originally, "imperfect" or "half-formed").

The other syllabary, *hiragana,* was developed at about the same time, also out of a direct connection to Chinese, but in an entirely different way. It evolved not through reading but through writing, as a kind of aesthetic endpoint to the art of calligraphy practiced among the nobility, who regularly wrote poems in Chinese. During the Heian period (ninth to twelfth centuries), when such writing reached its height, the tracing of certain characters, often used for their *on* (phonetic) readings, became highly stylized and finally reduced to a few flowing, cursive strokes. These were called *onna-moji* ("woman's writing"), since women were not allowed to study Chinese (though many did so) and thus were forced to write in Japanese through these aesthetically simplified *kanji.* By the end of the Heian epoch, *hiragana* (lit. "smooth" *kana*), woman's writing, became the syllabary used in combination with Chinese characters to form the main portion of the Japanese writing system that has carried down to the present. *Katakana,* meanwhile, attained a variety of uses, including phonetic spelling of Japanese words for emphasis (similar to Western italics), and more importantly, the phonetic rendering of foreign words, especially, beginning in the sixteenth century, those derived from Western languages.

Katakana was a creation of the ascetic world of the Buddhist priesthood. In actual form, it is simple, angular, unadorned. *Hiragana,* by contrast, sensual and curvilinear, originated in the rarefied cosmos of the Heian aristocracy, where luxury and decoration defined an everyday sensibility (see fig. 16). It was created not as a reading aid for "holy truth," but as a type of writing for those required to produce "art," in the form of sophisticated, stylized expression. Modern Japanese writing, therefore, in combining these two scripts with *kanji,* and its complex accumulation of sound and meaning variation, offers in its phonetic, semantic, and graphic qualities a visual representation of historical origins—indeed, of history itself, inasmuch as this history combines wholesale importation of another language with several styles of nativization.

Despite the development of the *kana* systems, however, Chinese remained the official language of scholarship and government for centuries thereafter. Most new words imported from abroad, whether from China or, later on, the West (beginning in the sixteenth century), were written phonetically in characters (including such terms as *pan,* for "bread"). Others, like *tabaco,* were pronounced as such but written semantically in Chinese (with the *kanji* for "smoke" and "grass"). Moreover, orthography remained entirely faithful to the Chinese standard, starting at the right and proceeding vertically. These and other uses of Chinese were given great stimulus in the early Edo period (seventeenth century), with the establishment of the Tokugawa shogunate and the adoption of neo-Confucianism (Chu Hsi school) as state-sanctioned doctrine. Chinese studies underwent something of a profound renewal as a result of this political-cultural move; Chinese was taught to the

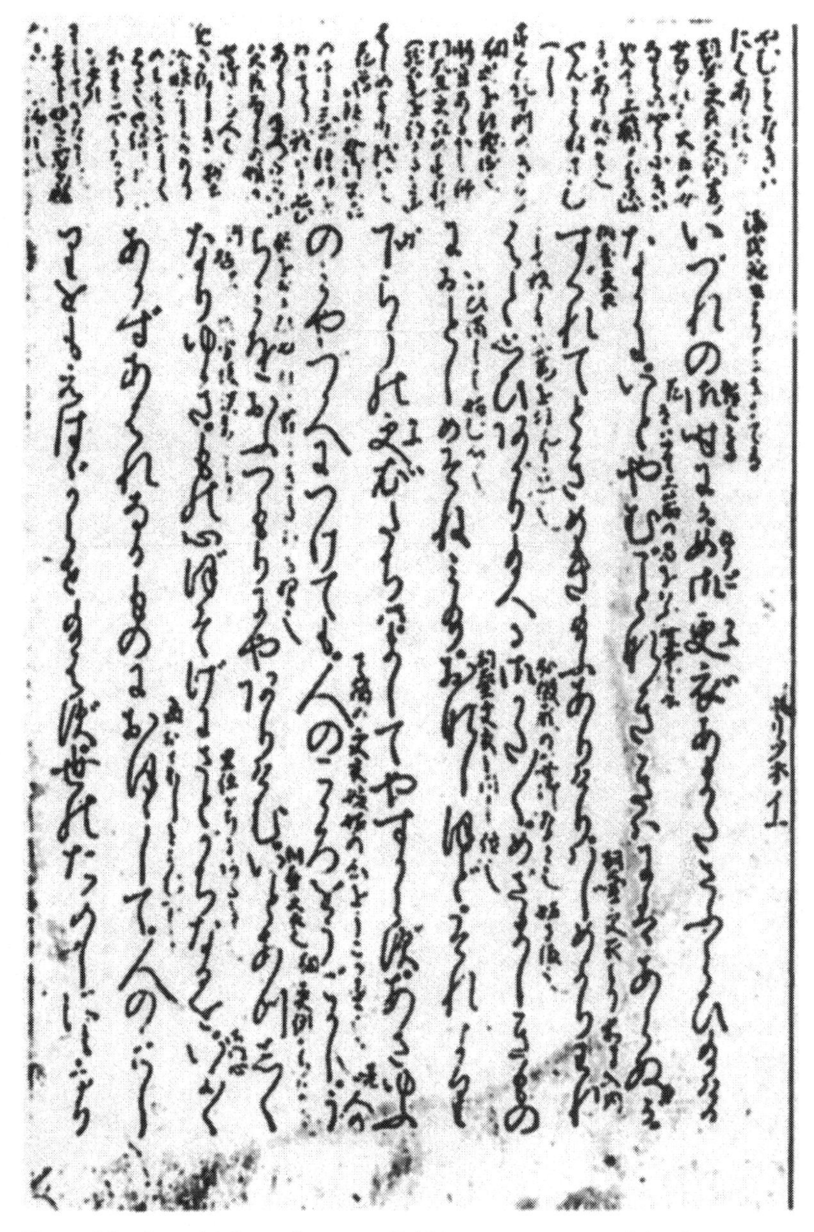

Figure 16. Late (eighteenth-century?) Tokugawa printing of the *Genji Monogatari* (Tale of Genji), showing *hiragana* as the main text (large script) and abundant glossing in mixed *kanji* and *hiragana* (smaller script), used in part to interpret the former. Reprinted, by permission from Miller 1967, plate 8.

samurai class, who constituted the top of the social hierarchy, as a foundation of their schooling. It was also promoted by the advent of printing, by private education (through which the lesser classes sought to compete with the samurai), by public lectures, and by the generally increased status of official scholars as leaders of authorized sentiment. Chinese remained the language of formal learning and social privilege in Japan up until as late as the nineteenth century. As such, it was bound to serve as both doorway and roof to a self-proclaimed vernacular revolution. This began in the eighteenth century, proceeded in several directions, mainly as forms of rebellion against the primacy of Chinese over Japanese language studies, and was not without ironies. One group of scholars and writers, for example, the so-called *kokugaku-sha* or "intellectual nationalists" who advocated for "pure Japanese," rejected all foreign influence and tried to reinstate ancient styles of writing used in the so-called Japanese classics of the early medieval period—the *Kokin-shu* and *Manyō-shu* (eighth century), for example, which had both depended upon writing but a half-step removed from Chinese itself. Among the rising merchant class, meanwhile, literary tastes ran toward the development of a number of popular genres of writing that employed more colloquial Japanese, e.g., folk tales, children's books, pornography, farce, and feuilleton fiction.

Such styles did not replace earlier methods of writing. In the manner of past developments, they simply added to the many overlapping, sometimes conflicting currents that had already come to make up Japanese language use (fig. 16), and they were followed by still other introductions and reintroductions that did the same. By the end of the Edo period (mid-nineteenth century), the writing and reading of the native language had been diversified into such a range of styles that literacy itself was often in a confused state: single characters and simple character combinations, for example, had acquired so many different readings (phonetic and semantic both) that they often could not be accurately deciphered without some sort of reading aid. Among the most difficult to decipher were personal names (this remains true today), which very often had to be accompanied in written works by *furigana,* small *hiragana* placed beside characters to indicate their correct sound (and therefore reading).

New Influences from Abroad

The supremacy of Chinese as the language of scholarship was challenged in another essential way. During the three centuries of the Edo period, Japan had been barred from open relations with other countries, especially those of the West. Yet this only helped create a new language of specific learning—restriction of European trade to Dutch merchants ensured that the Dutch tongue would become the inevitable medium of early contact with the occidental sciences and thus the vehicle of learning in this realm. By 1800, this was largely the case. Importation from the West of various technologies (telescopes, globes, guns, etc.) required translators versed in Dutch to study, make sense of, and apply them. The centuries of the Edo period had revealed, by trickles, the progress in technology being made by Europeans, and it had become clear to more than a few that profound changes would soon be necessary if Japan were to remain protected for any significant length of time. The visit of Commodore Perry's "black ships" (naval gunboats) in 1853, intended to force open Japanese trade to America, ended this hope and gave voice to another—direct competition with the West through absorption of its material powers. Following restoration of the imperial government in 1867, the country as a whole turned toward Europe for models of "modernization."

The linguistic effects of this historical shift were hardly surprising. When added to the forces of vernacularization already in motion, it spelled the downfall of Chinese, particularly in those areas related to natural philosophy. Western languages—Dutch early on, German and English somewhat later—came to rule the sciences, and in their immediate wake Japanese itself began to penetrate all levels and kinds of scholarship. The experience of Western languages, meanwhile, added a host of new, often confusing loan-words, sounds, and writing conventions (left to right) that convinced many of the need for some type of final standardization, and for uniting written and spoken Japanese. Between 1860 and 1900, a number of proposals were put forward. Some advocated using Western languages as a model, by which only *katakana* (the "Japanese alphabet") would be employed for writing; others recommended doing away entirely with all native forms and using the Roman alphabet instead. A more measured, and eventually successful response, put forth by the famed leader of the "Meiji Enlightenment," Fukuzawa Yukichi, was to limit the number of *kanji* for general public use and to standardize all readings according to a scheme that would incorporate both *on* and *kun* versions, along with any others that approached a similar level of usage. The rationality of such a proposal did not prevent it from being hotly debated for decades, held in limbo during the era of militarism in the early twentieth century, and then resurrected and finally adopted after the Second World War, when the Ministry of Education put into effect a series of sweeping changes, decided on some years before, that have acted (not without difficulty) to create the semblance of a modern, relatively stabilized and uniform written language.

Well before this, however, by the early twentieth century, written Japanese had become an amalgam of many influences. This was apparent even on the level of orthography: Japanese writers now frequently used Western-style punctuation, including commas, periods, and question and exclamation marks, and they could choose among three fairly common schemes of visual organization, including vertical from right to left (traditional Chinese style), horizontally from left to right (Western style), or horizontally, right to left (mixed style). According to Miller (1967, 133), by the late 1920s, Tokyo newspapers retained type for between 7,500 and 8,000 different *kanji,* of which about 5,000 were thought to be "familiar" to the "educated reader." Yet what this meant, with regard to any individual ideogram, was a "familiarity" far in excess of what was required of the average (or even above average) educated Chinese person. History had imposed a burdensome richness upon the same characters in Japanese, through the multiple phonetic and semantic attachments that had accumulated over time.

In the postwar era, the so-called *Tōyō-kanji* reforms, though implemented in schools throughout the nation, could not abolish with a wave of the official wand all of this traditional complexity. Even today, it is easy to find examples of writers using older (prewar) forms, including character readings and combinations, as well as various orthographies in literary, popular, and technical publications. When national interest shifted to the United States during the Occupation and English became a focus of considerable social attention and favoritism, a still newer complexity was added: the writing of Western loan-words in the Roman alphabet, especially if these words were in English. Today, this is particularly true in the sciences, where it is now a regular practice to insert various English nomenclatural items and citations in published articles in such fields as geology, astronomy, computer science, and biotechnology.

Taken as a whole, contemporary written Japanese represents a field where overlap occurs between a host of graphic-historical deposits. This is true, moreover, no matter what the subject material, no matter what the level of writing. Indeed, to say that the visual realities of this language, as written today, wear their history on their sleeve is to state only half the truth, for they also reveal much of the actual processes behind this history. Graphically, the different types of writing—*kanji,* the two *kana*s, the Roman alphabet—might be said to compete with each other at some level. Despite all its recent reforms, Japanese remains a battleground between ancients and moderns, a space where different eras, places, philosophies, and origins have all left their historical trace.

Scientific Discourse in Japanese: General Aspects

What has just been said about the Japanese language in general is perhaps even more true of scientific discourse than other forms of expression. To understand this, one must also understand something fundamental about this discourse.

Scientific Japanese differs from technical languages in Western countries in one very striking aspect. Because it is written in ideograms known to almost any college-educated person, nonscientists can readily decipher the basic meaning of many of the most complex terms. The layperson can, in a sense, read scientific jargon far more easily than his or her counterpart in the West (who is no longer regularly schooled in Latin and Greek). Two simple examples might suffice to show this. The terms "pyroclastic" and "protoxylem," no doubt obscure to the average humanities major in the United States or England, are written in Japanese with the ordinary characters for "broken by fire" (*kasai*) and "original living wood part" (*genseikibu*)—in this case, *kanji* that any seventh grader knows. Such legibility is by no means maintained for all technical terms, however. A great deal of contemporary scientific nomenclature is written in *katakana* and represents direct phonetic importation from Western languages, English in particular. A term such as "obduction," rendered into Japanese as *obudakusyon,* would undoubtedly make as little sense to educated nonscientists on either side of the Pacific. However, in certain fields where the use of *kanji* tends to dominate (e.g., botany, mathematics, and a sizable portion of geology and astronomy), general public access to technical language is extraordinarily high by the standards of occidental linguistic reality.

What, then, of style? What literary qualities might scientific discourse possess in Japanese? What principles of rhetoric might it uniquely depend upon? To answer such questions, even briefly (as will be done here), one must consider a principal division. The rhetoric of Japanese science, that is, like the Japanese language generally, needs to be viewed in terms of two basic dimensions, one stylistic-semantic in nature, the other graphical. In reality, these are often merged into a single metal, yet with several layers of attempted persuasion.

If one looks at style, minus the graphical dimension, the overall characterization is fairly simple. Given the history involved in its formation, Japanese scientific writing has qualities that are perhaps both unexpected and predictable: despite profound linguistic differences of many kinds, this writing is quite close to English in its basic aesthetic and rhetorical aspects, certainly much closer to English than to other Western languages, such as German, Dutch, French, or Spanish, or to Chinese. This is clearly the case in such aspects as tonal formality, general simplicity in sentence structure, use of first-person pronouns, manner of citation, article organization, and, most of all, the heavy reliance upon jargon as the syntactic center of expression. Of course, important differences exist: the general prolixity (repetition of individual words, terms, phrases) of Japanese imposes a redundancy that would be unacceptable in any occidental scientific language. Yet, except for such "climatic" differences, the similarities to English are striking. While one can often find in technical French or German examples of complicated literary technique, involving parallel constructions and the like, Japanese scientific writers commonly follow the English example in keeping things as direct and unadorned as possible. The following provides an example:

> Beppu Bay lies on the eastern margin of the Beppu-Shimagara graben(1) (Matsumoto, 1979) and Hōhi volcanic zone ([written in English]; Kamata, 1989b), which developed as a result of volcanic activity and graben(2) formation in Pliocene-present time (approximately 0–5 Ma [million years before present]). Within the Hōhi volcanic zone, andesitic volcanism was associated with normal faulting and caldera subsidence involving 2–3 km vertical subsidence of pre-Tertiary basement rocks. (Kamata 1993, 39)

Based on this fairly literal translation, it is evident that such writing has all the straightforward dullness and efficiency one expects from contemporary geologic discourse in English.

Yet when one begins to consider the graphical dimension, i.e., how the passage appears in Japanese, an entirely different and extremely rich rhetorical universe emerges. One sees, for instance, that the term "Hōhi volcanic zone" is written in English and thus stands out amidst the flow of *kanji* and *hiragana*. The effect is a type of subject emphasis that makes the mere use of underlining or italics in Western languages appear paltry by comparison. At the same time, all numbers in the article are also given in Western (Arabic) symbols, adding emphasis to the quantitative aspect of the treatment.

The term "graben," on the other hand, is written in not one but two separate ways, within the same sentence: the instance labeled (1) appears in *kanji* and represents the adopted Chinese term, which was absorbed during the late nineteenth century; the second instance (2) uses the phonetic symbols of *katakana,* thus indicating a distinctly Western origin, "graben" being originally derived in the early nineteenth century from the German for "ditch" or "trench" but also adopted in occidental geologic parlance in general (English, French, Spanish, Italian, etc.) to indicate a fault-bounded rectangular or rhomboidal area of sinking. The double writing of "graben" in the above Japanese passage is not done for effect only. Instead, *Shimagara-chikō* (*chikō* = graben) is part of a proper name, coined either late in the nineteenth century or early in the twentieth. To juxtapose it against *gurahben* in *katakana,* which is contemporary usage in geo-discourse, is to place different eras or tendencies of nomenclatural coinage—different cultural loyalties—immediately side by side.

Finally, the terms "Pliocene" and "Tertiary," both proper names of Western origin for specific periods of geologic time, are rendered in *kanji,* with the characters, respectively, for "bright new era" and "third period." The first of these translations represents something of a nativized elaboration: the Greek roots mean, simply, "more new." The Japanese, in contrast, has a distinct dynastic ring to it, reminiscent of "era of shining peace" (*Meiji-jidai*) or "bright harmony" (*Showa*). The second term, meanwhile, *daisan-ki,* is merely a direct translation from the English. Thus again, two very different sensibilities placed shoulder to shoulder, two different realms of choice revealing of historical influence.

Such complexity can be found everywhere hovering at the surface of scientific writing in Japanese. No doubt it is not perceived as such by its users (else, one wonders, how would any "science" get done?). As insiders, they have been trained to ignore or forget such realities, just as Western scientists only rarely question the genealogy of the language they employ. Yet there are a number of other observations that may be made along these lines. For Western readers, it is somewhat striking to find in Japanese scientific journals that illustrations, tables, charts, graphs, or other nontextual elements are presented entirely or largely in English. The effect, as one might expect, is both entirely remarkable and distinctly useful. Floating in a sea of otherwise incomprehensible symbols, there are here maps and diagrams made entirely legible for the Western eye, oases of valuable, summarizing information. Why this particular use of English? Since early in this century, Japanese scientists have been acutely aware of their linguistic isolation and have striven to internationalize the products of their labor. This isolation, however, also allows them an ingenuity. A standardized demand to compose illustrations in English effectively requires literacy in this language, thus access to technical writing from many other nations. It also recognizes (inadvertently?) that scientists everywhere often "read" the latest research in exactly this way, by skimming the visual information offered, even before perusing an abstract. (Indeed, the deeper internationalism of scientific discourse as a whole may lie not merely in certain modes of writing, but in styles of reading as well.)

In the face of all this, one perhaps feels an urge to declare the equal complexity of technical speech in European languages. Haven't terms here come from Latin, Greek, Arabic, English, French, German, Russian, the twelfth, seventeenth, nineteenth, and twentieth centuries, and so forth? Certainly they have, and this has many implications with regard to the content of science and the ability to decipher its historical evolution through a kind of textual archaeology. In the case of Japanese, however, it is the graphical element that once again makes for an important degree of difference (I do not say "exotic" difference). To better understand the Japanese case, one needs to imagine a wholly novel kind of English. This scientific English would contain Greek and Arabic terms written in the original Greek or Arabic sign systems. Other terms dating back to English's Anglo-Saxon or Frisian origins would also be written in the symbols of these languages. Moreover, one would need to conceive the simultaneous use of medieval and modern forms of certain words, each with its own graphical integrity preserved. In short, all derivations would have to be *kept visible.*

Japanese Science: The Cultural Context

The terrain of modern Japanese science has been created almost entirely since the late eighteenth century. This was a period that followed the lifting of a strict ban on Western books, a policy that had been in effect since 1630, when contact with Western nations as a whole was restricted (not closed off) by the early Shoguns. Most favored nation status was granted Holland and China, the former of which was allowed to maintain a trading station and factory on the island of Deshima (lit. "Departure Island"), in Nagasaki harbor. Any and all books unloaded here had to be inspected by official censors before even being allowed to touch mainland soil. Most Western works considered for admittance were those translated into Chinese, from which any mention of the Christian religion or of God had been extirpated. By the 1760s, some Dutch books on medicine and various almanacs were being imported as well, and among the items shipped to Japan were included many scientific instruments (esp. telescopes), which had the effect of quickening an interest in Western learning (MacLean 1974).

At the time, this was calculated to appeal to the existing tradition of natural philosophy in Japan. This tradition was concentrated in two main areas: botany (actually pharmacology, linked to herbal medicine) and astronomy (above all, calendar-making). Both subjects were deeply linked to the Confucian social order put in place by the Tokugawa shogunate, which imposed a rigid class structure with the intellectual professions dominated by physicians, Confucian scholars, and priests. All of these were hereditary in practice and all were enforced by the great resurgence of neo-Confucian study and scholarship the government supported. In a sense, this resurgence made Chinese scholarship a required foundation for any branch of intellectual endeavor, with the Chinese classics treated as the source of wisdom and a path to virtue. Pharmacology and medicine in general drew on detailed study of these works, the plants and animals described within them, along with their uses, as well as the theory and illustration of the human body. Similarly, astronomy depended on literary study of chronology related to the seasons, constellations, and planets, as expressed in these ancient texts. New reference works were also produced that catalogued these phenomena, commenting upon their occurrence and meaning within the "books of wisdom," and, somewhat later on, developing "theories" or "hypotheses" to explain them, usually in metaphysical fashion, based upon points of debated interpretation. All of this seems to have been generally, perhaps vaguely, known or made known to the Dutch traders, who tailored their imports of "science" accordingly.

The Chinese literary tradition, though preeminent, did not, however, enjoy a complete monopoly over intellectual life. Even by the late seventeenth century, the Japanese classics (poetry and prose alike) were also being invoked as important sources of information on native plants, descriptions of astronomical phenomena, and weather-related events. Dissatisfaction with the inaccuracies of the Chinese calendar in particular led to a desire for reform and also to a greater openness to considering the possible benefits of Western astronomical methods and ideas. With the lifting of the censorship ban on European books in 1720 by the Shogun Yoshimune (who himself seems to have been interested in the techniques of Western calendar-making), a slow turn began to take place away from China as the major seat of knowledge and toward Europe. This did not mean anything like a quick overturn of Chinese-inspired natural philosophy; on the contrary, even up to the early nineteenth century official scholarship and education remained wedded to the older classics as the major reservoir for higher forms of truth and study. For the sciences as well, these works provided a metaphysical, vitalist framework or filter through which occidental ideas were often absorbed, in somewhat reconstituted form. On several levels, science in Japan thus had literary origins. As a separate area of knowledge, "science" emerged from a series of canonical works whose understanding was restricted to an elite group of scholars.

In this respect, it is important to note that the vast majority of European books entering Japan before about 1770 were Chinese translations or commentaries in Chinese by Jesuit missionaries living in China. Most of these books were concerned with scientific subjects, for a distinct reason. Japanese censors had been ordered to examine all texts for any sign of religious reference; this meant, in effect, that those books most able to pass inspection were of a technical nature. From the 1770s onward, such books began to include volumes in Dutch in ever increasing numbers. These were examined, and sometimes requested, by a "college of interpreters" in Nagasaki who had been trained in the Dutch language at government command. An observer writing in 1776 noted that this "college" included between forty and fifty interpreters, assigned to assist the Dutch on Deshima, and encompassed journeymen translators, apprentices, and students, whose "conversation with the Europeans involves exclusively physique, medicine, and natural history, this to the annoyance of the Europeans" (MacLean 1974, 18). By this time, books on botany, astronomy, surgery and anatomy, mining, and navigation, plus many dictionaries, had been imported. The training these interpreters and translators received was often rudimentary at best, and doubtless unequal to their apparent enthusiasm, but it did help provide the basis for a new realm of scholarship, *rangaku* or "Dutch Studies," that proved extremely significant before the turn of the century.

After being inspected (i.e., read to the degree that was possible), books in Dutch were then shipped to the government library in Edo (modern-day Tokyo), where they were stored and kept out of general circulation. Some books might also have come from the private libraries of Dutch officials on Deshima. In any case, what Nakayama (1969) says of astronomy was true for other fields of Western science as well during this early period—namely, that they had little success in penetrating the existing neo-Confucian system of learning and inquiry. And yet, before long, it was this system itself that helped elevate the worth of technical knowledge from Europe, to such a degree that *rangaku* studies became the focus of nearly all interest in the West. Thus, ironically, even as late as the early nineteenth century, "the knowledge that the Japanese had managed to glean from and about the West was almost entirely confined to the natural sciences" (Blacker 1969, 14).

Tokugawa Natural Philosophy: Adapting the Chinese Example

Before the nineteenth century, the number of Japanese intellectuals even partly literate in a Western language was practically negligible compared with that part of the population who could read and write in Chinese. By force of circumstance, therefore, as well as tradition, Chinese natural philosophy continued to exert a major influence on Japanese thought during this time. Yet this philosophy was hardly a single, unified weave of thought; it had multiple strands, whose clash of color may have been just as important to Japanese adaptation as anything else.

One of these strands rose out of neo-Confucian beliefs, to which the Tokugawa shogunate gave government sanction. In Tokugawa Japan, neo-Confucianism meant the so-called reformist or Chu-Hsi school, whose founder had lived and written in the late twelfth century (Sung Dynasty). Within this philosophy, as adopted by the Japanese, there existed two fundamental terms for helping explain the nature of the universe and the place of human beings within it. One of these terms was *ri* (in Chinese, *li*), meant to indicate the ultimate ontological and moral principle that determines order in the universe. According to Chu-Hsi, the perception of *ri* within or behind the material world puts one in touch with the governing harmony of the cosmos and thus leads one to higher wisdom and ethical conduct. Posed alongside this, meanwhile, was *ki* (Chinese *ch'i*), whose description by Chu-Hsi is at best vague, having to do with primal energy of movement and appearance and whose definition—no doubt due to such vagueness—became a topic of much debate in the seventeenth and eighteenth centuries by Japanese scholars interested in the natural world (Saigusa 1962; Nakayama 1969). One of the first of these was Hayashi Razan (1583–1657), perhaps the most influential of the neo-Confucian scholars. For Razan, *ri* signified spirit, the fundamental essence; *ki,* on the other hand, was equivalent to the actual movement and workings of the material world. *Ki* was the materialization of *ri,* what brought it into the realm of action (Saigusa 1962, 53).

During the next century and a half, Razan's interpretation was modified and reconstituted by a number of thinkers. Yet it established a dualistic view that remained intact. More conservative scholars, holding to a literal reading of Chu-Hsi, tended to privilege the *ri* side of the equation, with its embodiment of virtue. Others, however, focused more on the methods by which this could be accomplished, and this is where *ki* tended to play a role. Chu-Hsi had laid out a number of aphorisms regarding this kind of inquiry, among the most important of which were (here translated into Japanese) the following: ***kakubutsu kyū-ri*** (investigate things and exhaust, or penetrate, the *ri*) and *sokubutsu kyū-ri* (probe into things and exhaust, or penetrate, the *ri*). In each case, the things (*butsu*) to be investigated or probed or penetrated were associated with the actual physical world. The goal was spiritual; the means, potentially at least, material. All of which constituted "a this-worldly mysticism with rationalistic implications" (Craig 1965, 139).

Such ideas, with their demand for greater specificity, were distinct adaptations. The Japanese were moving away from both the original metaphysics of Chu-Hsi and from the contemporary Chinese example, which adhered to ideas of *ch'i* as a grand cosmological force of energy and formation. Chu-Hsi's philosophy represented a comprehensive synthesis of the moral and the material; with regard to the heavens, it expounded a scholarly astronomy that involved no mathematics and no real observations. For practicing astronomers in China itself (who were employed by the court to predict and explain celestial events), this philosophy was useless and largely ignored. The importation of Chinese almanacs and other astronomical writings therefore meant that this split between "theory" and practice was transferred to Japan. Razan and others, in their interpretations of *ki*, were striving toward a new unity, based (necessarily) on greater emphasis given to the material side of the equation.

The Jesuits and China: Ironies and Advantages in Influence

A large percentage of the earliest books on Western science imported to Japan were those written in Chinese by Jesuit missionaries during the early 1600s. These books were intended to help secure a favorable position for the missionaries with China's emperor and his court, and to no small degree they did just that. For a full 150 years, the Jesuits were the primary—at times the only—translators of Western scientific thought to China. Obviously enough, this was a position of enormous historical import, which the missionaries well recognized. The famous Matteo Ricci, one of their earlier leaders, wrote more than thirty separate works dealing with occidental science, mainly astronomy, and helped establish a pattern followed throughout the seventeenth century that kept the Jesuits in partial favor.[3] But the missionaries, no less than

3. Extensive documentation of the life stories and Chinese writings of the Jesuit missionaries can be found in Bernard 1945.

their counterparts in Europe, could not escape the effects of the Counter-Reformation, with its condemnations of Copernicus and Galileo. However far from Rome to Peking, the distance was easily crossed by Western ships bringing news of the Church's injunction of 1616 against the teaching of heliocentrism. The results for Chinese science appear to have been something near to tragedy. In the words of Nathan Sivin:

> Jesuit missionaries, who alone were in a position to introduce contemporary scientific ideas into China before the nineteenth century, were not permitted to discuss the concept of a sun-centered planetary system after 1616. *Because* they wanted to honor Copernicus, they characterized his world system in misleading ways. When a Jesuit was free to correctly describe it in 1760, Chinese scientists rejected the heliocentric system because it contradicted earlier statements about Copernicus. No European writer resolved their doubts by admitting that some of the earlier assertions about Copernicus had been untrue. . . . To the very end of the Jesuit scientific effort in China, the rivalry between cosmologies was represented as between one astronomical innovator and another for the most convenient and accurate methods of calculation. . . . The character of early modern science was concealed from Chinese scientists, who depended on Jesuit writings. (1995, 4.1)

No work by Copernicus or Galileo, Kepler or Newton, Descartes or Huygens was ever directly translated into Chinese by the missionaries. Instead, they wrote their own treatises, strategically simplifying and deforming the actual record; in a very few cases they rendered works by Jesuit astronomers, such as Christopher Clavius. The greatest of "the moderns" was said to be Tycho Brahe, and the sum total of European astronomy, as put by one Jesuit author around 1640, "in its essentials has not gone beyond the bounds [set] by Ptolemy" (Sivin 1995, 4.21). The historical position of Copernicus, meanwhile, was changed into that of a later entry in a line of medieval thinkers whose work had been refined, updated, and rendered obsolete by Tycho. A century later, when Newton's great *Principia Mathematica* was being popularized throughout Europe and the Roman Church wore its old condemnation as a public symbol of intellectual defeat and backwardness, the Chinese remained convinced of a pre-Copernican universe. Through the power of their writings—through their decision *not* to translate, but to rewrite the texts of occidental astronomy—the Jesuits helped make the Earth stand still in China.

Japan's ban on Western books, which began in 1630, may thus have had a beneficial side to it. Certainly one of the main targets of censorship was the so-called Ricci corpus, which included nearly two dozen individual works. A blanket statement along these lines, however, would obviously be wrong: these works included valuable treatises on mathematics, astronomical instruments, cartography, and other subjects, as well as a partial translation (by Ricci himself) of Euclid's *Elements* (Nakayama 1969). No small portion of Western science did enter China, in accurate fashion, at the hands of the Jesuits.

As it happened, the Jesuits in Japan did not pursue an intellectual and pedagogic policy as did Ricci and his followers in China. Instead, they confined themselves to evangelical work. This left Japanese intellectuals with a choice between neo-Confucian natural history and a handful of Jesuit works on Western science that had arrived before the ban of 1630 and were not proscribed thereafter. The latter did not amount to very much, and Western science did not achieve a particularly significant following at this time. Chinese natural philosophy, with its focus on metaphysical interpretations of the natural world, remained entrenched. Yet, in the wake of thinkers such as Hayashi Razan, this philosophy continued to evolve in the direction of a more materialist sensibility. Indeed, a new focus on "things"—*butsu*—became a hallmark of the next century.

Miura Baien: The "Moderns" Come of Age

Such implications came to the fore in the era when Western science first began to be imported in significant quantity, after 1720. Ideas of *ki* began to be associated with the physical workings and actual phenomenology of the natural world. At this stage, that is, a purely abstract philosophy of Chinese origin was increasingly rejected. "Those who investigate natural things," wrote Miura Baien (1723–1789), "must not, like all prior followers of Confucius, remain dogmatic in their adherence to the words of the sages. For mastering truth, nature itself is the better teacher" (Saigusa 1962, 55). Baien represented the new type of *rangaku* scholar who emerged in the late eighteenth century, deeply schooled in the Chinese classics yet drawn to the originality of occidental science and its instruments and quite willing to use neo-Confucian philosophy to rationalize such interest. His particular interpretation of this philosophy is an adaptation: he had made a careful study of books on European science and came to view them as offering pragmatic "tools" for pursuing deeper questions. At the same time, he denounced or turned away from the speculative ruminations of more orthodox neo-Confucians, whom he saw as backward and closed (Nakayama 1964). Baien was thus one of the key "moderns" to argue against the "ancients." He backed his thinking with actual work, being among the very first to build and employ such instruments as a celestial globe, a telescope, and a microscope—all devices for taking the eye and mind deeper into the visual physicality of nature. In his scheme of thought, *ri* was tantamount to Plato's *noumena* or Kant's *a priori,* the ideal form of what "should be." *Ki* signified the reality, the force of what existed. To investigate it meant to study a range of other componental realities: *sei* (quality or character), *ryo* (mass), and *shitsu* (substance) among them.

Baien thus represents a crucial turning point in the development of modern Japanese science. In a sense, he stands close to its origin. His framing of concepts such as those just mentioned helped provide the very first vocabulary, relating to concepts of physical process and its analysis. Such basic terms as "matter" (*busshitsu*), "mass" (*shitsuryo*), and "body" (as used in physics, *buttai*) go directly back to him. They are terms that had a profound effect on the way nature was viewed, creating as they did a number of linguistically discrete object-areas for inquiry, distinct from cosmological ethics. Such terms, as it happens, were not wholly coined by Baien but instead adapted out of the Confucian tradition. *Shitsu,* for example, had long been used by Chinese scholars and philosophers to refer to the material quality of things, as a secondary manifestation of more fundamental metaphysical principles. It was a *kanji* caught up in the philosophical and religious doctrines of Chinese learning.

Hiraga Gennai: The Introduction of Electricity

If Baien marks the new philosophical attitude of the late eighteenth century, another writer, even more renowned at the time, represents its material ambitions. By any standard, Hiraga Gennai (1729–1780) was the most magnificent, diverse, accomplished, and eccentric dilettante of the age, a man who took inspiration from the West and transformed it into a veritable exhibition of specific achievements and failures. The catalogue of his activities runs through a dozen fields—playwright, novelist, mining innovator, mineralogist, potter, political writer, naturalist, pornographer, artist, and impresario. Different scholars have attributed to Hiraga different "major contributions": he was the first to apply Western painting techniques to Japanese subjects; he invented a form of asbestos shielding; he crossed previously unbroken lines between popular and literary authorship; he introduced electricity to Japan.

For our purposes, it is the last of these that merits particular attention. Hiraga had visited Nagasaki and the Dutch factory on Deshima and was taken with the array of things he saw there, including books. According to one report, he sold off all of his possessions, including his bedclothes, to buy a volume of Dutch zoology (in Dutch). Hiraga had earlier studied *honzo-gaku,* the field of botany and pharmacology as practiced by neo-Confucian scholars, involving exegesis of the Chinese classics and the discussion of plants therein. By the early 1750s, following his first trip to Nagasaki, Hiraga became wholly disenchanted with such study. Exposure to Dutch works seems to have revealed to him a vast world of worthy phenomena beyond the covers of ancient Chinese books. Sometime during the mid-1750s, he formed the idea of holding major annual expositions of Japanese (not Chinese) "products" for the public to view and wonder at. These "products" included medicines, native plants, rocks, minerals, types of wood, and a range of other items specific to the Japanese islands. Though the original idea was to confine these items to "natural productions," the catalogues for each exhibition make it clear that archaeological phenomena (e.g., old temple roof tiles) were also included (Maës 1970). The result was a type of grand *cabinet de curiosités* or *Schatzkammer,* a collection of physical curiosities meant to put on display the marvels of Japan, in no uncertain nationalistic tones.

Hiraga's specific goals are interesting to note. He wanted to overturn the existing bookish monopoly of the Chinese classics with regard to the pantheon of worthy objects; he desired to give new knowledge of Japan a concrete dimension; and he sought to engage the general public in such knowledge by showing it the spectacles that lay everywhere in evidence (Maës 1970). Hiraga called these exhibitions *bussan-kai*—a meeting for the display of natural products. One of these products that drew particular attention was medicinal plant material *honzo-gaku* experts had brought back from their explorations into the Japanese countryside, an activity, backed by the Edo government, that Hiraga claimed might reduce dependence on foreign medicines (notably Chinese).

Hiraga's penchant for public performance and for Western learning eventually came together in a very different type of spectacle. Apparently, he had by some means procured a damaged Dutch import, a simple electrostatic generator introduced sometime before 1771 in Nagasaki, as part of a collection of "strange objects," and he proceeded to work on it intermittently and with diligent curiosity over a period of years until he finally produced his own working version, of which he seems to have made no less than fifteen different copies. The original generator is attributed to the Dutch in accounts of the time and was said to serve medicinal purposes, as a means to apply heat or fire to damaged portions of the human body (Maës 1970, 152). To both the machine and the "artificial fire" it produced, Hiraga applied the term *erekiteru*—written phonetically in *kanji*—which he apparently adopted in shortened form from written documents. This seems to constitute the very first word for "electricity" used in Japan. It was an abbreviation of the Dutch term, *elektriciteit,* which in Japanese was written *erekiteristitato.* Hiraga's use of the shortened version defined a sort of linguistic adaptation Japanese scholars frequently employed with regard to longer and more complex terms in Dutch (this particular style of adaptation remains very much in use today).

The man who coined this term remained, to his dying day, more of an evangelist for knowledge than an investigator for it. In explaining "his" invention of *erekiteru,* Hiraga turned not to Western ideas but to Buddhist cosmology, specifically that of the Shingon sect. Within this scheme, the universe consisted of five elements—air, wind, earth, water, and fire—with the last an essential part of living things. *Erekiteru,* in essence, was the materialization of this, evoked by the machine. Such was the explanation given by Hiraga to Ishikura Shingozeamon, a famous contemporary Confucian scholar. Ironically perhaps, the term *erekiteru* did not survive the next century, but was instead replaced by the Chinese word *denki.* Hiraga's followers, against his own wishes, followed his example and used the relevant device mainly as a showpiece, rather than as an instrument capable of inspiring scientific inquiry. This "low" use of electrical force seems to have helped taint Hiraga's term in the eyes of later scholars. Before 1850, the word *denki,* written with the character for "lightning" and that for *ki* (*ch'i*), had been established in scientific circles.

Translating "Science"

This brings up the question of "science" itself. How was this word translated in the early modern period? As used by Baien, Hiraga, and other *rangaku* scholars, up until the mid-nineteenth century, this term was none other than *kyū-ri*—the "exhaustion or penetration of *ri,*" a primal principle set down by Chu-Hsi. Such is how the Dutch word *Natuurkunde* was translated for use by Japanese thinkers, i.e., through a Chinese philosophical word given Japanese pronunciation. Later on, between the final decades of the Edo era and the Meiji Restoration (1867), it was replaced by *kagaku* (lit. "course" or "branch" of "study"), a fairly literal rendering of the German *Wissenschaft.*

Kyū-ri, however, was never abandoned. Instead, it served for a time as the term for "physics," under the auspices of the belief, imported from the West, that this field lay at the base of all the other sciences and thus remained in league with traditional concepts of *ri.* Indeed, this connection has never wholly been given up, as the eventual term adopted for "physics" that remains in place today is *butsuri*—the *ri* of "material things," another word taken from Chinese (Nakayama 1992). Such complex blending of Chinese and Western sensibilities was an inevitable part of Japanese science in its early phases.

Such blending should not, however, be taken as a sign of compatibility or reconciliation. Thinkers such as Miura Baien reveal that a deep-seated tension had grown up within the *ri-ki* dialectic, one that increasingly divided virtue and knowledge from one another (Craig 1965). By the late eighteenth century, books in Dutch were being allowed into the country in ever increasing numbers, and the government's support of the *rangaku-sha* ("Dutch studies experts") had grown, partly based on the idea that the knowledge involved could be of important practical benefit. Visits by Russian and American warships during the early 1790s greatly added to this belief. Conservative scholars, on the other hand, felt that spiritual and moral power, the essence of Japan, would be enough to repel any invaders and that nearly all contact with Western knowledge was contaminating. The effect of these two camps of thought was effectively a freeing of "science" from "ethics" on the one hand, and "virtue" from "science" on the other. Western science came to be viewed more and more as a series of practical techniques, whose province was real-world power in a material sense.

Language and Loyalty: Issues in the Translation of Western Science

These brief comments on the intellectual origins of Japanese science return us to a central point: the history of early modern science in Japan is largely, if not entirely, a history of translation. With regard to scientific knowledge, Japan's isolation was then, and has always been, far more a matter of language than of geographical or political seclusion. The fundamental reasons for this are fairly obvious. Few, if any, European scholars were able to speak or read Japanese prior to the late nineteenth century. And even then, among those precious few who did learn it, the great majority were not scientists or engineers but instead people with artistic, commercial, religious, and diplomatic interests. Indeed, this has remained true up to the present. The requirement that Japanese scientists, amateur or professional, learn foreign languages in order to pursue their chosen field has been a constant element of training for two centuries (recall that a sizable number of technical journals in Japan are even today published in English). Well into the twentieth century, "science" in Japan meant translation in the larger sense.

At the time Western science was first beginning in Japan, the languages of scholarship were still Chinese or *kanbun,* a Chinese-Japanese hybrid. These were the systems of expression into which Western knowledge had to first pass. Inevitably, this meant a heavy Chinese/Confucian influence over the terminology coined at this time. It also meant that any break with the past, any final shift to a more modern view, would be dependent upon a change in linguistic fidelity. To modernize, Japanese science had to discover a new language of knowledge.

Beginning in the late eighteenth century, this new language was Dutch. It seems important to note that the term *rangaku,* which originally meant "study of the Dutch language," came to be used to denote the learning of *all* things Western, science in particular. By the early 1800s, works in Dutch had found a relatively secure place within certain portions of the scholarly community and were supplanting those in Chinese within certain scientific fields. There was a desire for more direct contact with Western learning, and for abandonment of outdated methods and beliefs. The feeling that European learning was mainly "technique" (*geijutsu*) and thus did not oppose neo-Confucian precepts allowed Japanese scholars some freedom to choose, linguistically and intellectually, between China and the West. One historian describes the situation in this way:

> Japan had been culturally dominated by China until at least the eighteenth century, but it was not politically dominated. The Japanese never lost their sense of national identity or their self-confidence. Compared with Chinese political satellites such as Korea and Annam, Japan was in a better position to modernize on its own initiative. The Japanese were free to choose either Western or Chinese science as they pleased. [Moreover] the existence in Japan of a plurality of influential philosophies also provided a sound basis for appreciating new ideas. There had been Buddhism, Confucianism, and Shintoism; why not another approach? (Nakayama 1969, 230)

Such a characterization is too simplistic, of course; the Japanese were never so "free" that they could cast off their entire intellectual heritage in natural philosophy with a wave of the hand. Various continuities were to remain, nearly to the end of the 1800s. Moreover, not all scholars agreed that the West should be so avidly embraced, particularly at the expense of former pieties. But despite these complexities, the general attitude that saw Western science as significantly superior to Chinese "technique" in a number of areas grew rapidly and impressively. Medicine (especially anatomy and surgery), navigation, and astronomy were three such areas widely recognized by the late eighteenth century, and military technology followed close behind, particularly after visits by Russian and American warships. The Tokugawa era, in general, was a period of growing nationalism in Japan, and this too turned many Japanese away from China, as we have seen in the case of Hiraga Gennai and Miura Baien. The National Learning movement, which gained steam throughout the late seventeenth and eighteenth centuries, adopted a platform of supplanting Buddhism, a Chinese import, with Shintoism, a native religion, and creating a Japanese canon of classics for study, which would be learned alongside the existing Chinese texts. Ideas for the founding of schools of national learning, such as those of Kamo Mabuchi or Motoori Norinaga, consistently vilified the "chaos" and "philosophical pretensions" of China, holding instead that it was Japan which governed itself according to the natural laws of heaven and earth (Keene 1968). To many Japanese, China signified a teacher that had gone astray, become arrogant. Allegiance to Chinese scholarship carried with it a profound debt, one that could only be eliminated by displacing China from the center of the intellectual/cultural universe. Such was the inevitable whetstone on which any nationalizing consciousness would come to sharpen itself.

This did not mean that Chinese natural philosophy was abandoned; on the contrary, this was felt to be part of Japan's own legacy, a necessary and nativized ingredient. It was rather that China itself, politically and culturally, had not lived up to the precepts and principles of the ethical cosmos: it had gone astray, fallen from the "right path" into "false government," "petty rationalizing," and the "addiction to sophistry." Western knowledge, on the other hand, brought with it no direct burdens of morality, no sense of obedience and rebellion. Neo-Confucian doctrine helped pose this knowledge as "mere technique," stripped of higher ethical-cosmological principles. This implied, however, that it could be absorbed without compromising loyalties. Even those who regularly abjured all things foreign, such as the influential and ultrapatriotic writer Hirata Atsutane (1776–1843), spoke up on behalf of European science:

> The Dutch have the excellent national characteristic of investigating matters with great patience until they can get to the very bottom. . . . Unlike China, Holland is a splendid country where they do not rely on superficial conjectures. . . . Their findings, which are the result of the efforts of hundreds of people studying scientific problems for a thousand or even two thousand years, have been incorporated in books which have been presented to Japan. (Tsunoda, de Bary, and Keene 1958, 41–42)

Here was a perception of Western science conceived in the terms most familiar to Japanese scholars—that is, a *literary* canon, compiled and handed down through the ages. Such "wisdom," the suggestion ran, could easily be used by Japan in its effort to grow strong and autonomous. But even when such writers refused to consider European learning, every mark they made against China was a mark in the long run for Western science. Every step backward for the former proved an advance for the latter. Western scientific learning, in other words, became a cultural-political lever to help Japan displace the image of a superior "China." Even at this early date, "the barbarian West" was a tool that aided Japan in its formative efforts to become an independent nation-state.

Growing debate, unrest, and philosophical conflict led the conservative Shogunate in 1790 to pass another ban on "heterodox teachings." Yet the overall effects of this attempt at censorship were small. Books from Holland continued to enter the country at an ever-increasing rate. Political and cultural orthodoxy remained highly porous and changeable. By the 1780s and 1790s, a growing number of physicians and interested samurai had taken up the study of the Dutch language. This had been accelerated by release to the general public of the first book translated from Dutch into Japanese, a medical work entitled *Tabulae Anatomicae,* originally written in 1731 by a German physician, which contained a host of anatomical drawings of great and fascinating detail that clearly added new dimensions of visibility to the existing Chinese-influenced anatomy. The resulting translation, the *Kaitai Shinsō,* was published in 1774 and "started a great wave of interest in Dutch learning of every description" (Keene 1968, 24). Though fears of the "barbarian menace" persisted in many quarters, this new interest, backed by the view of Western learning as "mere technique" and by desires to add to the national stock of practical knowledge, essentially guaranteed that a large number of new Dutch works would be allowed entrance.

In 1811, the government set up the Office for Translation of Foreign Books (Banshō Wage Gōyō). This was an attempt on the one hand to gain control over the content and spread of this new knowledge, but also an obvious acknowledgment of its growing reality and importance. The Office had a somewhat checkered career, being largely an administrative bureau early on, then, with another rearward shift in the political wind, a censoring board during the 1840s, finally becoming in 1855, after Perry's visit, the center of an official training institute that not only taught Dutch, but English, French, and German, as well as courses in most of the sciences. (No doubt its most significant influence in the early years was its publication of the first Dutch-Japanese dictionaries.)

By the first decade of the nineteenth century, Dutch had become the language of Western science. It remained so until the beginnings of the Meiji era in the late 1860s, when the shogunate was finally overthrown. By that time, Japanese scholars, translators, and officials had come to realize that their earlier gateway to Europe had in fact been something of a side entrance. English and American naval power, French chemistry and astronomy, German physics and medicine: all these became the sources of great fascination, fear, and envy. But before a new shift of linguistic interest could occur, a greater degree of openness to outside influence had to take place. The political and economic, not to say cultural, circumstances leading to such openness have been discussed too often for iteration here. A point to be made, however, is that official desires to acquire modern technology, and thus science, played their part in determining final decisions in this story.

Prior to the Meiji Revolution (1867), Japanese science remained largely noninstitutional. Except for the Office of Translation, *rangaku* was carried on primarily by individuals, even when they worked under government auspices. Despite mixed feelings regarding China, most were nonetheless loyal Chinese or neo-Confucian scholars. As late as the 1830s, it was common practice for *rangaku-sha* to make translations of Dutch works into *kanbun* or classical Chinese. Moreover, even the most xenophobic critics of China or aggressive advocates for the new National Learning (e.g., Kamo Mabuchi) would never have advocated something so radical as a purge of Chinese thought and its complete replacement by Western philosophy. Such thought, like the Chinese language itself, had long been part of Japan. Something akin to its denial would only come later, when science entered its first major phase of institutional growth.

Science in the Meiji Era

Regarding this phase, it would be too simple to maintain, as many historians have, that modern science began with the Meiji period in 1867, as by some divine shift in the intellectual wind. While it is certainly true that the government took matters firmly in hand at this point, viewing scientific knowledge as the key to modernizing the country, exporting students overseas and importing foreign teachers, this did not mark a complete break from the past. The translation of Western works was an activity long performed under government support and supervision. The bureau of translators in Nagasaki was a kind of institute for the introduction and transmission of Western learning, and, together with the book repository in Edo to which many translations were sent (often, as it were, in internal exile), was part of the larger governmental effort to take institutional control over science. So too was the *Kaiseishō*, or "Institute for Western Studies," which the government set up in 1855. These efforts were the brighter side to a long-standing attitude that had its darker moments in such events as the Siebold Incident (1828) and *Bansha no Goku* affair ("Jailing of the Office of Barbarian Studies"), in which important *rangaku* scholars were arrested, tortured, and forced into confessions that led to either execution or ritual suicide.[4] In these cases, the government's violent reaction was a result of its perception that strict policies of information control were being violated. It was a sign, once again, of the great power invested in Western knowledge, whether this power was thought to be contaminating or enlightening.

The official Meiji program, therefore, was merely an extension of this greater belief. If the logic of favoritism had shifted, to now rest entirely on the side of importing occidental knowledge, the fundamental rationale for such control hadn't: books, teachers, ideas, technologies—these were essential capital vis-à-vis national strength at home and around the world. Planning and maintaining control over the dissemination of this capital had been the government's chosen task from the very beginning. One measure of the long-term success of this attempt is exactly the central point of this essay: namely, that "science" was for so long equivalent to "translation."

Foreign Teachers, Native Students: Enlightenment Philosophy and Language Issues

The programs for importing foreign teachers (the *oyatoi*) and exporting talented students (*ryū-gakusei*) were begun during the 1870s and 1880s. Throughout most of this period, the motive to Westernize reigned supreme (though not without some local resistance), especially in technical, literary, and government circles (Gluck 1985). The 1870s

4. The Siebold Affair, in particular, is one that historians have pointed to as an indication of the repressive measures the government could take when it perceived the signs of potential threat. Philip Franz von Siebold was one of the first Europeans allowed to enter and teach in Japan. He arrived in 1823 from Germany, as physician to the Dutch factory, and was permitted to teach medicine and biology in Nagasaki until 1829. In the year 1828, the shogunal astronomer, Takahashi Kageyasu, was given the task of drawing up a map of Sakhalin Island, a job for which most information existed in Dutch. Takahashi contacted Siebold and made a trade for this information by showing the German several coastal survey maps of Japan itself. This was highly illegal, and when the government discovered it, a vicious reaction ensued. A large number of people only peripherally involved were sent to prison; Takahashi himself died there only a few months later, apparently after being tortured. Siebold was held for several months and then expelled to Europe. Finally, all of Siebold's students in Nagasaki were also arrested.

in particular were the heyday of the Japanese Enlightenment (*keimō*), spearheaded by thinkers and writers such as Fukuzawa Yukichi, who helped make commonplace the vilification of the old feudal system, and who proclaimed the modernism of adopting Western ways in all areas of life, from clothing to poetry. Watchwords of the day were "civilization" (*bunmei*, an old Chinese word) and "progress" (*shinpo*). But not these alone. Older terms employed by Confucian scholars were also taken up and redefined: *kyōgaku* ("empty learning") and *jitsugaku* ("practical learning"), the former being applied to scholarship without virtuous behavior and the latter intended as the opposite. Enlightenment writers, in a sense, turned the meaning of these terms against their originators. *Kyōgaku* came to indicate the attitude (Confucian, after all) that knowledge must always have a moral end; *jitsugaku*, meanwhile, referred to real-world learning that could be put to use in everyday life, toward the improvement of lives and industries, the building of a new, more forward-thinking nation. Fukuzawa was among the most ardent and effective proponents of this sensibility, and associated *jitsugaku* with Western science in particular. Indeed, as one historian of this era has noted,

> It was . . . due to lack of interest in the laws of nature, Fukuzawa was convinced, that Japan had failed to progress to the blessed state of civilisation. She had put far too much emphasis on one particular kind of knowledge, ethical knowledge, at the expense of ethically neutral scientific knowledge. She had believed that virtue was the sole element in civilisation, whereas it was abundantly clear that what caused "progress" . . . was not virtue but knowledge. There had been no very startling progress in virtue since ancient times, for moralists and saints had been unable to do more than merely comment on the principles laid down by Christ and Confucius. . . . Whereas in the sphere of [science], "we know a hundred things where the ancients knew one. We despise what they feared, mock what they marvelled at." (Blacker 1969, 54)

Notwithstanding the simplicity of such comparisons, Fukuzawa remained a guiding voice throughout the Meiji period, and his ideas took hold in most intellectual circles.

Overemphasis of such influence has been common, however, and the period has often been interpreted as one in which Japan sought to "refound itself on the basis of European culture" (Kobori 1964, 3). Yet the ideology of the moment was not so simple; there was no wiping the slate clean. As Carol Gluck has written, "in those years the apostles of civilization often initiated their exhortations with some variety of the phrase, 'for the sake of national strength and expanding national power'" (1985, 254). The impulse toward nationalism, which had long viewed occidental science as a needed element, never wavered. The move to Westernize only increased this impulse, providing it with a new set of practical agendas and methods. This is why "civilization" and "Enlightenment" could go hand-in-hand with a renewed official idolatry of the emperor, who stood as the personification of national unit—a unity that would triumph through time (history) and would one day (soon) come to justify Japan's own colonial "adventures." Nationalism, in other words, meant advancing the internal strength of the country by incorporating the best that other nations could provide, those that had already proved their worth in various areas of modern power.

According to Bartholomew (1993), a total of only about eight thousand teachers from Europe and the United States were brought into Japan between 1870 and about 1900. In contrast, tens of thousands of *ryū-gakusei* were sent overseas. This reflects an important policy of the government: to reduce, as quickly as possible, any direct dependence on foreign sources of knowledge and expertise. Pupils of the *oyatoi,* it was hoped, were to replace them, and in greater numbers. In 1872, the government initiated a plan to divide the country up into eight academic districts, each with an institute or university of higher learning headed by a European professor. The idea was to create a series of nodes, from which qualified students would radiate outward, permeating the body of the nation with the new knowledge. (Only three such schools were actually established, however—in Tokyo, Osaka, and Nagasaki.) Students returning from overseas, meanwhile, would both help in this process and extend well beyond it, enriching the government, the military, the new universities, and commercial enterprises with a wealth of expertise.

This dual effort to nativize Western science both from within and without must be understood as an overall rational plan, one based, profoundly, on the powers of translation. It was through translation—into Dutch, into Chinese, and into Japanese—that Western science had always been known. Now this tradition of contact would extend far outside the country, into the true centers where this science had been built, involving a host of new languages like English, French, German, Russian, Latin, and Greek. The Japanese experience of Western science now became polylingual, and therefore multisocial. Writers and officials may have spoken of it as a single entity: *rangaku* or *banshō* ("books of the barbarians") early on, and now, as *seiyō-gaku* ("studies from the western seas"). But science in Japan was more than ever, both linguistically and practically, an extremely heterogeneous thing. One should note, in this regard, that no part of the government's plan ever involved teaching foreign professors the Japanese language. Those *oyatoi* who did learn it were discouraged from using it in their professional work. Their books, articles, and lectures were translated by students, who therefore carried out the national plan for knowledge control by maintaining control over the development and use of scientific discourse itself. Westerners, that is, very rarely added to this discourse themselves in any direct way.

This matched, as well, a larger linguistic reality we spoke of earlier. By the late nineteenth century, the situation of written Japanese was enormously pluralistic, confused, a melange of formal and informal, ancient and modern, communal and personal forms. The 1870s and 1880s in particular were a time of great dissension and debate in official circles about what to do with the Japanese language. Many, such as Fukuzawa, viewed it as clumsy, feudal, and inefficient, wholly inadequate to the learning of Western ways, especially science. Proposals were made to do away with *kanji* altogether and to Romanize the language (i.e., use Western phonetic symbols) entirely. Conservatives argued instead that an effort had to be made to "purify" the mother tongue, either by transforming all Western and Chinese words into newer Japanese ones, or else by doing away with the "foreign" altogether (Miller 1967). In the end, the conflict retreated to a matter of emphasis. There was, after all, nearly complete agreement that Western scientific knowledge held a key to the country's future. The question was how best to nativize this source—what the proper balance should be between adoption and adaptation, between the "Westernization" of Japan and the "Japanization" of the West. The argument about language was an argument about both practical use and political/cultural symbolism. The "West," however—and therefore "science"—was never a single entity. It was divided into a series of roughly planned favoritisms.

Changing Loyalties: From Europe to America

While foreign teachers and even *ryū-gakusei* were often selected haphazardly (see Bartholomew 1989), great care was taken in choosing certain "target" nations. Indeed, the preferences decided upon by the Japanese government present a fairly accurate portrait of the state of Western science at the time. Nakayama (1977), among others, has delineated these preferences. They include:

Germany: physics, astronomy, geology, chemistry, zoology, botany, medicine, pharmacology, educational system, political science, economics;

France: zoology, botany, astronomy, mathematics, physics, chemistry, architecture, diplomacy, public welfare;
Great Britain: machinery, geology and mining, chemistry, steel making, architecture, shipbuilding, cattle farming;
U.S.A.: mathematics, chemistry, general science, civil engineering, industrial law, agriculture, cattle farming, mining;
Holland: irrigation, architecture, shipbuilding, political science, economics.

Japan recognized Germany and France as leaders in basic scientific research. In the first part of the nineteenth century, French chemistry (Lavoisier) and natural history (Buffon, Cuvier) had formed the basis for these disciplines in Japan. During the early Meiji years, chemistry and physics were actually taught in all three major languages: German, French, and English. Favoritism granted to France, however, declined after the Franco-Prussian war. British and Scottish technicians were prized in mining, railways, and roads. It was the Germans, however, who came to be seen, especially from the 1880s onward, as the essential model in nearly all areas, in medicine particularly, and for university teaching in basic science. More German *oyatoi* were employed than all other nationals combined. During the 1880s, German models were used by the Japanese government to reform the entire system of higher education, and many political institutions as well. Whereas in the 1870s only 27 percent of the *ryū-gakusei* chose Germany as their country for study, this number climbed to 69 percent by the 1890s and 74 percent by the first decade of the twentieth century (Bartholomew 1989, 71).

As a whole, therefore, Japanese science was largely conducted in the languages of several Western countries. Japan's new scientists frequently taught, wrote papers and monographs, even conducted conferences in German, English, or French, depending on which nation had been adopted for a particular field. Again, this did not represent a new development, but merely continued an ancient tradition related to the import of specialized knowledge that had begun with Chinese and later moved on to include Dutch during the early years of *rangaku*. Now, however, it meant that different disciplines—or even different groups within single disciplines—quite literally did not talk to each other. There was an often radical absence of interplay between fields. Not until the turn of the century did Japanese itself finally begin to replace foreign languages as the medium for scientific learning and inquiry.

This shift might have required yet more time, if not for an important development. This was the formation of scientific societies, like the

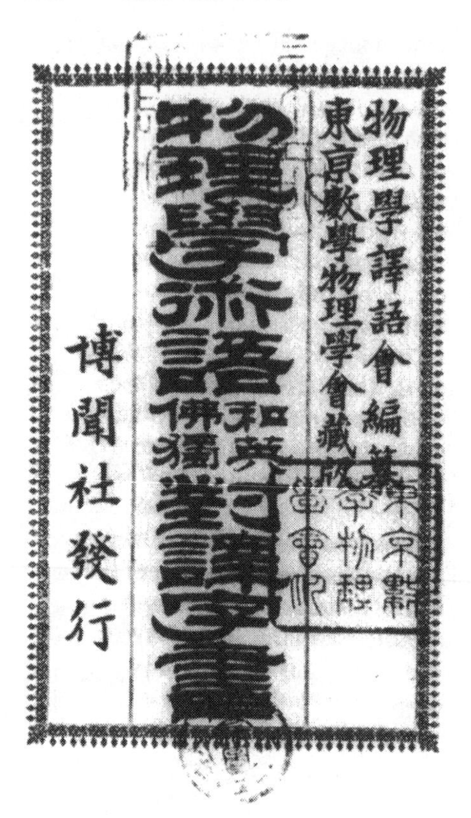

Figure 17. Cover page for a dictionary of physics terms in Japanese, English, French, and German. Published in 1888, this small text of eighty-eight pages took more than thirty scientists several years to compile (Watanabe 1990). It represents a crucial step in the history of Japanese physics, as it helped compile and standardize for the first time terminology from a wide range of translated source material. Reprinted, by permission, from Watanabe 1990, 14.

Tokyo Chemical Society and Physico-Mathematical Society, which undertook conscious efforts to compile dictionaries and to standardize important terms (fig. 17). These societies took on such linguistic tasks as their first duty. The large number of newly translated works, including textbooks, often applied different nomenclatural systems to the same phenomena (e.g., names of the elements), and this contributed to a general and widespread confusion. Titles derived from French, for example, with regard to chemistry, differed strongly from those taken from the German, which had far fewer words cognate with Latin. Such confusion, therefore, partook of the larger absence of linguistic standards in society. Different nomenclatures in science paralleled the perplexing number of writing systems in literature (*kanbun, manyō-gana,* which had been revived, Chinese itself, colloquial Japanese, plus several orthographies). Just as some Japanese researchers might prefer German or English terms (transliterated into Japanese), or else a mixture of these, so did some novelists of the day choose a classical style of writing, others the vernacular, still others a complex alloy of the two. A perplexing number of writing systems in the larger society was thus matched by a perplexing number of ways to write individual ideas or realities in science. Science, indeed, was an integral part of the larger literary moment. Standardization began, in effect, when professional groups formed and began to impose order by using the ambient chaos as a reason to promulgate a particular linguistic program. In 1888, for example, under the direction of physicist Yamagawa Kenjiro, a dictionary of physics terminology in Japanese, English, French, and German was published (see Watanabe 1990), providing Japanese physicists with an unprecedented document for unifying their field and developing pedagogic standards. Such developments made it finally possible for Japanese scientists to lecture and write in their native tongue. Moreover, this took place against a background of increasing demands for "Japanization" in the last years of the nineteenth century, when reaction began to set in against the previous decades of zealous Westernization, a wave that moved forward on claims that Japan had become largely self-sufficient in technical power and should no longer admit subservience to the West (Gluck 1985). The new technical power of the state had been proven, in no uncertain terms, by Japanese wartime victories over both China (1894–1895) and Russia (1904–1905), both however largely a result of Japanese naval strength built and equipped with much help from Britain (Japan's major ally in Europe at the turn of the century).

The new nationalism, on the other hand, did not at all mean the end of translation as a central part of Japanese science. Students were no less required to study German or English than in prior decades. Models for research, involving new government and industrial laboratories, continued to come from Europe, and now also from America. The first phase of linguistic standardization had ended by about 1900 and, despite noble efforts, had not been as inclusive as hoped, in part due to lingering factionalism. The pace of research in the West, meanwhile, was continually a topic of mixed news, excitement, and anxiety. With its fledgling institutional structures and comparatively small research community, how could Japan possibly keep up with what was being done in physics in Germany or engineering in America, let alone compete on an even level? The realities of such inequality ensured an ever-increasing flow of new terms from abroad into Japanese. But it also maintained the necessity for Japanese scientists to know, and be able to work in, foreign languages. By the 1920s, when the National Research Council of Japan set about founding a set of major research journals in the fundamental fields of astronomy, physics, chemistry, geology, zoology, and botany, it decided to do so on the basis of the apparent linguistic future of science itself. That is, the Council decided that most problems of readership would be avoided and international circulation enhanced if such periodicals were all published in English. Thus were born the *Japanese Journal of Chemistry* (1919), the *Japanese Journal of Geology and Geography* (1922), the *Japanese Journal of Astronomy and Geophysics* (1920), and so forth.

Notably, German was not chosen for this effort. During and after World War I, Germany lost its favored nation status in science. The war dramatically interrupted intellectual and commercial exchanges between the two countries (Japan, for example, imported a great deal of its pharmaceuticals, medicines, and industrial chemicals from Germany) and revealed a depth of dependence that did not sit well with official hopes for Japanese self-sufficiency and, only a few years later, imperial expansion. The war also helped make clear to the Japanese government and Japanese scientists and engineers that a larger portion of the industrialized world spoke English than any other language, that America in particular had rapidly modernized itself into an industrial and military power of ever-expanding influence, and that, despite its apparent lead in such fields as physics, chemistry, and medicine, Germany had been the undisputed loser in the conflict and was having significant trouble recovering.

In the 1920s and 1930s, meanwhile, scientific research in Japan was basically taken in hand and advanced according to military and colonial goals. The government, army, and private industry all worked together to effect the "Manchurian incident" of 1931, which resulted in a munitions boom that staved off the effects of the Great Depression and caused an estimated 50 percent of the country's engineers to be employed in military research, with a substantial portion of the remainder, and many basic scientists as well, involved in indirectly related activities (Nakayama 1977; Hiroshige 1973). It was during this period in particular that "big science" became central to Japanese capitalism and Japanese imperialism, and began to truly achieve, if selectively and for troubling goals, a level commensurate with scientific-technological work in Europe and America. Needless to say, this continued during the years of mobilization leading up to World War II. Throughout the period, America was perceived as both a looming nemesis and an enviable international power.

After Japan's defeat in World War II, scientific work was taken away from the military by American occupation forces and turned over to a series of government-run research councils and agencies, directly overseen by scientists imported from America (Yoshikawa and Kauffman 1994). The building of a new institutional infrastructure for Japanese science was viewed as a crucial element in the reshaping of the nation into a democratic state. On the Western model, scientific work was transferred from the military sector to private industry and the universities and, somewhat later, to new government labs. Japanese science now turned its full attention to America, as a source of both inspiration and appropriation. The details of this shift and its institutional effects, however, are less important here than is the fact that this allegiance, in general terms, has remained intact for more than fifty years, a length of time nearly equivalent to the whole of the Meiji era. During this time, the United States might be said to have become in the late twentieth century what Holland had represented earlier. Certainly *rangaku* has found its correlative in a variety of species of *bei-gaku* (*bei,* or "rice," being the traditional designation for America).

The twentieth century, then, has seen a progressive changeover in the favored language for Western science. As early as the mid-1920s, German had given way to English, a pattern that continued into World War II, despite hostilities, and that has culminated in the postwar period. As of the late 1940s, this turn toward English had to be seen against the larger backdrop of preferences granted the United States and the English language within Japanese society generally—preferences that arose from a complex range of motives. Among scientists, the turn to English has involved a psychology of purpose linked to intellectual need as well as to professional and personal status. To a large extent, of course, Japanese scientists share this need with researchers everywhere, given the status of English worldwide in the sciences. Since a sizable majority of articles in international journals have come to be published in English, since technical conferences are more often held in this language than in any other, and since, in the case of Japan particularly, the closest diplomatic, industrial, and academic ties have been with the United States, it is inevitable that some command of English would be viewed as a required part of contemporary science. Added to this is the larger social image English has acquired in Japan, an image associated with being urbane, cosmopolitan, "in touch," unconfined by traditions viewed as valuable and unique but also provincial, overly "Japanese." Again, there are parallels here with other nations and cultures. But in the case of Japan, the war, occupation, and continued physical isolation of the country add singular burdens, and a note of augmented anxiety, to desires to learn English. In any case, scientists in Japan are part of this greater cultural turn. Their need for English literacy should not be seen as purely professional, but instead as part of a larger historical moment.

The central point is this: that Japanese scientists today, no less than a century ago, are very often engaged in efforts of translation—reading, writing, and listening to a foreign language or else consuming information that has been transferred into their native tongue from elsewhere. While these activities no longer account for the whole or the major part of science in Japan, they are still central, unavoidable. Indeed, if the predominance of English worldwide has forced the translator's role upon scientists nearly everywhere, this represents in Japan merely the continuation of an age-old reality, reaching back centuries.

6 Japanese Science in the Making

Of Texts and Translators

Early Beginnings of Western Translation

Between 1770 and 1850, a great deal of Western science entered Japan via the Dutch language. Copernican theory, Newtonian physics, the biology of Linnaeus, Lavoisier's chemistry, the astronomy of Laplace and Lalande, much of Western medicine—all these and more were introduced and taken up, mainly by physicians and scholarly samurai. Dutch, of course, was not the language of origin for most of this knowledge. To most Japanese at the time, however, habituated as they were to thinking of Holland as the epitome of Europe, works written in Dutch were seen as primary sources. To read these works in the "original" was tantamount to being in contact with scientific knowledge at its generative base. The reality, however, was that many of the books imported from Holland were in fact translations of translations (etc.). Moreover, due to the tenuous political climate with regard to Western ideas, *rangaku* scholars often felt it expedient or necessary (for their own survival) to make certain changes in these texts, and in their own translations of the texts into Japanese.

Motoko Ryoei and Heliocentrism

These characteristics are well demonstrated by an early work that helped introduce the heliocentric view to Japan. This text, by one of the most famous translators of the early period, Motoki Ryoei (1735–1794), was titled *Oranda Chikyū Zusetsu*, "An Illustrated Explanation of the Dutch View of the Earth" (1772). Ryoei's source text was a translation of a translation, a Dutch version (dated 1745) of a French original, *Atlas de la Navigation et du Commerce qui se fait dans Toutes les Parties du Monde*, written in 1715 by Louis Renard (Nakayama 1969). But whereas the Dutch work—titled *Atlas van Zeevaert en Koophandel door de Geheele Weereldt* ("Atlas of Navigation and Commerce for the Entire World") had included a number of important maps as well as a seaman's guide to their use, these very central parts of the work were entirely omitted from Motoki's Japanese rendering. Not only was the result incorrectly (if revealingly) titled the "Dutch View of the Earth," it was also lacking in any "illustrated" dimension whatsoever. It is more than likely that Motoki left out the maps in part due to fears of government censorship and reprisal. Their absence was a direct expression, one could say, of the official desire for the continued isolation of the country—for the effective deletion of Japan from maps of this type.

What of Motoki's choice of terminology? The word he selected for "heliocentric" was entirely expressive of the time and, one might say, the general situation of knowledge. This term, *taiyō kyū-ri,* was wholly in keeping with neo-Confucian natural philosophy, denoting "the exhaustion of *ri* with regard to the sun." Motoki accepted for use the traditional term for "sun" (*taiyō*), this being the ancient Chinese combination of the characters for "thick" and "yang principle" (positive, male, daytime, etc.). A revolutionary new theory was thus introduced through entirely traditional linguistic means. This was revealed in another way too. According to Nakayama (1992), Motoki's manuscripts show that, prior to publication, he actually used the indigenous Japanese phonetic alphabet, *katakana,* to write the names of Greek and European astronomers mentioned in the text, as well as the names of geographical places in Europe, Africa, and the Middle East. Yet, when it came time to publish his work, he felt obligated to change all of these foreignisms into phonetic *kanji,* in obedience to convention. Scholars of the day were still scholars of Chinese before they were scholars of Dutch. Indeed, traditional neo-Confucian language seems to have been a means by which *rangaku-sha* could introduce entirely new systems of thought under cover.

Rangaku translators such as Motoki began the process of rendering Western science into Japanese by using existing Chinese words or by borrowing them from Chinese dictionaries and lexicons, often of ancient vintage (the Sung dynasty of the tenth to thirteenth centuries, viewed as a high point of mainland civilization, was a favorite reference point). With time and confidence, however, this method declined in favor of another. Rather quickly, scholars found that even the total Chinese vocabulary was inadequate to the task before them. No existing terms could be found to accommodate, even remotely, some of the concepts and nomenclatures of Western science. The Japanese translators therefore began inventing their own character combinations. The pace for this was set by one of Motoki's students, Shizuki Tadao (1760–

1806), one of the most important figures in the entire history of modern Japanese science.

Shizuki Tadao: Newton and the Language of Physics

It was Shizuki who brought Newtonian theory and vocabulary to Japan, in effect founding the language and therefore the concepts of modern physical science. This he did in a work of translation, *Rekishō Shinsho* ("New Writings on Calendrical Phenomena"), a title that reads as a masterpiece of nomenclatural disguise. The work appeared in three volumes between 1798 and 1802, the result of no less than two decades' labor. Shizuki seems to have used several texts as sources for his final work, but his principal original was a Dutch text, *Inleidinge tot de Waare Natuuren Sterrekunde* ("Introduction to the True Natural Philosophy and Astronomy," 1741) by Johan Lulofs. This work had been translated from the English writer John Keill's *Introductiones ad veram Physicam et veram Astronomiam* ("Introduction to the True Physics and True Astronomy," 1739), a widely read popularization of Newtonian ideas that, as it happened, Keill had himself originally published in English decades before (1720–1721) in two separate volumes (Knight 1972). The popularity of these works convinced their author to render them into Latin in order to reach a wider audience among the educated of Europe; the irony here is that Keill's ambitions were nicely fulfilled not by acceptance of his Latin version but by the translation of *this* work into the various European vernaculars, such as Lulofs did. Thus, even with regard to its influence in Europe, Newton's work had a complex textual history, having undergone multiple levels and stages of translation.

Lulofs's rendition of Keill, like most Dutch translations, was a mixture of fidelity and augmentation, the translator having added a number of clarifying notes and explanations. Shizuki's version, however, required added levels of adaptation. Many of the terms Newton used were so utterly foreign as to have no easy counterpart in Chinese or Japanese. They did not merely have to be invented; they had to be conceived. Yet, they also had to be inserted into the larger linguistic frame of the time. Their success, in a sense, depended on their acting as a mediating substance posed halfway between two very different cultural-intellectual systems.

Perhaps as a precaution, Shizuki chose a title that invoked the ancient Chinese term for "calendar-making" as his word for "astronomy" in its most general sense. Other, more specialized terms derived from Chinese classical literature existed at this time too: *tenmon* (lit. "documents on the heavens"), which usually referred to the study of existing writings and was also employed for astrology; *tengaku* ("study of the heavens") and *seigaku* ("study of the stars"), which implied the use of direct observation; *seisho* ("calendar-writing"); *kenkon* ("heaven and Earth" or, alternatively, "emperor and empress"), a metaphysical term signifying the cosmos as a creation of opposites. Finally, the word *tensetsu* ("explanation of the heavens"), derived from an earlier work, *Oranda Tensetsu* ("Dutch Explanations of the Heavens," 1796, by Shiba Kokan), was also in use when the *Rekishō Shinso* was written. Shizuki used nearly all of these at different times in his text.

The first two volumes of the *Rekishō Shinso,* in fact, were written in classical Chinese. But in the last volume, and in his final rendering of the entire work, Shizuki wrote in Japanese, and a certain change in emphasis occurred, generally speaking, away from *tenmon* toward *tengaku,* i.e., from "astronomy" as literary study of the Chinese classics and Chinese calendar to "astronomy" as a more direct examination of celestial objects themselves. By this time, astronomy had come to be called by its now standardized name, *tenmongaku:* a compromise struck between ancients and moderns.

In his use of such terms, Shizuki shows himself a kind of battleground between these two poles. As an official interpreter in the government's employ, he felt his translation had to obey certain conventions—indeed, as a loyal neo-Confucian *rangaku-shi,* he believed strongly in the value of Chinese thought and natural philosophy and attempted, in a number of places, to reconcile Western ideas with this philosophy by tying Newtonian principles back to the primal text of the *I-ching* (Book of Changes). The heliocentric view, for example, he attempted to justify on the basis of well-worn ethical principles:

> [T]here always exists a governing center in everything. For an individual, the heart; for a household, the father; for a province, the government; for the whole country, the imperial court; and for the whole universe, the sun. Therefore, to conduct oneself well, to practice filial piety . . . to serve one's lord well, and to respond to the immeasurable order of heaven; [these] are the ways to tune one's heart to the heart of the sun. (Nakayama 1969, 185)

Thus did Newton make his entrance into Japan in the service of Confucius.

But Shizuki was a modern, too, quite consciously so. In his day, there were no Japanese or Chinese equivalents for terms such as force, gravity, velocity, elasticity, attraction. Shizuki absolutely avoided simple phonetic translation into Chinese sound characters. Instead, he invented semantic equivalents whose *kanji* always attempted to offer a clear, precise meaning. His purpose seems to have been in part a didactic one, i.e., to provide a usable, practical base for understanding the new science of Newtonian physics. It is obvious, in any case, that he labored greatly and with impressive creativity in this area, producing a true nomenclatural system that might well be compared with that of Lavoisier for chemical entities.

All of the terms just mentioned—force, gravity, velocity, elasticity, attraction—and many others, Shizuki coined according to a simple formula. This involved taking the respective characters for movement, weight, speed, stretch, and pull, and combining them with the character for strength or power (*chikara*). Thus "force" became "movement-power" (power to create movement), "gravity" was rendered as "weight-power," and so forth. By doing this, Shizuki gave each term a concrete, easily understood meaning. He created a direct link to everyday experience that the English and Latin originals lacked. Newton himself once described "force" as mysterious and inexplicable, and several of his other concepts, such as "mass" and "velocity," as mathematical abstractions. In a sense, Shizuki made the Newtonian vocabulary more Baconian. He brought it more within the realm of perception—"weight-power" or "pulling power," for example, being far more tangible in terms of associated imagery than "gravity" or "attraction." Some of his other coinages bear this out as well. "Vacuum," for instance, he rendered with the ideograms for "true emptiness" (*shinkyū*). The focus of an ellipse he denoted with the character for "navel" (*heso*). For ellipse itself, he rejected the older Chinese term used by Japanese mathematicians, *soku-en* ("circle on its side"), and replaced it with *da-en* ("oval-circle"). On the other hand, he accepted the traditional Chinese word for "eclipse," with its allegorical, poetic combination of the characters for "sun" or "moon" with that for "eat." Finally, the Newtonian "corpuscle" he wrote as *bunshi* (lit. "part-small"), from the characters for "segment" or "division" and "child" or "offspring." This, as it turned out, was an epochal choice; it was adopted, first, as an equivalent for "atom" with the introduction of Daltonian theory, only to be re-coined at the end of the nineteenth century by the Tokyo Chemical Society as the word for "molecule," which it remains today (Sugawara, Kunimitsu, and Itakura 1986).

Finally, Shizuki coined a range of still more difficult terms, such as "centrifugal" and "centripetal force." Following his own lead, he designated "force" in these cases with the character *chikara* (strength/power), which therefore became a suffix. This, one might note, along with the use of *-shi* for "particle," set a pattern that has been followed down to the present day in coining any related terms for basic natural forces and particles (e.g., electromagnetic force, electron, proton, etc.). To render "centripetal" and "centrifugal," Shizuki broke with his preference for concreteness and instead seems to have advanced into the realm of the poetic. The former he wrote with the ideograms *kyūshin* (literally "want" or "request" and "heart" or "center"), i.e., "seeking the center." Centrifugal, meanwhile, was translated as *enshin* ("recede" and "heart"). Shizuki could have easily found more simple and unsuggestive ways to write these same words, as he did for other portions of the Newtonian vocabulary, but he chose instead to be more "Chinese" in his inventions here. This method may relate to his own larger attempt to interpret Newtonian ideas according to the metaphysics of *ki*. As indicated by the brief quotation above, notions of "the center," "inwardness," and "outwardness" were critical to the neo-Confucian cosmos within which Shizuki conceived his own theories of reconciliation. In any case, all of these terms remain unchanged today as the basic vocabulary of physics. Together with Miura Baien, Shizuki coined for posterity the fundamental language of physical science for Japanese society.

Scholars who have written about Shizuki Tadao, and there have been many (see, for example, Ohmori 1964a and b; Nakayama 1969; and references therein), commonly relegate him to a "premodern" category, according to the long-standing bias that modern science did not begin in Japan until the Meiji era (see Low 1989 for an excellent discussion of this problem). That such an idea is overly narrow can be gathered immediately from the above. Despite allegiances to traditional Chinese thought and language, Shizuki clearly inaugurated a critical segment of modern scientific discourse—as any adequate translator of Newton would inevitably have done—laying down vocabulary and patterns of linguistic formation that have remained foundational ever since. With Shizuki Tadao, the Japanese tradition in scientific language begins to depart significantly from the Chinese model, while still including it. It inaugurates a type of quiet "revolution" whereby the past, in terms of discourse, becomes less a single dominating influence than an ingredient in a complex evolution of nativization that would soon involve other elements as well. The language of Newton in Japanese, therefore, presents a condensation of history at a moment of transition, when fundamental loyalties regarding culture and thought were in motion. To the knowing eye or tuned ear, the daily use of this language today rehearses this history like an unending ballet. The dance of its movements, its conflicts, have been reduced to the residue of shadows, perhaps. But they are there nonetheless, if only behind the thickening veil of "standardization."

Darwin, "Evolution," and "Survival"

We mentioned earlier that few of the *oyatoi* ever lectured or wrote in Japanese. Linguistically, on a professional level, they were never allowed

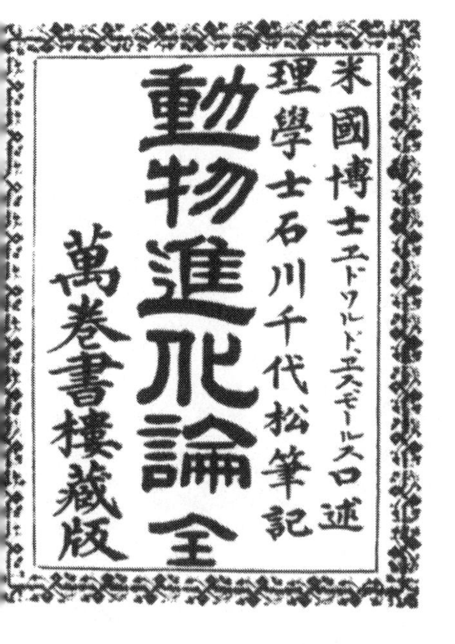

米國博士エドワルド・エス・モールス口述
理學士石川千代松筆記
動物進化論 全
萬巻書樓藏版

英國 チャーレスダーヰン著
日本 文學士 立花銑三郎譯
生物始源 一名種源論
明治二十九年二月
經濟雜誌社出版

Figure 18. Cover pages of the first editions of two important works on evolution translated into Japanese. On the left is the title page for *Dōbutsu Shinkaron* (Animal Evolution, 1883), a translation of the lectures of Edward Morse by his student Ishikawa Chiyomatsu. On the right is the cover to Tachibana Tetsusaburo's translation of Darwin's *Origin of Species*, titled *Seibutsu Shigen* and published in 1896. It is interesting to note that while Ishikawa is listed as a "physicist" (or "doctor of physical science"), Tachibana is a "doctor of Japanese letters." Moreover, the publisher of the Darwin translation is not a technical press, but instead the *Keizai Zasshi-sha*, an economics journal. This reflects, in some part, the earlier influence of Herbert Spencer, whose doctrines had become well known two decades before. Reprinted, by permission, from Watanabe 1990, 51,70.

to "leave home." In the case of the American zoologist, Edward Morse, his writings were handled by one of his more gifted students, Ishikawa Chiyomatsu, who in 1883 helped fully introduce Darwin to Japan by collecting a series of Morse's lectures into a book he titled *Dōbutsu Shinka-ron,* "Theory on the Evolution of Animals" (fig. 18; see Watanabe 1990). The term for "evolution," *shinka,* Ishikawa either coined himself or adopted from one of several earlier works published in the 1870s on Darwinian ideas, most notably Izawa Shuji's 1879 translation of Thomas Huxley's *Lectures on the Origin of Species* (1862). In either case, it did not exist in Japanese scientific discourse before the 1870s.

It was, however, an excellent choice: comprising the characters for "advancement" and "change," it had come to enjoy a wide currency outside science prior to its adoption there, denoting "progress" as the idea of the "forward movement of society toward civilization," especially under the auspices of Western science. It was a term that could be said to have embodied two sensibilities at once: Darwin's own Victorian view of evolution as a process of continual improvement and, more immediately, the ideology of the Japanese Enlightenment, with its call to a civilizing nationalism. The political side to Darwinian language, therefore, was aptly retained, to serve the purposes of Japanese self-imagery.

Interestingly enough, however, such nationalism helped dictate that Darwin's own texts would be among the last works to be translated in this area. Darwin himself, as an author, would be the last to speak for his own ideas in Japan. Instead, it was Herbert Spencer, missionary of social Darwinism, who came to be adopted as the apostle of evolution theory. Indeed, by the time the first translation of *Origin of Species* appeared in 1889, at least twenty renditions of Spencer's work were already in circulation. Moreover, Darwin's book was translated not by a biologist, geologist, or natural historian, but instead by a literary scholar, Tachibana Sensaburo (see fig. 18). Before this, only four works on the actual science of evolutionary theory had been published, two from books by Huxley. Even within scientific circles, the little of Darwin that did get through during this period was doubly "translated," being imported through the writings of his own English interpreters. Finally, even in Ishikawa's rendering of Morse's lectures, *Dōbutsu Shinkaron,* there is a continual tendency throughout to take examples from the animal and plant kingdoms and apply them to human situations (Watanabe 1990).[1]

Spencer had been introduced to Japan through American contacts. If Morse was perhaps the more important of these, another was the famous philosopher, art critic, and oriental scholar Ernest Fenellosa, one of the very few *oyatoi* who did write and publish in Japanese. Japanese authors, meanwhile, took up Spencerian ideas and disseminated them broadly in copious writings, at times with government support. During the 1880s and 1890s, as the intellectual atmosphere of Japan grew increasingly conservative and nationalistic, many thinkers, officials, and students found

1. Watanabe (1990, 69) provides an interesting graph showing numerical percentages of articles dealing with the theory of evolution for the natural sciences, social sciences, and humanities during the 1880s. The graph is specific to a single journal, *Tōyō Gakugei Zasshi,* a major scientific publication, comparable perhaps to the *American Journal of Science* in the nineteenth century. The data show that evolutionary theory had a much greater occurrence in social science writings (26 percent) than in the natural sciences (5 percent) or the humanities (1 percent).

themselves drawn to the concept of a struggle between nations, with "higher" species eventually winning out over "lower" ones (Nagazumi 1983). Indeed, the theory had no small attraction for those who argued against further Westernization. Before 1900, the theory of evolution tended to operate politically both within and outside of science.

Before the century was out the language of evolution took on a more striking cast. Terms such as *skinka* reflect the era of their origin, the early Meiji period of hope and "progress." But by the late 1880s and early 1890s, the pitch of nationalism had shifted to more reactionary concerns about national moral standards, loyalty among the people, and at the same time, about national destiny in terms of empire (*teikoku*). Western nations were being viewed more in oppositional terms, again as colonial aggressors, and as destructive models for Japanese character and virtue. There was a strong resurgence of Confucian ethics, especially evident for example in the Imperial Rescript on Education (1890), which linked "virtue" directly with such things as obedience to authority, national sacrifice, and belief in the emperor's divine status. In this atmosphere, Spencer himself was translated in sometimes hyper-Spencerian terms. Kato Hiroyuki, in his *Jinken Shinsetsu* (New Doctrine of Human Rights, 1882), provides one of the best and, at the time, most influential examples. Not satisfied with a literal rendering of Darwin's famous phrase, "survival of the fittest," so central to Spencer's own philosophy, Kato felt compelled to evoke more of what he perceived to be its deeper significance, and wrote it thus: *yūshō reppai*—"victory of the superior and defeat of the inferior." This, he asserted, was "the law of heaven," governing the world of plants and animals as well as that of human beings and the cultures they build (see Watanabe 1990, 71–74). For a brief time, Kato's "victory of the superior . . ." was actually adopted into biological discourse. Though largely abandoned before the second decade of the twentieth century, replaced by a much milder alternative, *tekisha seizon* ("survival of the most suitable"), Kato's phrase was nonetheless revived during the era of rising militarism in the 1920s and 1930s, when eugenics came to Japan. Today, it is no longer used on any sort of regular basis and has been veritably driven from the language of evolutionary biology. Yet it has not disappeared; it is still listed in dictionaries, without comment, as an equivalent to Spencer's seemingly immortal phrase.

Origins of Modern Chemistry in Japan: A Complex Litmus of History

The development of modern chemistry in Japan, including the periodic table as written in Japanese, presents an almost unparalleled example of the collision of cultural and political influences in the realm of scientific language. It is an example that both parallels and goes well beyond the case of Newtonian or Darwinian language. This is because the relevant nomenclature evolved over a much longer period of time and involved a much greater array of sources. These included a native tradition of element names, in part derived from China: substances such as iron, gold, silver, and so forth had long been known by their Chinese or Japanese titles. Then, when the element concept was introduced, these substances were effectively segregated into a new category and their traditional names cast into question. Added to this "native" group, most of whose titles were eventually retained, there was an increasing number of new elemental substances from the West imported under a range of names that, through the course of the nineteenth and early twentieth centuries, reveal a change in loyalty to Dutch, Latin, German, and finally English sources. Reconciling past, present, and future in this area proved to be a task that encouraged no end of new and competing nomenclatural systems. Different researchers proposed a succession of these systems between the 1830s and 1890s. How Western, or Japanese, or Chinese, or mixed, a particular system might be depended on the intentions of its creator, intentions that nearly always came back to some type of larger historical motive.

Western chemistry was first introduced into Japan in the early nineteenth century through native physicians interested in pharmacology. These men were all trained in traditional forms of Chinese herbal-based medicine and were therefore all scholars of Chinese literature. Their interest in Western medicine, in fact, was partly a direct outgrowth of this background. Books on European plants and pharmacology led them, rather quickly and with excellent logic, to works on botany and then to the importance of chemical understanding with regard to the analysis and manufacture of medicines. The most important early works in Japan were published between 1820 and 1850. These have been discussed in some detail by a number of scholars (Sugawara and Itakura 1990a and b; Tanaka 1964, 1965, 1967, 1976; Doke 1973). They include, in particular, three books by *rangaku* scholars. The earliest of these was *Ensei Iho Meibutsu Ko,* which can be translated as "A Reference on the Specialty Products of Far Western Medicine," published in 1822 by Udagawa Genshin and his son-in-law Yoan. A second notable book, by Rinso Aochi, was titled *Kikai Kanran,* a poetic Chinese name meaning literally "A Gaze over the Broad Waves of Air and Sea" (but more often translated by scholars today as "A Study of Nature"). Finally, there was Udagawa Yoan's own epochal *Seimi Kaisō,* best rendered as "Principles of Western Chemistry"—*seimi* being a phonetic transcription of the Dutch "chemie"—published in no less than twenty-one volumes between 1837 and 1847.

The most influential of these was surely the *Seimi Kaisō*. Indeed, this multivolume work is widely acknowledged to be one of the foundational writings of modern Japanese science, not least with regard to chemical language. A leading historian of Japanese chemistry, Tanaka Minoru, describes its importance in no uncertain terms:

> The most impressive characteristic of this work is that its author, through largely independent effort, succeeded in translating into Japanese nearly all of [Lavoisier's] chemical nomenclature and terminology. No less was he successful in rendering into easily understandable speech and clear expression the precise description of scientific conclusions, chemical analyses and experiments, the characteristics of substances, etc. Present-day chemical nomenclature in Japan largely owes its existence to the work of Udagawa. Until the fall of the feudal government in 1867, the *Seimi Kaisō* played a decisive role in the spread of scientific chemical knowledge in Japan. (Tanaka 1976, 97)

Udagawa Yoan was a complex and brilliant man, a wide-ranging polymath of enormous curiosity and intellectual vitality. Not only did he write on chemistry; he also produced works about Western botany, zoology, history, geography, music, linguistics, and mathematics. While still very young, he was adopted into the Udagawa family, well known both as Chinese-school physicians to the Tsuyama clan and as outstanding *rangaku* scholars and translators. His adoptive father, who had translated as many as thirty books on Western anatomy, urged Yoan to first study Chinese philosophy and composition, the *kogaku-ha* or "school of ancient studies," which the young man eagerly did. "If you lack the ability to compose a Chinese sentence," he was told by the elder Udagawa, "you cannot achieve medical learning either. . . . But neither forget that translation is important work and worthy for a man to sacrifice his whole life to" (Doke 1973, 104). Again, there was no contradiction inherent in such advice: being a physician meant comprehending deeply the sacred works of Chinese medicine, but at this point in history it had also come to mean the capability to read the increasing number of translations of Western medical texts too, many of which were only available in Chinese or in *kanbun* (to read which essentially required literacy in Chinese). For those such as Udagawa Genshin, the "ability to compose a Chinese sentence" seemed to have had a double meaning as well. It meant, on the one hand, the old rhetorical view of knowledge, based on repetition and imitation. Yet it also meant an ability to add to the stock of knowledge itself, by translating and interpreting works of Western science into the language of Japanese scholarship. Whether or not Yoan understood such advice in this way, he came to practice it, and at a very early age. Before he was even twenty he had largely mastered the Dutch language and had begun writing technical articles (in Japanese) on various botanical and chemical substances.

One of these works, a small booklet on botany, represents a very remarkable, even startling merger of Eastern and Western influences, and the creativity such a merger could produce. Entitled *Seisetsu Botanika-Kyo* (1822), the book was quite literally a "Sutra on Western Botany" and a critical work for the introduction of Linnaean botany into Japan. It is written in the literary form of a traditional Buddhist sutra, comprising a total of only seventy-five lines, each made up of seventeen *kanji,* arranged to produce a chant-like rhyme with decided oral qualities. Brief as it was, the work could only touch upon the Linnaean system; but it was able to show how all plants could be named according to a specific system that built titles out of descriptive elements. Thus did concepts of Western botanical theory enter Japan via a scheme of rhythmic chanting dating back to the Sung dynasty and possibly before. Where Shizuki Tadao had placed Newton in the arms of Confucius, here was a still more intriguing example of Western thought in Eastern form.

Udagawa's term for "botany" in this work, *Botanika,* was a phonetic transliteration of the Latin word *botanica.* Only a few years later, he stopped using this term and changed it to *shoku-gaku,* "plant studies," as shown in a subsequent work titled *Shoku-gaku Dokugo* ("Confessions on Botany," 1825). This nativization to Japanese, however, despite its simplicity and directness, did not become common usage. Instead, it was replaced by the word *shokubutsu-gaku,* literally "study of plant-things," adopted from a Chinese work of 1857 that gained great currency in Japan due to its combining of European and Chinese pharmacology (Nakayama 1992). In this particular case, terminology evolved away from, rather than toward, a Western standard.

Seimi Kaisō: A Masterwork of Adaptation

The *Seimi Kaisō,* Yoan's masterwork, is a compendium of translations from various Dutch books, commentaries on these, and the author's own experiments. In his preface, Yoan states that the earlier *Ensei Iho Meibutsuko,* which he helped write with his father-in-law, was for specialists and thus not appropriate as a textbook or learning aid. It was "too profound and difficult for beginners, so I decided to write my own book *Seimi Kaisō* in more plain and accessible terms using more fully the elucidative descriptions in simpler Western chemistry books [as well as] my own chemical knowledge accumulated through various experiments." (Doke 1973, 113). His intention, clearly, was a pedagogic one, to make Western science more available to educated Japanese. The books he chose to translate from were mainly introductory in nature (see Tanaka 1976 for a discussion of these, of which there appear to be at least twenty). The most important seems to have been a Dutch translation of a German version (*Chemie fur Dilettanten,* 1803) of William Henry's *Epitome of Chemistry* (1803), itself largely a simplified English rendering of Lavoisier's *Traité Elementaire de Chimie* (1789). These works had all been popular in Europe, aimed at a lay audience and beginning university students. Yoan's choice was therefore well-considered. Moreover, it is crucial that he wrote the *Seimi Kaisō* not in Chinese, as was previously the rule, but in his native tongue.

Udagawa set out a clear definition of "element" as a "primitive indivisible substance" and proposed a system of names. Part of this system, and the term for "element" itself, he borrowed from the *Ensei Ihō,* on which he had collaborated. The word for "element" was *genso,* employing the characters for "origin"/"beginning" and "simple"/"essence." The latter of these ideograms, *-so,* he used as a suffix for a number of other elements for which no names yet existed, e.g., *suiso,* or "water essence," for hydrogen; *tanso,* or "coal-essence," for carbon.

The first volume of the *Seimi Kaisō*, devoted to a discussion of element theory, lists fifty-eight names, in dictionary order of their first sound. A fair number of these, if they corresponded to well-known metals or other substances, were allowed their traditional titles. This meant they were written in *kanji,* having been adopted from Chinese over the centuries, but had since come to acquire Japanese pronunciations. Other elements, meanwhile, were taken directly from both Latin and Dutch and translated phonetically into *kanji* via the *on* (that is, Chinese sound) system. As was the case with other scholars, Udagawa seems to have first written these names in the native phonetic alphabet, *katakana,* in his handwritten manuscripts, but then changed to the conventional *kan-on* system for publication, adding miniature *katakana* alongside them to indicate proper pronunciation (Sugawara and Itakura 1989). At times, this led to interesting "errors." For example, since Chinese had no hard "k" sound, a name like chlorine could only be approximated, as *su-ro-rin.* In contrast, three elements in particular—platinum, bismuth, and molybdenum—the author translated from the Dutch strictly on the basis of meaning (which had a basis in the vernacular alchemical tradition). Platinum, in Dutch *witgout* (literally "white-gold"), was rendered by Udagawa with the two appropriate characters, as *hakkin.* Molybdenum, *waterlood* ("water-lead"), became *sui-en*. In addition, Udagawa included several non-elements, later deleted, that had been proposed by Western chemists, notably Berzelius. All of these were named in Japanese with the *-so* suffix, as in *onso* for "warmth-essence."

Udagawa's term for element, *genso,* and his use of the *-so* suffix—which later became standard usage for Japanese chemical nomenclature—were modeled directly on Dutch words, which in turn had been derived from German. In particular, the Dutch/German ending *-stoff* was the direct inspiration for Yoan's scheme of translation. This ending is used to write "element," as *Grondstoff* (Dutch) or *Urstoff* (German), and also appears at the end of many element names, such as *Koolstof/Kohlenstoff* for carbon. This is significant to the whole of Japanese chemical discourse not simply because of its influence but because of its larger meaning in terms of cultural inheritance. Udagawa's scheme represents, in effect, a historical choice between two competing linguistic traditions of Europe. One of these systems was that of Lavoisier, who had revolutionized chemical nomenclature by creating a naming scheme patterned on Linnaeus. In this system, the critical activity of each elemental substance was embodied in its title, via Greek or Latin roots. The gas whose central principle was to combine with other substances in order to make acids (the "dephlogisticated air" of Priestly or "vital air" of others) could thus be termed *oxy-gene,* meaning "acid-producer" (*oxy- from the Greek,* oxus, for "sharp," "bitter"). To Lavoisier, as to many other thinkers of the French Enlightenment, language was an analytical instrument or method that could be used to both contain and employ the natural order for human use and understanding. His systems of names, for the elements, acids, and bases alike, was soon seen to have great power and was adopted before long both in France and England, and subsequently America. Though Dalton's atomic theory came to amend much of Lavoisier's work, the latter's nomenclature remained in use and has obviously carried down to the present.

In contrast to this scheme, however, was the German chemical and mining tradition, which had its own names for substances that later came to be known as elements. This tradition had buried within it an older, mainly alchemical view of matter as containing various "essences." Unlike Lavoisier's concept of *substance chimique,* which focused on a capacity for interaction, the German view tended to see the chemical universe in terms of "pure" and "impure" matter constantly engaged in revealing its primal truths through various principles of activity. To Lavoisier's *oxygene* or *hydrogene* ("water-producer"), eighteenth-century German chemists coined the terms *Sauerstoff* ("sour-essence") and *Wasserstoff* ("water-essence"). No primal suffix, defining of the final substance of "element" itself, ever existed in the French scheme. A concept such as *-stoff* had no place in a nomenclature built out of qualities and exchanges. In the early nineteenth century, moreover, as the Lavoisieran system gained currency in many parts of Europe, German chemists refused to alter their nomenclature. No doubt this was linked, in some part, to Germany's strong and historical antipathy toward France, which, under Napoleon, had defeated and humiliated both Austria and Prussia and had helped bring about nationalizing reforms in all areas of German intellectual life. In any case, the long-term result was that the Germans, Dutch, and other northern European countries never abandoned their own native systems of chemical nomenclature.

As a *rangaku* scholar, Udagawa received the inheritance of the Dutch/German tradition by fiat. This had been ordained in the early seventeenth century by the shogunate's choice of Holland as Japan's only major trading partner in the West. In coming to Japan, European science was forced to pass through the filter of Dutch linguistic realities. In the case of chemistry, this meant a quiet yet momentous acceptance of Lavoisier adapted to a Teutonic onomastic scheme.

According to Sugawara and Itakura (1989), a total of twenty-three elements in the *Seimi Kaisō* were translated into native Japanese without using the Chinese sound system. These included: (1) traditional Japanese names with one character (in most cases originally derived from Chinese); (2) traditional names with two or more characters (some of Chinese derivation, some not); (3) literal translations from the Dutch (e.g., platinum, as discussed above); and (4) translations using the *-so* suffix. The remaining thirty-five were given in *kanji* translated by the *on* sound system, having been taken phonetically from either Latin or, less often, Dutch. In terms of sheer quantity, then, the nomenclature of the elements at its earliest stage of formation seems most obedient to Chinese influence, which at this time (the early nineteenth century) still commanded the language of scholarship as a whole. But a crucial change had occurred: not only had a host of new substances been added to the traditional chemical pantheon, but the most basic term in all of the new chemistry, *genso,* had been rendered in a manner loyal to Western terminology. This was the historical signal of the direction the coming vernacular revolution would follow within Japanese science.

Kikai Kanran: An Alternative Scheme

To better understand this, it is important to note that Udagawa's system of names was in competition with another scheme. Rinso's *Kikai Kan-*

ran, published in 1826, was a popularizing attempt at explaining basic Western concepts of physical and chemical phenomena. Written in Chinese, but with a lucid style that reveals solid understanding of the science involved, the book became a standard reference for *rangaku* scholars, just as Udagawa's did (Tanaka 1967). In large part, it was a translation of an elementary Dutch textbook, *Natuurkundig Schoolboek* ("Textbook in Natural Philosophy"), written by Johannes Buijs for use in upper-level grammar schools of the day. It was later translated *in toto* into Japanese and expanded with a commentary by Kawamoto Komin, as *Kikai Kanran Kōgi* ("Commentary on the *Kikai Kanran,*" 1851).

With regard to his naming system for the elements, Rinso also adopted the German/Dutch *-stoff* tradition but translated it differently, rendering the term "element" itself with the characters *gen* and *shitsu,* meaning "original matter/quality." Rinso thus used *-shitsu* as a suffix equivalent to Udagawa's *-so;* for example, hydrogen became *suishitsu,* or "water-quality." As noted above, *shitsu* had a long history of use within Confucian metaphysical philosophy, indicating something close to a merger between material and spiritual natures. In using it, the author was proposing a nomenclatural system that adopted Western terminology into traditional Chinese forms.

Despite the popularity of *Kikai Kanran* and its support by influential writers such as Kawamoto, Rinso's system did not flourish. Indeed, not only was Udagawa's system preferred; by the 1860s it had been extended to replace a number of traditional, single-character Japanese names, now written with the *-so* suffix too. Still further, during the 1870s another important translator and writer of chemical works, Ichikawa Morisaburo, proposed that all nonmetallic elements be put in this form. This proposal, however, was not adopted. In fact, it proved to be merely one in a veritable explosion of new naming schemes proposed in the politically chaotic years between 1860 and 1880, when Japanese society as a whole and the scholarly community par excellence underwent a great deal of change and confusion regarding the place of, and proper balance between, traditional vs. modern (Western) ways.

Naming the Elements: Competing Schemes in the late 1800s

Indeed, it is striking to see how nomenclatural systems within chemistry reflect the patterns of larger ideological issues during this time. These systems were many, overlapping, and, for the most part, temporary. Each suggests a particular type of vision for Japanese chemistry, a canonizing of fidelities to certain features of the past or hopes for the future. When seen from this viewpoint these systems clearly divide into five basic categories, characterized by: (1) Chinese influence; (2) traditional Japanese influence; (3) Chinese and Japanese influences combined; (4) Western influence; and (5) western and Chinese influence combined. These different systems are not always easily separable; two or more can sometimes be found within a single work by a single author, applied to different groups of elements. Individual chemists, that is, often reveal more than one precinct of loyalty.

Chinese influence can be discerned in two particular systems proposed during the 1860s and 1870s.[2] One system, found in the work of Mizaki Shosuke, made use of the suffixes *-ch'i* (ki in Japanese)—the same character so central to neo-Confucian yin/yang natural philosophy—and also *-shō*, meaning "spirit," "energy," or "purity," another metaphysical association. It appears, in fact, that Mizaki consciously replaced every occurrence of *-so* in Udagawa's *Seimi Kaisō* with *ki* (Sugawara and Itakura 1990a). Another system, offered by Kiyohara Michio in his translation of a Chinese work (*Kagaku Shōdan*, 1873), stated that each element should be given its own single ideogram, which would often involve creating new characters by combining two existing *kanji* according to a prearranged system of prefixes indicating "solid" (using the character for "rock"), "non-solid" (using the *kanji* for water), or "metal" (using the character for "gold" or "metal"). Both systems had their "standard" Chinese ingredients. Where one adopted a conventional terminology (with all its many associations), the other took advantage of a creative logic inherent in the writing system itself, something that had been done often through the centuries by Chinese and Japanese authors alike. Either way, both proposals sought to filter Western ideas into Japan through Chinese linguistic tradition. It therefore revealed a combination of loyalties, yet with favoritism granted China via language.

Meanwhile, an entirely different type of linguistic conservatism, expressive of nationalistic feeling, came in the form of a curious and entirely unique scheme offered by Kiyomizu Usaburo in 1874. This work was *Mono-wari no Hashigo* ("Steps to a Division of Things"), whose title, written entirely in *hiragana* and meaning something like "Analytical Procedures," amply betrayed its author's intentions by using an ancient poetic word (*mono-wari*) that otherwise had little business (and no use) within the confines of science. Kiyomizu's idea, in fact, was to

2. The idea of "influence" here needs to be qualified. The element concept, after all, seems not to have been accepted in China until the 1870s (Sugawara and Itakura 1990b), well after it had become current in Japan.

replace nearly all element names with new titles coined in the manner of classical Japanese literature—the Japanese of the *Manyōshu* and other ancient poetry and prose collections. "Element" itself would be written as *oh-ne,* with the characters for "great" and "root." Individual elements would use *-ne* as a suffix and would employ classical words, written in *hiragana,* as a prefix. Thus, oxygen would be rendered as *sui-ne* (*sui* being equivalent to *suppai*), "bitter-root," or more poetically, "the root of all bitterness." Some elements would keep their traditional titles, and the previously unknown rare earth metals would even be given their Western sounds. But all would be written in the native syllabary of *hiragana,* and thus the "feel" of the classical tradition would be maintained as a central sentiment.

The mixing of Chinese and Japanese influences, on the other hand, came in a predictable form. In the 1860s, for example, just prior to the Meiji Restoration, a writer named Ueno Hikoma published a work in Chinese entitled *Seimi Kyoku Hikkei* ("Official Manual of Chemistry," 1862) in which he suggested that all substances with long-standing Japanese titles retain these and all others be rendered from Latin into *kan-on* (Chinese sound system). This made for more than a few clumsy titles: even simple names such as chromium had to be written with as many as five or six characters, i.e., a character each for the sounds *ku-ro-mi-u-ma.* The emphasis on Latin, meanwhile, imposed a few interesting alterations of its own, e.g., oxygen being turned into "oxygenium," written with a total of six *kanji.* Rendered in this way, the Western origins of these names were effectively erased. To read them, moreover, meant that one had to be a Chinese scholar, as in prior times, still a reality for most chemists in the 1860s but less and less so thereafter. Finally, despite its title, Ueno's book ignored the Udagawa nomenclature, which had already become the most commonly used by that time. Ueno's ambition was to impose a new "official" scheme, more in line with traditional scholarship.

In the end, however, the most complex and varied category of nomenclatural proposals were those that looked ardently westward. This was perhaps to be expected, but the diversity of ideas is still striking. Some of these ideas came as direct consequences to the new and sometimes radical proposals for linguistic reform then current in Japanese society generally. Kawano Tadashi's *Seimi Benran* ("Handbook of Chemistry," 1856), for example, proposed writing all the names of the elements in the Roman alphabet. This would have necessitated choosing among different versions for individual elements, of course (e.g., among English, French, Dutch, and German names), but in any case was tantamount to swallowing the European/American system whole, in its own native state. In contrast, Ichikawa Morisaburo suggested extending Udagawa's *-so* nomenclature to include a much larger number of substances, thus in effect more deeply penetrating chemical language with the Teutonic notion of "essence." From the late 1870s on, a growing fidelity to German science inspired the convention of changing any older Dutch names into German and replacing certain Latin ones with their German equivalents as well. Names in English began to be used about this time too, at first in equal number to the German but then less and less so. The main conflict, at least in the early stages, was whether to express the chosen titles in *kan-on* or in *katakana.* The latter had been used sporadically for a few elements since the 1820s, but never consistently. It was Ichikawa, following the example of none other than Mizaki Shosuke, who finally helped set the pattern in the 1870s (Sugawara 1984).

Mizaki, who died at the age of thirty-six in 1874, in fact was extremely active as a translator of chemical works in the last several years before his death. In his books, one sees the old and the new visions of chemistry, as embodied in nomenclature, combined to a remarkable degree. Mizaki had studied in Nagasaki under a well-known Dutch science teacher, Frederik Gratama, whose lectures he translated into several books that then became widely used. His most influential work, however, was published after he had left Nagasaki and had been recruited by the government to take up a professorship of physical science at Tokyo University Medical School. This work was a translation of a Dutch rendering of the German chemist C. R. Fresenius's *Anleitung zur Qualitativen Chemischen Analyse* (1869), which had become the most well-known and standard introductory textbook on chemical analysis throughout Europe. In bringing this work into Japanese, Mizaki helped introduce a number of important new concepts, including Dalton's atomic theory and Avogadro's molecular hypothesis. But it was his handling of chemical nomenclature that caught the immediate attention of his contemporaries. Mizaki divided the elements into two basic groups: those with their traditional Japanese names (e.g., the common metals), and those imported from Europe, which he translated from Latin by means of *katakana.* Next to this list, he gave equivalents in English, in *kana,* and also in *kan-on,* for which he chose the Chinese sound character combinations most often used by chemists of the day. This system of both juxtaposing and intermingling Eastern and Western names he had used before, in his translations of Gratama's lectures (especially *Rika Shinsetsu,* "New Treatise on Physical Science," 1870), a work from which Ichikawa Morisaburo apparently drew inspiration for his own *Rika Nikki* ("Diary of Physical Science," 1872). Mizaki's list became a standard part of chemical textbooks, where it served as an essential reference. Mizaki's translation of Fresenius's textbook also came to be widely adopted as a basic work throughout Japan and helped spark debate over the term "chemistry" itself. Indeed, both Mizaki and Ichikawa had ignored the older *seimi,* derived from the Dutch, in favor of the newer word *kagaku* ("study of change"), a Chinese loan-word given Japanese pronunciation.

Mizaki and Ichikawa represented a newer generation of translator-scientists. They too, like Udagawa and other early chemists, were from families in which the father had traditionally been an official physician of some sort. Yet, as young men came of age in the years of the Meiji Enlightenment, their careers diverged from those of their predecessors, in fact and in spirit. They were recruited into government service, not merely as translators but as teachers too, as replacements for the *oyatoi* in the new scientific institutes of Osaka and Tokyo. They were regularly exposed to a greater range of linguistic materials than were Udagawa and Rinso, and their fidelity both to European science and to Japanese comprehension of it—often at the expense of Chinese traditions, shown by their adoption of Western nomenclature in the form of *katakana*—was a temporal mark of their higher "civilization." Those who opposed them, and who remained adherents to the *kan-on* system, even into the 1880s, were generally older men who had stayed in the medical profession, and who, in addition to employing Western therapeutic techniques, continued the traditions of herbal treatment and pharmacology linked both practically and philosophically to the linguistics of Chinese convention. This division within the chemical community, however mild it might have been, was nonetheless a direct expression of more eager conflict in the power relations of Japanese scholarship as a whole.

From roughly this point on, the story of chemical language in Japan begins to shift. The use of the Chinese sound system (*kan-on*) declines rapidly, while more and more German and English words enter in, reflecting the larger change of political and cultural loyalties regarding knowledge. Declining use of Chinese was true throughout Japan, a result of the nationalist movement that promoted a Japanese identity based on things Japanese, especially the language. At the level of education, the government had taken over the job of building schools and instituting curricula that tended to favor the learning of the Japanese classics and, in terms of "things foreign," Western languages, especially German and English. By this time, Germany and England were viewed as the most powerful European states, both intellectually and militarily. The final humiliation of China, in Japanese eyes, was its quick defeat by

Table 1 Linguistic Origins of Element Names in Japanese

	1886	1891	1900
Japanese	16	23	23
Common to Latin, English, German	24	26	31
English only	13	12	0
German only	11	13	18

Japan's own modernized military in the 1894-95 Sino-Japanese War. In the period between about 1880 and 1900, China came to lose nearly all prestige as a crucial cultural reference point. Full vernacularization of Japanese intellectual life, like the introduction of European science, now took place under the canopy of freedom from the "old" and embracing of the "new."

Beginning in the 1880s, Japanese chemists, like society generally, struggled to gain control over their own discourse. This struggle is reflected in three major position papers presented between 1886 and 1900 by the nation's leading professional chemical organization, the Tokyo Chemical Society (*Tokyo Kagaku Kai*). In these papers, the Chinese sound system is essentially abandoned as a method for writing element names. For those titles expressed in *kanji,* Japanese names are used, including Udagawa's system utilizing the suffix *-so,* which was further extended to the halogen family as a whole (Sugawara and Itakura 1990a). All other names are written in *katakana* and are derived from one of three sources: European titles common to Latin, German, and English; names used in English only; and names in German only. Reviewing the three standardizing proposals in some detail, Sugawara and Itakura (1990a) have noted certain changes in the emphasis given these various categories over time (table 1).

The changes shown are a mirror of the larger political circumstance of Japanese chemistry at the time. Under the *oyatoi* system, two groups of chemists had been established in Japan: one under English-speaking teachers from Britain and America, the other instructed by those from German-speaking countries. Because of this social reality, the Tokyo Chemical Society felt it had to forge a compromise. Yet the most striking aspect of the numbers given in table 1 is the disappearance of names in English and the increase of those in German (plus, one might add, the complete absence of any Dutch names). More generally in chemical discourse, one sees at this time the first use of the Roman alphabet to express chemical equations and also adoption of the German system for naming common chemicals: for example, "silicate" being written *keisan-en,* literally "jade-bitter-salt," a direct transcription from the German *kieselsaures Salz;* or "sodium bicarbonate"; *jū tansan-natrium,* a literal rendering of *doppelkohlensaures Natron,* "double-carbon-bitter sodium." The growing favor enjoyed by Germany in the basic sciences, physics and chemistry most of all, is here in evidence. Meiji nationalism had found an excellent model in Bismarck's Prussia, with its strictly centralized system of education, its use of intellectuals as civil servants, its unparalleled chemical industry, and its imperial discourse of spiritual force and national destiny. Germany, after all, had turned the century to its advantage: defeated and humiliated by France early on, it had turned the tables entirely after 1871, becoming the new premier military, intellectual, and cultural power in Europe. This story, in however stylized or subtle a form, could not but draw Japanese admiration and in some part serve as a background narrative for official hopes of Japan's own apotheosis.

The German model did not last, but fell to pieces in the new century. Even before 1900, Germany's participation in the "Triple Intervention" along with Russia and France, by which Japan was ordered to return some of its conquered territory to China after the 1894–95 war, left a bitter aftertaste. During World War I, Germany stopped supplying Japan with valuable chemical exports, causing portions of the national economy to suffer great losses (Bartholomew 1993). In the Pacific, Japan sided with Britain, its long-term ally in naval matters, and (in a small way) helped destroy German influence there. More generally, the effects of World War I were, so to speak, devastating on the German example. Bismarck's vision had collapsed into defeat and poverty. From that point on, England and, somewhat later, the United States became Japan's models, and English began to be perceived as the more important language with regard to international science and thus progressively replaced German in the periodic table of *katakana* names. Though some noises were made during the xenophobic era of the 1930s about eliminating *bango* ("barbarian speech") from all areas of Japanese scholarship, this was never taken wholly seriously by the government or the scientific community. In the early post-World War II period, linguistic reform, which sought to simplify and standardize the entire language for a variety of reasons, did away with many *kanji* that were either complex to write or had very limited use. In the sciences, this meant the replacement of certain characters with their sound equivalents in *katakana.* This process, along with increasing interest in the United States, led some chemists to advocate replacing all traditional Japanese names with English ones, but this proposal was not adopted. In 1949 and 1955, the Ministry of Education issued a series of dictionaries and other publications intended to standardize scientific language, as taught in the schools. In the sections on chemistry, several element names (fluorine, arsenic, and sulfur) were changed: in 1949, they were stripped of their original characters and given simpler ones (those included in the new *Tōyō Kanji* list); then, in 1955, they were finally rendered into *kana* (Sugawara and Itakura 1990b). Dictionaries of that time, particularly in specialist areas, often speak in their prefaces about a need to "streamline" or "advance" education and knowledge at all levels in order to "contribute toward a progress in national strength vis a vis Europe and America" (Mamiya 1952, v). These last small changes in the names of the elements are perhaps only a final footprint to show that the rhetoric that lay at the base of modern science in Japan had never wavered.

Epilogue

The history of Japanese science, its international dimension, is to be found in the language with which this science is written today. This is naturally true for science in any tongue; rhetoric and nomenclature here, as elsewhere, contain the linguistic deposits of centuries. But in the case of Japanese, the literacy needed to decipher these deposits is one that must untangle political, literary, philosophical, and scientific developments, all within a larger linguistic context whose evolution reveals a profound shift in the most basic of cultural loyalties.

But there is something else here, too: the internationalism of Japanese science, as expressed in its dependence upon translation, shows up in the very reality of how the history of this science is now being written, by Japanese themselves. Anyone interested in this area of study, that is, will be forced to read articles and books by Japanese authors written in four or five different languages: Japanese, Chinese, English, German, French. Each of these languages, moreover, tends to be divided along traditional lines regarding the field(s) in which they had the most influence. The many essays and articles written by Tanaka Minoru on the history of Japanese chemistry, for example, are nearly all in German; those by Nakayama Shigeru on the history of astronomy are given in English.

Where, in the West, might this sort of thing happen? Where might one find the correlative of a well-known Japanese author writing in German on the development of Chinese natural philosophy (Saigusa 1962)? Or surveying in French the diffusion of European science in the orient (Kobori 1964)? Where is it felt to be required in America or Europe to adopt a voice utterly different from one's own in order to reach an international audience? In more recent years, there has been a tendency for most Japanese historical journals to publish fewer articles in French and German and an increasing number in English. This, too, obviously enough, matches the larger and more long-term tendency within Japanese science as a whole to adopt English as its international model. Such is the state of Japanese scholarship, which therefore wears its history on its sleeve. In seeking to enter this realm, then, one must also don this garment and be swept up in its colors, patterns, and fashions. Translation, in one form or another, remains at the heart of the content and the experience of Japanese science, whether for those who practice it or for those who study it.

PART III
The Contemporary Context

REALITIES OF CHANGE AND DIFFERENCE

7 Issues and Examples for the Study of Scientific Translation Today

A Preamble

Examined in light of its historical complexities and importance, translation reveals itself to be a formative influence in the making of scientific knowledge. This has been shown, decisively I hope, with regard to science past. The case of astronomy brought the reader from the depths of antiquity to the brink of the Renaissance, while the case of Japanese science moved things forward in time from the late medieval period into the twentieth century. This leaves open the question of contemporary science.

Given the enormous diversity of scientific expression today worldwide, this would appear a rich, if somewhat daunting, arena for inquiry. Indeed, the case of Japanese science suggests as much by itself. What follows, therefore, is a brief selection of issues and topics involving several different languages and scientific cultures. It is not my goal to be comprehensive in any sense; rather, I hope to suggest the validity of using a variety of approaches. The power of translation, including its ability to impose important transformations, has not diminished in the current era, despite the continuing jargonization and standardization of scientific discourse. This, I hope, will become clear in the sections below.

First, however, it is necessary to take note of a particular issue. This has to do with the deeply embedded belief that, due to the nature of scientific discourse today, its transfer across languages counts as an unimportant literary event, a matter of simply passing coins from one hand to another. This belief has its foundation in another article of faith: the presumed universalism of scientific discourse. Perhaps the greatest myth of all has been the notion that a discourse of knowledge could be created that, used "naturally" in everyday situations—in writing, speaking, and listening—and penetrating all the world's major languages, would emerge essentially untouched by any of them. For if science is truly a universal form of discourse, then all questions of translation (save those of accuracy) become trivial. If, on the other hand, scientific speech is today more like literature or philosophy in being unable to escape certain dependencies upon localized linguistic phenomena, then the complexities of transfer across languages and cultures remain. Indeed, if scientific discourse undergoes even a *few* substantive changes across linguistic boundaries, translation must be accepted as imposing its own determinations on technical knowledge in the present.

In order to help settle this question, we will consider two test cases with regard to the issue of universalism. The first is the question of mathematical texts; the second, that of English as the current-day *lingua franca* for scientific work.

Mathematics as Literature

It would seem justified to propose that mathematical works, due to the nature of the symbolic systems involved, represent the extreme case of universal expression in science (and perhaps in any field). Such texts, that is, appear invulnerable to significant change between languages. Mathematical discourse is the purest form of scientific logic, occupying a space above the more troubled topology of normal human speech.

In reality, however, such is not the case. Even the most densely mathematical research takes place within a linguistic context. This can easily be confirmed by a glance through journals, monographs, and textbooks in such fields as theoretical physics, biostatistics, physical chemistry, and cosmology, as well as any branch of mathematics itself. Equations, formulas, propositions, measurements, and alphanumerical or geometrical expressions of all kinds are to be found nested within written explanation and discussion. This is inevitable: as yet, mathematical articulation does not approach a fully self-sufficient system of communication (among human beings, at least).

To help convince the reader of this, I have chosen, at random, a brief section from a recent research report in mathematical physics having to do with ordered vector spaces (Busch 1998, 2):

> **1.1.** In a fundamental investigation [2] Ernst Ruch proposed an interesting extension of Felix Klein's geometry programme for affine geometries based on normed real vector spaces V. Defining the figure of an (*oriented*) *angle* as an ordered pair $([x], [y])$ of directions (rays, $[x] := \{\lambda x | \lambda > 0\}$ for $x \neq 0$), two angles are called *norm-equivalent*, $([x], [y]) \sim ([x'], [y'])$, if there exists a linear map that sends $[x] \cup [y]$ onto $[x'] \cup [y']$ such that corresponding pairs of points are equidistant: $\|\alpha x_0 - \beta y_0\| = \|\alpha x_0' - \beta y_0'\|$ for all $\alpha, \beta, \in \mathbb{R}^+$. (Here $x_0 := x/\|x\|$, etc.).

The *direction distance* $d[x/y]$ (from x to y) is defined as the family of distances $\|\alpha x_0 - \beta y_0\|$. This is succinctly summarized in terms of the map

$$(1) d: ([x],[y]) \mapsto d[x/y],\ d[x/y]: \mathbb{R}^+ \times \mathbb{R}^+ \to \mathbb{R}^+, (\alpha, \beta) \mapsto \|\alpha x_0 - \beta y_0\|,$$

The direction distance induces an ordering of the classes of norm-equivalent angles via

$$(2)\quad d[x/y] > d[x'/y'] : \Longleftrightarrow \forall \alpha\beta \in \mathbb{R}^+ : \|\alpha x_0 - \beta y_0\| \geq \|\alpha x_0' - \beta y_0'\|.$$

It is evident that written English plays a crucial role here. Without it, the symbolic system in use would be virtually meaningless, since each fundamental element must first be defined linguistically before being set into motion by the various equations. Each symbolic definition or qualification, moreover, is like an ingot within the flow of the narrative: it requires the written text to move it forward, toward a larger assembly into a "map" or formula.

The precise roles of verbal expression in mathematical reports, the contours of rhetoric and organization, may well define a unique precinct in scientific discourse, one that has not yet been probed very deeply. The existence of these roles, however, can hardly be doubted. In the end, given this central dependence on written language, mathematics-based science is subject to most or all of the main considerations presented in this book regarding translational change. However direct and straightforward the argument of the above passage may seem, there exists considerable room for change or deformity during transfer between languages. Subtle differences in the expressive content of such terms as "normed real vector" or "family of distances," connotative shifts in the rendering of words like "interesting," "sends," "succinctly," and "induces," are all likely, even inevitable. This is what experience, and theory, teach. There can be no such thing as absolute one-to-one correspondence in terminologies across languages. Much of what follows below should help to demonstrate this.

English as the Universal Discourse of Science: The Case for Qualification

If science be universal among the world's languages, both in its general and its specific standards of expression, then it should certainly display little or no important variation in any single language. To test the truth of this, it would seem necessary to tackle one such language in particular—the most widely used, for example. Indeed, one of the most generally held perceptions about science today is that it has finally discovered one, international speech. This speech, of course, is English, whose global spread in the precincts of technical subject matter increases with each passing year. No doubt the notion that English has become a "new Latin" for the world's scientific output has an element of irrefutable truth to it, as measured, for example, by sheer quantity of publications, by the use of English in technology and technology transfer, by the most favored foreign language studied by science and engineering students worldwide, by the most frequent source language for scientific translations, or by any number of other measures (Strevens 1992). It has been estimated, moreover, that users of English worldwide for all purposes total two billion or more, barely a fifth of these being native speakers (Crystal 1985).

Do these people, scientists and engineers among them, all speak the same English? Can it be said that American and British forms of the language dominate everywhere, without exception? One thinks immediately, perhaps, of such Commonwealth nations as Canada, Australia, New Zealand, and Malta. But what of India, Hong Kong, Singapore, Nigeria, and Pakistan? Are the forms of English essentially identical in these societies as well? This would be unlikely, given the many varieties that obviously exist with regard to other major colonial languages—for example, French (Haiti, West Africa, Quebec, Algeria) or Spanish (Mexico, Spanish Sahara, Argentina, the Philippines). As it happens, one no longer speaks of "world English," but instead of "world Englishes"[1]:

> Wherever there is an English-using community, sufficiently large and sufficiently stable as a community, there may arise a localized form of English (LFE), identifiable and definable through its distinctive mixture of features of grammar, lexis, pronunciation, discourse, and style. . . . LFEs vary greatly and in numerous ways. Some are complex, with many subdivisions: "Indian English" is the name commonly applied without distinction to a great many subdivisions of the LFE in India. . . . "Philippine English" also displays a considerable range of internal variations. "Sierra Leone English" has fewer variations, but the special history of Sierra Leone as a community . . . means that while Indian English and Sierra Leone English are both LFEs, their detailed anatomies, so to speak, are very different. (Strevens 1992, 34)

During the past two decades, it has been recognized that standard American or British versions of the language represent but a single genus of what has become a much larger linguistic family. Depending

1. Witness, for example, such journals as *World Englishes* and *English Worldwide: A Journal of Varieties of English*.

on its range of uses, the LFE can remain a foreign language (China) or serve as a true second language (India), in which relevant forms of communication take place among educated elites. In either case, accepted standards of grammar, punctuation, sentence structure, syntax, word usage—in short, all the major ingredients of writing—differ significantly from those of Britain or America. The lingering problem about whether such standards produce only "errors," "misuse," and "deviancy," meanwhile, is not a linguistic question, but a cultural one.[2] No self-respecting linguist, for example would consider Quebecois or Haitian French a "degraded" form of the French language. Such nativization is the fate of any living international form of speech.

What, then, are the implications of "world Englishes" for science? Several points might be made. First, not all technical fields have favored English equally, whether as a foreign or a second language. Several studies have indicated that international scientific publications tend to favor English in physics, biology, and engineering especially (roughly 80 percent and higher, as measured by major abstracting and bibliographic services), while chemistry and mathematics lag somewhat behind, being often written in the local language (average 55 percent—Michel 1982; Swales 1985). Nations such as Germany, France, Russia, and Japan retain an acknowledged lead in certain fields, and researchers in those fields continue to publish in their respective native tongues.

Second, and more to the point, one finds that the grammatical, syntactical, and lexical norms for using English in scientific writing do in fact vary significantly from one LFE to the next, and this shows up in journal and book publications. Comparing papers or abstracts that have appeared in journals published in China, Malaysia, and India, for example, will show that one is not dealing with a standardized scientific discourse at all, but instead with something approaching a series of technical dialects, whose contents are adapted in overt and subtle ways to the conventions of each particular linguistic community. This reality, too, varies among disciplines: different scientific fields exhibit different levels of nomenclatural standardization even in English. And this sort of thing is bound to be compounded by a range of other factors specific to each cultural setting, including the comparative level of "modernization" among different fields and the complicated mix of older and more modern terminology that exists in certain ancient disciplines (e.g., botany or astronomy). In any case, different Englishes, to an important extent, make for different sciences—if by "science" we are to under-

2. This issue has been much discussed in the relevant literature. I refer the reader to the various essays and references in Kachru 1992.

stand a domain of knowledge and labor grounded in the written and spoken word.

Scientific English in India: Some Examples

To make such claims, however, is to call for evidence. The following example is from the work of a well-known Indian research chemist, and will help to illustrate the point nicely:

> In India many plants are multipurpose ones and their various parts are used. For example, the mango tree yields a delicious fruit, its wood is of high value and even its leaves were at one time used for feeding cows and for obtaining a dye called "Indian yellow" from their urine. Many such cases may be mentioned. All are familiar with the principle of cross breeding in order to obtain better results. In the case of sugar cane the crossing of the noble ones with reed canes resulted in better varieties. . . . It was at one time visualised that crossing of potato and tomato would result in potomato yielding edible tubers and edible fruits; this has not been so successful. But in Nature there are many plants of this multipurpose type. *Pachyrrhizus erosus* called in India "Sankh alu" comes under this category. *Pachyrrhizus erosus* is cultivated throughout India, but not known in a wild state. The tuberous root resembles a turnip in taste and consistence. . . . Eight compounds in all have been isolated from [the beans of this plant] which thus constitute remarkable source of chemicals of complex type. (Seshadri 1968, 16–17)

Many aspects of this passage will sound odd or aberrant to a native speaker of American or British scientific English. In addition to the tone, there appear to be grammatical lapses, missing articles, redundancies, clumsy transitions, even wrong word choices (e.g., "consistence" instead of "consistency"). The author also brings up matters that would be deemed unnecessary, distracting, or even nonscientific—for example, the opening details regarding the mango tree (which lack any direct chemical references) or the "taste and consistence" of *Pachyrrhizus erosus*.

Yet a survey of Indian chemical writing in English would show that these apparent slips constitute proper usage. They are not merely acceptable, but demanded. In particular, the appeal to the experience and knowledge of a specifically Indian audience is quite striking in many of the details, as well as in the author's mention of the Hindi title for the subject species. Doubtless the religious and cultural connotations associated with "Indian yellow" dye, with its ancient uses in both sacred and secular textiles, would be largely lost on a Western audience. Yet a close reading of the text shows that such connotations are central to the argument being offered, which is concerned with the chemical analysis and agricultural improvement of native plant species. In the terms set out by B. J. Kachru, such standards constitute not deviations but differences, fully legitimized within the context of contemporary Indian science.

As a type of experiment, one might contemplate what would be required to make the science of such a writing acceptable to a North American or British journal. The following offers one possibility:

> There exist in India a large number of plants with multiple uses. The mango tree, for example, yields fruit, wood, and, via cow urine, the dye "Indian yellow." Such uses can be enhanced through cross breeding, as demonstrated by the case of drought- and cold-resistant sugar cane. Another example is *Pachyrrhizus erosus,* known only in its cultivated form. To date, eight crystalline compounds have been isolated from the seeds of this plant. . . .

The differences between the two passages should not require detailed comment. They are enough to suggest that even the so-called "cognitive content" is not the same in each case.

Lest this appear an isolated example, let me provide two other samples, both from one of the most active areas of contemporary scientific publication in India: geology. The first passage comes from a major research paper on the nature of the crust in northwestern India, the subject of significant international interest in the geological community due to ongoing continental collision in the region. Here are the opening passages of the article:

> The Precambrian terrains of the shield areas record complicated geological and tectonic development history. In the Indian subcontinent, the cratonic areas of the major segments in south, central, western and eastern areas have developed through crustal interaction processes whose responses are discernible in the geological and geophysical attributes of the rock components of each segment. In deciphering the nature of these crustal segments, it has become necessary to relate surface geological features with those of deeper crustal parts. The information pertaining to the latter is derived in recent years through deeper geophysical probes. With this idea in view, a project was launched by the Department of Science and Technology, New Delhi, on deep continental studies in relation to Rajasthan Precambrian shield. In order to have a clear focus on the problems and their resolution the studies were conducted along a tran-

> sect . . . that cuts across most of the important lithological and tectonic units of Rajasthan Precambrians. (Sinha-Roy, Malhotra, and Guha 1995, 63–64)

Again, the most efficient way to demonstrate the differences such a passage exemplifies from American geological parlance is to rewrite it in the latter:

> Precambrian shield terrains record complex histories of geological events. In the Indian subcontinent, major cratonal segments have developed through crustal interactions recorded in the geological and geophysical properties of their respective rocks. Deciphering the nature of these segments requires that surface geology be interpreted in light of deep crustal features recently identified from geophysical surveys. With this goal in mind, a project focusing on the Rajasthan shield was launched by the Department of Science and Technology, New Delhi. In an attempt to resolve as many important questions as possible, a transect (Nagaur-Jhalwar transect, or N-J) was conducted across the most significant lithological units and tectonic boundaries of the Rajasthan Precambrian terrain.

Much of the abbreviation seen in the edited version is due to the deletion of words that would be considered redundant, unnecessary, or confusing in American (or British) geological discourse. For example, the second sentence in the passage takes out the word "areas" (confusing), the directional terms "south," "west," "east," and "central" (no specific referent), and the repetition of "segment" at the end (redundant). There are also terminological shifts: geophysical "probes" become "surveys"; crustal "parts" is changed to "features"; tectonic "units" is altered to "boundaries"; and so on. Idioms, too, are adjusted: "with this idea in view" is edited to "with this goal in mind." The point of all these "translations"—and others besides—is not merely to make the passage more precise in common usage and mode of intention familiar to Western researchers. There is much more that is "lost," both on an epistemological and an aesthetic level, than simple ambiguity.

To take a single example, changing "probe" to "survey" causes deletion of an entire realm of connotative urgency. As a noun, "probe" retains a great deal of the implied action of its corresponding verb; it suggests a penetrating, anxious, even meddling effort that draws upon the author's own personal exploratory desire and that pierces deep into its chosen medium. Nothing of this confessional aspect exists in the far more sedate and remote "survey," which connotes a more removed and diffident, surface position. Indeed, what has happened in this exchange is a profound distancing of the author from his activity in a manner that neatly mirrors the larger historical movement within scientific discourse itself, from a language once dense with the experiential "I" to a speech where objects and processes become the subject of nearly every sentence.

Such differences in nomenclature extend in other directions as well. One finds in Indian geological discourse terms of art that do not exist in American or British scientific parlance but that constitute recombinant forms from Anglo-discourse. Examples include words like "lithoassociation" and "lithounit," which adapt the prefix "litho" (as in lithotype, lithofacies, lithostratigraphy, etc.) to new uses. But there are also adaptations from other sciences, or from mathematics, that are unique. One recent article, for instance, speaks of a specific stratigraphic boundary as a time-transgressive "plane" (Bhat and Chatterji 1979, 197) instead of a "surface" (such boundaries being almost never planar in their morphology). Another uses the word "sedimentaries"—as in, "The sedimentaries varying in age between 1700 and 700 Ma crop out in seven separate basins" (Chakraborty et al. 1996, 101). No such word exists in Anglo-geology; similar terms are "sedimentary rocks," "strata," or (somewhat inaccurately but nonetheless commonly) "sediments." In other cases, words found in Western geologic speech are used in somewhat different ways, as in "dominant geologic expressions" (in place of "formations" or "phenomena"). One rather striking example occurs in an article by T. M. Mahadevan entitled "Deep Continental Structure of North-Western and Central Indian Peninsular Shield—A Review." The concluding section of this paper bears the title "Directions of New Thrust"—which can only be read as a type of pun by a Western geologist, since the term "thrust" denotes a specific type of geologic structure (which is obviously not intended by the author). But it is the last paragraph of this section that most deserves citation for its revelations of word choice and usage:

> While adequate expertise is available in the country in achieving the above objectives, there is a need to generate high resolution geological and geophysical data that will enhance greatly the values of modeling. Investments in these investigations/studies are necessary to evolve not only wholesome models of crustal structure and evolution but also to provide the much-needed new innovative approaches to mineral exploration and planning seismic environmental safety. (Mahadevan Kochi 1995, 30)

"Values," "investments," "wholesome": these are not the stuff of which American-British geology is made. Nor is the term "seismic environmental safety" one that Anglo researchers would either recognize or comprehend immediately in its precise meaning and implications. Indeed, the technical phrase in Western English is "seismic risk," which relates directly to the probability of a seismic event (earthquake) occurring. "Seismic safety," by contrast, has no such associations to the actual study of seismology; it is more in the way of an inventive converse, dictated perhaps by the logic of linguistic opposites but lacking in corresponding scientific content. It *does,* however, at least suggest the possibility of such content being added, e.g., using "seismic safety" (or some similar word—"security" or "immunity," for example) to define those specific periods when the probability of an earthquake event is near minimum.

Indian geology therefore, and Indian science in general, are built upon a species of English with distinctive characteristics of its own, thus with individual contents as well. Geology in Indian English, as a shared knowledge occurring in a mixture of standardized and more fluid linguistic forms, cannot honestly be made equivalent to geology in American or British (or Hong Kong) English.

Cultural and linguistic realities impose a certain inevitable *resistance* to universal forms of discourse, science included. "English" does not belong to its native speakers. The more it has become a truly international language, the more diversely it has been adapted, and therefore the less universal in its standards it has become.

Internet Science and the Globalization of English

Increasing use of the Internet by scientists around the world is having important effects upon the relative dominance of English. In the late 1990s, many fields began to establish online journals, preprint archives, conference proceedings, major databases of all types, educational resources, and more. The overwhelming majority of this information is in English. To no small degree, this reflects the simple fact that the Internet was first developed in the United States and was already fairly well advanced and in use by academic and industrial scientists by the mid- to late 1980s. During the succeeding decade, it was mainly the native English-speaking nations—the United States, Australia, Britain, and Canada—whose research communities began to take greater advantage of the new medium and who have since moved furthest toward integrating it into daily scientific activity. With expansion of the World Wide Web in the 1990s, many sites can be found in other languages as well: German, French, and Spanish above all, with a growing number in Japanese and Chinese. Yet a majority of these offer English versions of their content, and it has even become somewhat common for these sites to first appear onscreen in English and then to provide links to versions in other languages. Online conferences, poster sessions, indexes, ejournals, and other resources are also in English, as are email exchanges on the international level. It is clear, in other words, that the Internet, as a new medium that promises to move rapidly, if indeed variably, to the center of scientific communication throughout the world, is also a powerful agent in the globalization of scientific English.

What this means, in practical terms, is that more than ever scientists are forced to be translators. Yet because of this very fact, one finds no single standard of Internet English in use; instead a considerable diversity of Englishes crowd the halls of professional science. Not only has the Internet provided new opportunities for various scientific Englishes to thrive, possibly even expand; it has also encouraged the creation of new, perhaps temporary "dialects," as the products of forced translation. Common errors made, for example, by French or German native speakers have become regular ingredients in Internet scientific prose. This may sound trivial in view of the greater powers for globalization, and it is almost certainly true that these sorts of "dialects" will pass out of existence before long. Yet it is crucial to realize that they, too, like the plurality of Englishes in general, represent part of the inevitable linguistic resistance to absolute standardization.

Geology in French and English

What, then, about changes that might occur in single fields across languages? Can these, too, be documented by relatively close readings? Taking on such questions allows one to break completely free of Bloomfield's ferric judgment regarding the presumed triviality of interlinguistic changes in science. But to begin in this way is, again, to call for evidence.

The following are two excerpts from recent research papers, both written by prominent investigators in African geology. These passages concern the formation of the African continent, which took place in the Early Paleozoic time period (600–450 million years ago). I present here the original text, and then my translation from the French executed in a direct, fairly literal manner, without overly scientizing or otherwise paraphrasing any elements in the original:

> L'histoire de l'Afrique au Paléozoique se déroule toute entière dans le cadre du super-continent de Gondwana. Celui ci, on le sait, s'était con-

> stitué comme tel aux environs de 600 MA, au cours de l'événement thermo-tectonique panafricain. Il ne se disloquera qu'au cours du Mésozoique, 450 MA plus tard. On doit, par conséquent, penser l'histoire de l'Afrique, durant cette période, comme déterminée à la fois par [quelques] facteurs principaux: l'héritage panafricain, la destruction des chaines formées, leur péneplanation, puis le "souvenir" de ces chaines. (Fabre 1987, 2–3)
>
> [Translation] The history of Africa in the Paleozoic unfolds entirely within the frame of the supercontinent of Gondwana. The latter, one knows, was formed somewhere in the vicinity of 600 Ma [million years ago], during the pan-African thermotectonic event. It broke apart only during the Mesozoic, around 450 Ma. One should, as a result, conceive the history of Africa during this period as determined simultaneously by [several] principal factors: the legacy of the pan-African event, the destruction of the mountain chains that had been formed, their peneplanation, and the "memory" of these chains.

Before comparing this with a similar passage in English, it is worth pausing to consider a few aspects of the above. Despite an abundance of technical terminology, there are decided literary qualities to the writing, evident in such evocative (counter-jargon?) words as "legacy," "destruction," and "memory." There is variation in sentence length, a conscious turn to metaphor (even if caged between quotations), a shifting of tone revealing of sensitivity to drama and effect. With this in mind, then, one might peruse a correlative passage in English:

> A review of the Palaeozoic stratigraphic record of western Gondwana reveals that for the purposes of rigorous reconstruction the information from many localities is sparse, and some of the relevant interpretations appear to be inconsistent or contradictory. Palaeozoic sedimentary rocks of western Gondwana are preserved on the broad northern and narrow southern margins of Africa. In describing these deposits we refer to basins principally as the geographical location of the record, without implying a particular Palaeozoic tectonic framework. (Van Houten and Hargraves 1985, 345)

Geoscientists in Britain today would recognize these lines as an example of good, even eloquent scientific writing. Nothing suggestive, overtly metaphoric, or otherwise "superfluous" is included. The tone is instrumental; the grammar crisp; the choice of terms and syntax functional. The narrative moves along smoothly, without any obvious literary edges to momentarily distract the reader. The intent is clearly to convey the maximum amount of technical matter with a minimum of "style."

Thus, if the French selection has a quality of writerliness to it—an embedded consciousness of aesthetic use—the English does too, but the standard is a far more restrained one. Such a standard has its direct and inevitable effects upon "content." The French excerpt, one sees, is by no means thin in this department; on the contrary, measured line for line, it is equal or nearly equal in density of sheer information. Yet it contains connotative material entirely absent from the second passage, material inseparable from the flow and logic of the argument. This might be shown, once again, by rewriting the French example according to rhetorical standards of scientific English, the result being something like this:

> The Paleozoic history of Africa took place within the context of the Gondwana supercontinent, formed approximately 600 Ma during the pan-African thermotectonic event and subsequently rifted apart around 450 million years later in the Mesozoic. The geologic evolution of Africa during this period should be viewed in terms of several principal factors. These include: (1) the tectonic imprint of the pan-African event; (2) the erosion and peneplanation of related orogenic systems; and (3) the influence of these systems on subsequent events.

The type of translation performed here alters the rhetorical strategy of the original French. This strategy, based on a mixture of statement and suggestion, attempts to convince the reader by means of persuasion and eloquence, more than outright declaration. The science in the rewritten text, meanwhile, has shifted toward a more bald statement of facts. The geology at hand is no longer a study of lithologic legacies, destructions, and memories—terms that call directly upon the philosophical basis underlying the discipline as a whole—but instead an examination of events and influences.

There are other, more subtle differences. The word *histoire* in the French example is used as it is in common speech, unadorned by technical adjectives and specialist phrasing, such as "Paleozoic history." This permits a much broader reading of the word, which encompasses both "history" in the general sense and "story" in the narrative sense, as something put together, even fictionalized, and then recounted by a particular speaker interested in engaging a group of listeners. This sense of putting together Africa, as a creation of two types of "history," also has a richness of meaning for geology as a science. It reveals something of the contemporary analytical intentions toward Africa, considered a unique entity with a unique "fit" in the larger assembly of the southern continents, the so-called supercontinent known as Gondwanaland that prevailed through Paleozoic time. "L'histoire de l'Afrique" implies a community of readers gathered around the writing and reading of this entity, the telling of its creation by contemporary narrative, the interpretive exploits it has attracted. Geologic history and the data that have built it both become the field of a hopeful yet indeterminate recital, one that admits its own tentative nature.

Missing from both English examples, therefore, is this potential window onto the larger scriptural, sculptural content of geologic science. An objection comes up at this point: does the average scientist really perceive this sort of content while perusing journal articles? The answer is that he or she probably doesn't, but also that this question is largely irrelevant. Practices of reading, whether in science or in literature, never exhaust the content of their objects. Such a degree of consumption may be, in a sense, a professed goal of technical literature, but, as the entire field of science discourse studies shows, it is impossible to achieve. Indeed, it would be foolish at this point to deny the many insights that have flowed from such studies in recent years, and more generally, the realization that scientific writing opens itself, in fact, to a host of analytical approaches no less diverse than those applied to literary work, including hermeneutics, genre analysis, reader-response criticism, semiotics, and so forth. Beneath its modernist gleam of efficiency, the scientific text is rich in rhetorical moves, slips of logic, suggestive terminologies, broken transitions, calls upon the larger society, and, as we have seen, philosophical and aesthetic manipulations.

To strike the point with new force, we might examine technical abstracts and their translations. These represent a forum where the writing is *intended* to be as functional as possible and where the ideal of a one-to-one correspondence between tongues is a clear goal. Many international journals today include an abstract written in the native language followed by a version in English and possibly others in French, Spanish, or German. Such examples therefore provide an excellent "control group" for examining changes in stylistic and content standards within individual fields. Note, therefore, the following excerpts from abstracts in another recent paper on north African geology, this time concerning central Tunisia:

> En effet, quel que soit le site initial des séries les plus compétentes, celles-ci ont finalement tendance à occuper une position structurale haute en fin de déformation. (Turki et al. 1988, 399)

> [Translation] Therefore, whatever the initial setting of the most competent strata, these have a tendency, finally, to occupy a high structural position at the end of deformation.

> [Article version in English] Therefore, despite the initial configuration, the most competent strata will reach the highest structural position at the end of deformation. (Turki et al. 1988, 399)

The more literary qualities of the French are again apparent. Note, for example, how "ont finalement tendance à occuper" is rendered into the flat "will reach." Beyond this, one sees that "quel que soit," a subjunctive phrase meant to suggest uncertainty, is changed to "despite," whereas "une position structurale haute" becomes "the highest structural position," actually a mistranslation.

Within even a single sentence, significant changes of meaning and style once more emerge. The English, without apology, tends toward attempted factualism. The French, meanwhile, is more indirect, cautious, more in league with the provisional conclusion than with the unassailable fact. Again, there are differences in basic philosophy implied: factualism assumes consensus; provisionalism looks for ongoing discussion and revision. Both philosophies are dynamic, productive, essential; but the outlook of each, and the means of its delivery, are different enough to convince one that here are different epistemological dialects, not a single, unified "science."

Aspects of Science in Non-Western Languages: Chinese Citations

The above examples suggest an early answer to the question of science and cultural-linguistic borders. But to deepen the evidence one level further, let us enter the realm of actual translation, through a side door however.

Citations are a crucial element of contemporary science. Their use in research papers, reports, monographs, and other documents has much to do with conventions of authority, the "past," collegiality, and originality. Among other things, they help establish an author's level of expertise in his or her particular subject and provide him or her with a means of displaying respect. What happens, then, when styles of scientific citation endemic to one cultural-linguistic community are transferred to another? How might such a transfer appear? Obviously, were science truly universal in its discourse, the answer would be simple: nothing of any importance would take place.

It has been confirmed, by extended study, that plagiarism among foreign students trying to learn English is significantly higher in the natural sciences than in the humanities (Braine 1995). Moreover, in a majority of cases, these attempts succeed, and science students who refuse to adopt this means wholeheartedly and only paraphrase their sources very often pay "a terrible price." Too often, that is, their adaptations have ended up "producing paraphrases that obscured or altered the meaning of the sources" and that yielded results "incomprehensible" or even "opposite" to what was actually intended (Braine 1995, 127). Plagiarism in this setting appears to reflect a number of factors: the complexity of scientific nomenclature in English; the demand for precision in word usage; and the pressures in academia to perform, even for those who have not yet mastered the language.

As it happens, charges of plagiarism have apparently been leveled at Chinese writers in particular. This suggests a possible cultural component, one related to rhetorical traditions in Chinese prose writing that historically may have had important influences on scientific discourse. The problem, having attracted some notice and concern, has been investigated by a number of scholars interested in pedagogy and cross-cultural language learning (Bloch and Chi 1995; Taylor and Chen 1989). Given the central role of "authority" in any act of textual borrowing, such studies have compared the ways in which Chinese and American academic authors make use of others' work in their own writings, including their use of citations. Over and above the various statistical analyses performed, several main conclusions emerge from these studies:

- Scientists writing in English show a far stronger preference for citing sources less than five years in age, while Chinese scientists often cite works in the five- to fifteen-year range, and show more than double the number of references to works published more than fifteen years earlier.
- Compared with Anglo-American researchers, Chinese scientists display less willingness to make critical citations; that is, to criticize their colleagues publicly in papers and monographs, even when they disagree with them. Such unwillingness is greatest in article introductions, where critical citation tends to be highest in English texts. In the main body of a paper, meanwhile, critical citation does exist, but at a significantly reduced level compared with English.
- Citations used as background to a particular argument are often more heavily concentrated near the end of an article. This is because Chinese authors frequently employ a somewhat indirect form of organization, such that the main argument is itself postponed until the latter half of the paper.

• Introductory portions to a significant number of scientific articles contain citations to such authors as Marx, Engels, Lenin, and Mao Tse Tung.

These kinds of observations obviously reflect important elements in the production of scientific texts by Chinese researchers. The quality of indirectness; the avoidance, until credibility has been established, of the signs of open conflict; the accepted demand for an expression of political correctness—such factors indicate a very different rhetorical framework for "science" than exists in North America or Europe. To the eyes of the American researcher, for example, technical writing in China must seem to wear its social and political context far more heavily on its sleeve, and in far more lurid colors, than anything s/he is accustomed to accepting as legitimate science. Much fruitful time and space might be spent linking the above observations to questions about the role of Confucian philosophy, traditions of hierarchical pedagogy, respect for precedent, and the influences of the Cultural Revolution. The essential point, however, is that the results of any such investigation would produce conclusions not merely about authorship but about the specific character of Chinese science.

Conclusion

If one can contemplate science as comprising a range of activities, all with a deep linguistic dimension, then the history of language—and of languages—becomes integral to the history of science. This means, of course, the history of translation too, for, as the brief examples above suggest, and as the previous chapters have tried to demonstrate, the movement of scientific knowledge across cultural-linguistic borders has always involved substantive change—the creation of new vocabularies; the deletion and addition of epistemological matter; alterations in logic and organization; major shifts in the rhetoric of persuasion; even such deep-seated philosophical differences as the declaration of "facts" vs. the suggestion of factual possibilities.

To say that these areas have little or nothing to do with something called the "true content" of science would be naive. Knowledge cannot escape its forms; and forms, in turn, must be recognized to have their determinations and involvements. The former positivist belief in a cognitive nucleus that would always remain pure and untouched by linguistic realities was, and is (where it lingers), founded on a distrust of language, a vision of its mutinous inadequacies. Thus only an artificial (or artificially constrained) discourse could ever hope to realize the dream of a one-to-one correspondence between phrase and fact, word and intended meaning.

Today one knows, as the Greeks knew, that there are no simple divisions between linguistic and epistemological domains. Indeed, the study of knowledge today returns, like a thirsty wanderer, to the flowing well of language studies, for whether one turns toward literature or toward science, one encounters complex and multilayered forms of human speech. Can science be called an "artificial discourse" in any important sense? The answer must be a resounding "no": millions of people speak, write, and read the multiple versions of this discourse every day, in the laboratory, field, hallway, classroom, and bathroom. Science represents a vast collection of living languages and thus, in a sense, "sciences" as well. Translation has been the historical process to ensure the truth of this, both in the modern era and long before. In rendering technical knowledge mobile between peoples and through the centuries, translation has been a crucial force behind both the creation and the continual refertilization of science.

8 Conclusion

Gained in Translation

The disparities in knowledge that beset the present era appear problematical only in the light of an ancient dream. Such is the dream of the unity of discourses, a vision of a universal *lingua scientia* able to bridge all differences, collect all efforts of intelligence, give order and direction to all uses of the human mind. This dream, though venerable, is not merely a province for the specialist in dead ideas. Throughout nearly the whole of what is commonly called the "modern era," beginning in the seventeenth century and continuing at least to some later portion of the twentieth century, this reverie has sought its fulfillment in the universality of natural science. It is a vision that does not weaken or fade easily; nor, perhaps, should it. Within the past few decades, it has risen to some local peak again and again in ideas proposing vast new syntheses among the sciences themselves, whether related to computer-inspired prophecy ("complexity theory") or subatomic physics. If the heat of enthusiasm has sometimes singed the sense of arguments in favor of this dream of a final unity, this is not the primary objection that might be raised. For if it can be more coolly admitted that what we call "science," today and in the past, is predominantly a reality of language—knowledge generated, shared, and used through media of written and spoken communication—then this dream must be relegated to the status of a dignified and useful hallucination.

It can be rightly said—and recent scholarship on the sciences has done much to demonstrate this—that no area of knowledge, however grand or specific its precinct, is entirely separable from its forms. There is no literature without styles and techniques of expression, no art without modes of visual representation, no science without the varied discourses of literacy and orality that give it reality in the world of human understanding. The history of this truth, as the present book has attempted to show, is even larger and more complex than has commonly been allowed. It is a history not merely of rhetoric, intellectual culture, and all the intricacies involved, but of entire linguistic communities as well and, still more, of the changing interplay and exchange between them over time. The growth of scientific disciplines has been no less dependent upon translation than has literature or philosophy or religion. Indeed, in a number of major cultures, such as India, China, and Japan, modern science is indivisible from translation, in fact *began* as translation, and as such was a matter of collision between languages, confusion and competition among individual technical discourses, without any hope of universality. This is well documented in the case of Japan: between 1750 and 1860, translation constituted the vast bulk of scientific work, and it was wholly expected that most scientists would be translators first, before they were experimenters, field workers, or theoreticians. During this period, the idea of "contribution" in Japanese science meant, above all, increasing the library of its textual resources. The most crucial work involved expanding the written foundations of scientific thought by incorporating works from abroad. And what this produced was a series of individual discourses bearing contents derived from traditional neo-Confucian natural philosophy, from Chinese science, from Western science, and from native Japanese extrapolations of each of these sources. Moreover, separate fields underwent separate linguistic-epistemological evolution as a direct result of shifting political-cultural loyalties among Japanese scientists to specific occidental nations, e.g., from Holland in the eighteenth century, to France and Germany in the nineteenth century, to England and the United States in the twentieth century.

For the greater part of its development, the content of work we now call "scientific" was extraordinarily diverse. Aspects of literature, philosophy, and religion (among other areas) were all woven deeply into the tapestry of such fundamental texts as Ptolemy's *Almagest,* Copernicus' *De Revolutionibus,* and Lavoisier's *Traité de Chimie.* The specific weave of these contents, as well as the threads themselves, tended to undergo change, sometimes of a radical, even impossible nature, between languages: Ptolemy being given a monotheistic cast in Arabic, or Newton acquiring Confucian overtones in Japanese, are but two examples of this phenomenon. The movement of scientific knowledge across cultural-linguistic borders has always involved substantive change—the creation of new vocabularies; the deletion and addition of epistemological matter; corruptions of names, dates, positions; alterations in logic and organization; shifts in the style of persuasion. The phrase "lost in translation" appears an impoverished cliché beside such historical realities. The truth is that enormous material has been gained in the actual transfer of knowledge from one people to another.

Questions of Difference and Universalism: A Brief Survey of Ideas and Issues

The role of translation in the making of science, whether ancient, medieval, or modern, is immeasurable, to be sure, but it is also continual proof of difference. The examples given in this book all make it clear that the idea of scientific discourse as a universal form of expression is at the least problematic. This being said, no such conclusion has the power to cast doubt upon the fact of an international scientific community, bound in part by conventions of speech, writing, and exchange. This community is real, indeed more real today than ever before, and so are its codes of expression. Yet the codes and conventions of this internationalism are hardly the whole story. A physicist in Boston may sense that he shares far more with his professional colleagues in Moscow or Beijing than with humanists or laypeople down the street. But then, as the saying goes, there is the rest of life. And the rest of life, with all its depth of sensibility, background, daily cognitive and emotional momenta, is inextricably bound up with language. No professional discourse, however powerful, has force sufficient to defeat all the localizing effects an individual language imposes, inevitably, upon its speakers and the material they produce within its medium.

Such a conclusion, however, goes against a strong tradition of writing and thinking about science. While we have noted this previously, it remains to demonstrate here the shape of the universalist tradition and what it implies. I offer only a brief survey of related ideas, in order to suggest how deeply entrenched and widely spread these notions have remained. Such a survey might begin with the Spanish philosopher, José Ortega y Gasset, who wrote an essay entitled "The Misery and the Splendor of Translation" (*La Miseria y el esplendor de la traduccion*) that has been often quoted and that deals directly with science:

> if we ask ourselves the reason certain scientific books are easier to translate, we will soon realize that in these the author himself has begun by translating from the authentic tongue in which he "lives, moves and has his being" into a pseudolanguage formed by technical terms, linguistically artificial words which he himself must define in his book. . . . It is a *Volapuk,* an Esperanto established by a deliberate convention between those who cultivate that discipline. That is why these books are easier to

> translate from one language to another. Actually, *in every country these are written almost entirely in the same language.* (Gasset 1992, 95; italics in original)

In short, because science is not really a language, but instead a kind of code, the translation of scientific works is, similarly, not really translation either, but something else—the production of replicas. For Gasset, this is what defines the universalism of scientific discourse. "Authentic" translation in science, the author says, takes place only in the moment when the original scientific text is written; this is where the burden of interpretation and transfer rests.

At the time Gasset composed his essay (the 1930s), and for more than a decade thereafter, language study in America was largely dominated by the work of Leonard Bloomfield, who wrote perhaps the only extended treatise on scientific language by a professional linguist. Bloomfield begins his volume by emphasizing:

> A scientific utterance differs from an ordinary utterance in this, that the conventional response does not allow of variations based upon goings-on inside the person of the speaker or of the hearer . . . the scientist's "concepts," "notions," "ideas," stomach-aches, and what not play no part in the communication, and what is spoken is accepted only at face value. (Bloomfield 1987, 208)

By this reckoning, science is a form of speech emptied of all psychology. This is what gives it the power of ideal passage between speaker and listener, this lack of subjective adventure. What might this imply with regard to translation? Bloomfield is unequivocal: when it comes to science "the differences between languages (as English, French, German), far-reaching and deep-seated as they are, constitute merely a part of the communicative dross. We say that scientific discourse is translatable, and mean by this that . . . the difference between languages . . . has no scientific effect" (Bloomfield 1939, 4). This is the crux of the belief in the universalism of scientific discourse: technical speech remains independent of whatever language or individual it might appear in—or, put differently, it takes possession of its speakers, mainly for purposes of transmission, just as the Greek Muse was thought to fill the vessel of its poet, like a votive, empty and waiting.

Related ideas have leaked into the writings of scholars in many fields. A recent and powerful formulation, sophisticated in contour, is offered by Louis Kelly, one of the most well-regarded translation scholars writing at present. Combining insights from linguistics, literary theory, and philosophy, Kelly has provided a theoretical history of translation yet to be surpassed both in scope and influence (Kelly 1978). Kelly's views on technical translation are apparent in several passages, including the following:

> If language is taken as "object-centered" or purely informational, the translator will suppress any awareness of aims other than symbol, and will therefore be able to find some reason for literal handling. This technical approach is not restricted to technical translation . . . assumptions of the essential objectivity of language signs have been bolstered by the theological argumentation of Jews and Christians, by the artistic theories of the Romantics, by the needs of scientists. . . . (Kelly 1978, 220)

There is a crucial point here whose historical accuracy should not be doubted. The mind-set of the translator interested in absolute fidelity springs from a type of linguistic fundamentalism that has its ultimate roots in religious adherence to the literal truth of the word. But Kelly is mistaken in something else. Such fundamentalism, as it happens, grew not from an "object-centered" view of language, but from a mystical one. Beliefs that invested final sacredness in the word conceived it as a form of divine presence fired into linguistic form or a primal model of rhetorical power and eloquence. St. Jerome and the late Syrian translators of the gospels provide an excellent example of the former, as we have noted, while the early Roman translators of Greek literature clearly exemplify the latter idea.

The true connection between sacred texts, Romantic concepts of the purity of art, and the translation of scientific knowledge has to do with the notion of an unalterable content, whose universality is presumably guaranteed by some type of natural or theological (in any case, higher, innate) law. In the case of science, this law regarding the "objectivity of language signs" supposedly emerges from the nature of scientific discourse itself, which is exhausted at the surface not as "information" but instead as "knowledge"—concepts, hypotheses, and interpretations, as well as facts and the *faits divers* of observation. Kelly's notion that translators of science are essentially forced to suppress all but the most literal readings underlies another claim, which adds clarity to the real issue at hand: "Given the traditional focus of translation theory on intent, theory-building has been the province of the man of letters. Acceptance of multiplicity of intent brings questioning; from questions come judgments, theory; from theory, experimentation. Hence the wide range of technique in creative translation" (Kelly 1978, 221). It has been one of the primary goals of the present volume to reveal how well this type of statement is matched to the realities of scientific translation through history—no more perhaps, but certainly no less, than to literature.

If literary intellectuals have tended to look upon the translation of scientific works as a matter of replication, how might technical translators themselves speak of their craft? One might assume a wholly different sort of attitude prevails; far more complex and laudatory in nature. Yet such is not the case. I offer two examples as a representative sampling. The first of these is drawn from a 1993 symposium held at the Centre of Advanced Study in Linguistics, Osmania University (Hyderabad, India), where a paper on the topic of "Problems in Translating Scientific and Technical Texts" had the following perceptions to offer:

> What makes the translation of scientific and technical texts significantly different from the translation of literary texts is the probability of one-to-one correspondence for the concepts and terms. . . . There can never arise the possibility of occurrences like Edgar Allen Poe's *Nevermore* which carries different semantic implications within a span of 110 lines. The question of interpretation never arises. . . . (Chandra Rose 1994, 63)

The notion of universality here gains a more specific dimension in terms of the belief in a near-perfect linguistic correspondence between languages.

It would be an easy matter to produce more examples of this kind, *copia redundat*. To spare the reader the burden of such proof, I will merely touch upon one final area. The field of machine translation, or "MT" as it is called, which had its first great flowering in the 1950s, arose from the concept of an "automatic translator" that had already been manifest in prototypes first patented by French and Russian engineers during the 1930s (Hutchins 1986). By the late 1950s, however, the computer had arrived, and so had Noam Chomsky's *Syntactic Structures* (1957), changing the face of American linguistics forever. The combination of advanced technology and an algorithmic "transformational grammar," it was thought, could be used to found a true "science of translation" (Nida 1964; Newmeyer 1980). That Chomsky himself demurred on all counts (linguistic universals, he said, were structural and formalistic, not syntactic or semantic, and as such could not hope to show the way to a correspondence theory of translation) did not at all deter MT enthusiasts, buoyed by the idea that it was wholly reasonable

> to expect that [in word-by-word automatic translation] patterns of the source language and patterns of the target language can be put into correspondence in such a way that certain kinds of meaning will be preserved, and that simple recipes for pattern transformation will emerge as well; ideally, the style of the translations obtained by the resulting process

> should be of a caliber comparable to the best that human translators produce. (Oettinger 1959, 250)

When it came to literature, or any sort of written discourse rich in complex and idiomatic usage, this type of transfer would never in a thousand years yield a convincing or useful result. Such effort would produce gibberish, comedy, or both. "Literature," however, would be destroyed. Not so science, however. The question was again asked: "Is it justifiable to call the product of such an automatic dictionary a translation?" And the response: "The experimental results suggest that for scientific texts the answer is yes." Moreover, "a monolingual reader, expert in the subject matter of the text being translated, should find it possible, in most instances, to extract the essential content of the original from this crude translation, often more accurately than a bilingual layman" (Oettinger 1959, 257–58).

Practically speaking, this statement is revealing, if mistaken (even a few mistranslated terms, if they be of sufficient importance, can render a scientific text tentative, as any working translator knows). As with Gasset and Bloomfield, the scientist is presumed to be, *de facto*, a type of rare, international species defined by the discourse s/he employs, a discourse beyond the provincialisms of individual speech communities. In the end, the early proponents of MT, and some of their contemporary progeny (though most now maintain a more humble position), have claimed a view of scientific language not so different from that held by literary intellectuals. Hope for a mechanical Rosetta Stone has produced strange bedfellows. The history of technology, indeed, shows that the newest machines have often awakened the most ancient of dreams. But the fact remains that it is a philosophy of language—not a love of apparatus—that will always underlie the ambition to automate the translation process.

One final caveat requires mention at this point. If scientific discourse cannot be considered universal in any absolute, or even functional, sense, where does this leave science as knowledge? Must we conceive scientific understanding—say, the concepts, facts, data, and so forth of molecular genetics or planetary astronomy—as also "doomed" to a significant degree of localism? The short response here (the only one we have space for) is that the question remains enormously complex and cannot be answered in any easy fashion. To state that "knowledge depends upon its forms" is not to make claim that these are identical. Whatever else might be said, the objective component to scientific knowledge and its technological products must be recognized as very large; indeed, its power to create concepts and machines that can be shared by any linguistic community lies near the base of its influence on social structures and practices in the modern world. The fundamental principles of plate tectonics consist of the same basic ideas in China as in Britain. A laser works identically in Japan and in Poland or Canada, though the precise expression of how it does so will differ. The question thus becomes one of the precise, inevitably variable relationship(s) between knowledge and expression. This is a theoretical problem, to be sure, but also a cognitive one. Posing the existence of an unalterable "content" of science will not suffice, not unless this "content" be redefined in every specific instance.

Translation and Scientific Advance

Once the idea of strict universality in scientific discourse is abandoned, the door opens to many important historical themes. The role that translation has played in the history of science has many facets, only some of which have been pointed out in this book. Science has indeed undergone—*is* undergoing—alteration between languages and peoples, and the details of such mutation reveal motifs that are indeed worthy of sustained investigation. First among these, perhaps, is the theme that before the seventeenth century—and in some countries (notably Japan) well afterward—scientific knowledge was almost entirely textual in its embodiments, a matter of truths and ideas held within the bounds of scrolls, codices, and books. This is not to deny completely the importance of oral dimensions; teaching and public lecturing, in particular, remained critical modes of passage. But these modes were themselves rooted in explicating and summarizing—orally duplicating, in other words—specific primary texts (and their handbook epitomes), whether those of Aristotle or Avicenna.

One encounters this time and again in ancient writings on scientific subjects: "wisdom" or *ḥakīm* resided above all in the long-ago hardened word of the sage, which could be advanced or continued, even corrected, by more "modern" hands, but not entirely surpassed. Despite all the amendments made to Ptolemy by Islamic astronomers, for instance, deeply felt veneration for the *Almagest,* the *Handy Tables,* or the *Planetary Hypothesis* never withered or weakened. Correcting such texts was tantamount to fulfilling their magnificence, not overturning it. Such primal texts were therefore more than mere obelisks of knowledge; they were like great mountains, immobile yet continually rising, changeable yet constant in their presence, to which scholarship must journey and return.

Translation thus brings an old topic—the process of scientific advance—into something of a new light. For example, it is apparent that the foundations of modern astronomy in the occident were not merely the result of a process of recycled knowledge and innovation, augmented by the revolutionary additions of individual luminaries. As a textual reality during this period, astronomy depended for its expansion upon efforts with the power to create new texts, new readers, new modes of writing and documentation. No greater efforts of this type might be found than those involved with major episodes of translation and retranslation. For early medieval Arabic science and late medieval European science, this meant a veritable flood of newly introduced materials, a transfer of enormous textual wealth that led directly to the wholesale reorientation of existing astronomical thought. It also meant the intermittent re-rendering of certain crucial texts that were therefore able to attract, train, and vitalize the thought of newer generations of scholars.

Such a process does not (re)make the history of astronomy into a series of saltatory "leaps," major discontinuities, separated by eras of relative stability when little translation took place. On the contrary, within manuscript culture, discontinuity was a constant, often unpredictable result. Scribal work performed by moderately or poorly paid copyists, whose education was often spotty and whose ambitions were less than pure, could always introduce corruptions, deletions, errors, and a dozen other changes into a particular work. These types of changes also arose from translation. In this case, however—and no doubt more generally too—they should not be seen en masse as unfortunate mishaps that impede good scholarly work or that possess qualities of interest merely as curiosities. Rather, they are one of the marks of adaptation; they are what has inevitably happened to textual works in the act of being transferred from one people and culture to another. There are, of course, many other types of adaptations that we have discussed, the additions, rewriting, omissions, glosses, and so forth that translators have so often performed upon the works of their choice. Most of these alterations, in short, were the result of purposeful recasting. Even more, indeed far more, than in literature, translators in the history of science have been editors (at times merciless ones). This is just as true for such "literalists" as Gerard of Cremona as for those who, like Cicero, followed a sense-for-sense approach.

Indeed, the particular case of Roman science illustrates the power of translation as a selective force for creating new cultural materials. Essential to this force is the ability to ignore or marginalize certain possibilities for knowledge, and to highlight others. Roman translators made the choice—compelled by the limits of their own background, by their cultural chauvinism, and by their sense of what would be useful and meaningful for the empire—to overlook the vast majority of Greek mathematical science. As a result, this science was kept out of Latin, and thereby out of the reach of the early Middle Ages. The repercussions of this for Europe should need little comment.

Displacement as a Cultural Process

Another theme that cuts across many case histories is that of displacement, as a frequent ingredient in the process of nativizing translated materials. By "displacement" is meant the collection of efforts within the receiving culture to erase or diminish any overt signs of debt to the source culture—a central aspect of nearly all the major episodes of translation we have examined.

Such displacement takes place on a number of mundane levels, through certain forms of adaptation. Glossing; reorganization; substitution of indigenous names, titles, or terms; and other substantive changes within a text constitute one such form. As in the case of Roman science, these can have the power to transform an alien knowledge into a known terminology, to conquer and colonize its very alienness while allowing it the appearance of the wholly new and desirable. The larger effect in this case was to effectively destroy the "original." So successful was Cicero, for example, at rendering the Greek constellations of Aratus into a homespun Latin that little need was thereafter felt to make contact with Aratus himself. Certainly this was the case with Arabic versions of Greek and Syriac works, and also, perhaps more pointedly, with late medieval Latin renderings of Greco-Arabic science. The truth of how Arabic writings, as *the* crucial influence on the formation of early European science, gradually disappeared from the immediate scene of obligated acknowledgment, to be replaced by the far more distant and "safe" Greeks, is a tale worthy of far more detailed consideration than has been given here.

Other forms by which epistemological debts have been canceled have included the summarizing of a famous work into a handbook or other epitome, which version then comes to replace the original, or else breaking the work up into a series of segments for inclusion in a number of new works on more specific subjects. The writing of commentaries must also be counted among the styles of displacement, though with caution. For commentaries have often been fertile parasites, invigorating the host (at least for a time) with augmented fame and importance. In some cases, the concepts of "sages" or "the ancients" have had some effect in this regard too. Attributing specific works that have already passed through several languages back to their original authors, in hallowed fashion, has had the effect of deleting any obvious traces of this passage, or of any debts to it. Islamic intellectuals reading Ptolemy in the eleventh or twelfth century can be forgiven for perhaps not knowing that the words before them had once passed through Syriac, Persian, and possibly Hindu.

The various changes from paraphrase to literalism, or vice versa, within receiving cultures have also played a part in this process of cultural displacement. Both approaches, that is, have been used to devalue older versions of a particular work, to create new translations of it, and thus effectively to build up a series of renderings that stand between the scholars of a particular time and the "original" text in the earlier language. The perceived need to "correct," "refine," and make more "accurate" an earlier version, to return to and take possession of a particular work in the name of a later era of "better" translators, often marked the effort of one generation to delegitimize or modify the achievements of another. In such cases, perceptions of inaccuracy in earlier translations may well be partly the result of linguistic change within the receiving language: no tongue or discourse stands still for very long, no matter how standardized, and especially if subjected to the sustained heat of one or another ideology. This is where the process of nativization involves internal displacement.

The Instability of Texts

It is popular, in the era of electronic textuality, to speak of the printed word as "fixed" and "stable." Once stained upon the page and offered to readers widely beyond the writer's and printer's reach, words are thought to be perdurable and to be able to persist in dependable form. The "culture of print" is often said to be a culture of continuance and preservation, at once funereal and reassuring.

That this is a myth of recent origin has been pointed out by more than one author (Johns 1998; O'Donnell 1998). The tale of the once and future Aristotle might stand as powerful evidence in support of such dissent. Books, in particular, have never been permanent commodities, except perhaps as literal objects, mummified in museum casements. New editions of old works, whether novels, poems, scholarly treatises, or travelogues, are constantly appearing with significant additions and deletions: "It is surprising what variations can occur between one printed edition of the same book and another, and if the work is a classic, often printed by different houses over a long period of time, a burgeoning of variant readings can arise comparable to those in works copied in manuscript" (O'Donnell 1998, 44). The unending struggles to establish what are called "authoritative editions" represent nothing if not recognition of the book's true instability—a sign of the continual need felt by successive generations to lay individual claim upon "the classics" through the imprimatur of selective textual editing. It is an irony of textual history that the term "authoritative" has long meant "temporary authority over" a particular work.

The realities of instability go much further and deeper. Almost any work written prior to the late fifteenth century (the advent of printing)—and many that appeared later on as well—today exists as a literal construction built from various portions of different manuscript editions. This is especially true of classical Greek and Roman works, *none* of which exists in anything even resembling its original form, instead having been assembled in selective, often uncertain fashion from medieval or early printed versions. We have noted this in the case of Ptolemy especially, and Aristotle too, but it is equally true of every well-known author of antiquity. Indeed, it is sometimes very instructive to read the introductions to the works of such thinkers, where discussion is made of the editions used. Here, for instance, is how the great Latinist Howard Rackham opens the Loeb Classical Library version of Pliny the Elder's *Naturalis Historia:*

> A large number of manuscript copies of Pliny's *Natural History* have been preserved; the oldest date back to the 9th or possibly the 8th century A.D. Attempts have been made by scholars to class them in order of merit, but it cannot be said that even those that appear to be comparatively more correct carry any paramount authority, or indeed show much agreement on doubtful points, while the mass of scientific detail and terminology . . . has necessarily afforded numerous opportunities for copyists' errors and for the conjectural emendation of the learned. Many of the textual problems raised are manifestly insoluble. (1942, xii)

For the interested reader, there are treatments that deal directly with the practice of textual criticism. A brief and excellent summary of the field, admitting of its many uncertainties, can be found in the well-known work *Scribes and Scholars* on the transmission of Greek and Latin literature. It is a revealing point, emphasized by the authors of this book, that much of the tracing of manuscript influence has been based on the identification and correlation of errors, corruptions, and missing passages.

Assembling the works of Plato, Cicero, Pliny *et alia* from manuscripts that are a minimum of seven hundred years removed from the lifetimes of these thinkers has a particular meaning. What one studies today, whether out loud in the overheated classroom or in the monastic shade of a library, is not Plato or Pliny, the Greek or Roman author, but instead "Plato" or "Pliny," a complex textual creation. There is no way to determine the exact relationship between these two types of entities. They are separated by far too many intermediaries, too many of whom have left their own touches before disappearing forever. In a quite literal sense, there can be no such thing as an "authoritative edition" in this realm. The history of passage for any particular work cannot be defeated by the attempt to freeze a particular, assembled version of it. There is always the possibility for other, perhaps even more Frankensteinian constructions. I do not say "reconstruction," for the fact is that the manufacture has no hope of bringing back something pure and original. History, time, and use have annihilated this entity; what lives is an assortment of fragments of its transmission.

There are other instabilities in both handwritten and printed textual culture. It has been frequently pointed out that the appearance of the page, with its breaks between words, its paragraph structure, its use of punctuation, and other aspects, represents a collection of very recent innovations, each of which has nonetheless undergone continual evolution (Martin 1988). Even the Victorians, as an example, used punctuation a good deal differently than we do today, and editions of many classical works published during the mid-nineteenth century reveal the effects of this. A century earlier, and such aspects as spelling, grammar, and even syntax were not always what they are today. The reality is that print has provided the conditions for both standardization *and* continual change—for historically momentary editions of authority that then pass away before their successors.

Yet the greatest proof of the text's instability is surely the very topic of this book, translation. Great authors, from Homer to Kafka, from Euclid to Linnaeus, have been rendered endlessly protean by the realities of this process, and seem fated always to be so. Nietzsche was not the first to observe how much is revealed of a particular age by the manner in which its translators remake famous works. It was another German author, however, who gave this idea a certain formative spin.

Probably no writer on translation has been as widely quoted, eagerly commented upon, and routinely recast as the literary critic and essayist Walter Benjamin. This renown is all the more impressive in that it stems from a single brief writing of barely a dozen pages, produced in 1934. Benjamin composed his elegant and enigmatic "Task of the Translator" (Die Aufgabe des Übersetzers) as a preface to his own German translations of Baudelaire. In it, he touches upon many fundamental questions regarding the nature and meaning of the translator's activity, and it is this concern that gives his essay such broad relevance. One of his key concepts, one that has deservedly drawn a great deal of attention among scholars of literature and language, is that of the "afterlife" (*Überleben*) of a work, brought to reality by the act and result of translation. Textual objects have continuity through time, space, fame, and adversity; they share with organic phenomena something essential—not mere survival alone, but a persistence that involves growth, change, loss, renewal, and the constant threat of death, disappearance.

Translation is the process through which written works acquire history. And it is not too much of an imaginative stretch or sacrifice to accept Benjamin's prescription that this be taken in literal fashion. The genre of literary biography takes the author as its subject and endpoint and, often enough, would have us believe in the person as an expression of his or her own deserved fame. But honesty forces one to admit that, in fact, it is not the writer but his or her texts that endure and have influence, indeed that are sustained wholly independent of their creator. The relationship, Benjamin himself might conclude, is a vampiric one: only by drawing upon the true, temporal life of the work is the writer able still to walk among us after death. Within the greater realm of literary studies, however, there is no such thing as a genre of biography reserved for the life of texts—yet such would obviously make sense, and not merely for works of poetry or fiction. How valuable might it be, for the sake of historical knowledge, to trace the dissemination of Newton's *Principia Mathematica,* its uptake and incorporation by later scientists and mathematicians, its popularizations, and yes, its translation into various vernacular languages, European and non-European?

Scholars have used Benjamin's idea of "afterlife" in a number of fertile ways. One of these has been to try to "explode the binary opposition between 'original' and 'translation,' which underwrites the translator's invisibility"(Venuti 1992, 6).[1] The translation, by this token, is no longer a replica or replacement, but a developmental stage, a further step in the growth of the expressive life to which the first text gives birth. Certain poststructuralist authors, Jacques Derrida and Paul de Man among them, have taken this further, deriving from Benjamin the idea that translation's deeper significance is how it reveals "instability" or "incompleteness" or "mobility" as an inherent property of every original (Derrida 1985; de Man 1986). Within the context of history, such pronouncements mostly recast the truth that translation always in-

1. Lawrence Venuti has been the most tireless and fertile author on this issue, having traced its history from the sixteenth century to the present in the literary realm. See Venuti 1995.

volves interpretation of a foregoing work, and therefore a certain level of dissatisfaction and filling-in. Still, the value of this type of thinking is that it helps sustain questioning of the tired notions of "originality" and "imitation." It helps make plain that the long-term survival of many famous works has come not from intrinsic merit alone (or at all), but from the canonizing power translations bestow upon them—a historical reality, which therefore adds productive doubt to the notion of a sourcework's absolute, unchanging singularity.

Benjamin believes in the fundamental importance of the word, as the basic unit of the translator's action. This idea seems almost quasi-mystical until he provides a simple, explosive example:

> While all individual elements of foreign languages—words, sentences, structure—are mutually exclusive, yet these languages augment one another in their very intentions. . . . In the words *Brot* and *pain* the same object [bread] is indeed intended, but the style or mode (*die Art*) of this intended meaning is different. In other words, it is in the style or mode of meaning that both words signify something different to a German and a Frenchman, such that these words are not interchangeable but, in the end, seek to exclude one another even while, taken on absolute terms of meaning, they signify the same, identical object. . . . With each individual language, intended meaning is never encountered in a state of relative independence, as with a single word or sentence, but is comprehended in a state of constant flux, until able to emerge as pure language from the harmony between all modes of intention. (Benjamin 1977, 54–55)

The author's terms here are extremely suggestive. They imply that there is no possible earthly equivalence between the words of one language and those of another, no matter how concrete and mundane the object "intended." Such equivalence, after all, would require that the entire context of each language be absent—all the individualizing factors (sound, syntax, history, sensibility) that make a particular speech expressive of a particular people at a particular time. Meaning, Benjamin seems to say, is never frozen, hardened for all time, least of all in nomenclature. Words are always part of a larger ocean whose precise deposits are "in flux," and this applies to every inlet, bay, and estuary of speech in daily use by human beings. Something approaching equivalence only emerges in a transcendental semantic space where one perceives the contours of how languages are driven by similar expressive needs and impulses, how they are reconciled by a "kinship" embedded within the basic human instinct to signify.

It is here, at this seemingly airy juncture, that the full force of Benjamin's ideas can be brought to bear upon questions of science and translation. One can hardly resist the urge, for example, to contemplate replacing Benjamin's *Brot* and *pain* with words like *Sauerstoff* and *oxygène* or *Trägheit* and *inertie*. Are we here in the presence of precise equivalents? Do these terms, which surely signify the same basic phenomena, do so in exactly the same way, with the same exact mixture of denotative and connotative significance, the same sensibility of linguistic presence, in all cases of writing, in all situations of expression? It may be common to maintain that they do; but this requires that all of these questions, so pressing and valued in language study today, be altogether ignored or disavowed in the case of science only. The lack of exact equivalence between words, however, is a microcosm of that same lack between "original" (or first text) and "translation," which Benjamin elsewhere describes as "fragments of a single vessel" (1977, 59). Every translation, whether in literature or in science, is an entirely new "fragment" of the larger possibility of what might be said. Such a "fragment" is a new work, a new "literary event" with a future and history of its own, including (in many cases) an ultimate influence that may match or even surpass that of the first text. There is a fundamental shift in interest, therefore, that Benjamin seems to ask for, a shift from the "original" to the future "originals" that emerge from it and that acquire their own power of life and effect through time. In the end, therefore, Benjamin's ideas never leave the arena of translation in the context of history.

Authors and Originals

Translation means the possibility of many texts, multiple versions, an ever expanding array of works, both among and within languages. This returns us to the idea of the "original," the primal source text, whose importance—or even existence—has been an object of epistemological debate in recent years. Another perspective on this concept, inspired to some degree by Walter Benjamin's notion of textual "survival" but more deeply informed by the type of histories examined here, is that the "original," as a topic for concern, can be considered either unnecessary or irrelevant. In part, this is because, in many cases, anything even resembling a primal source no longer physically exists. More importantly, however, it is because the reality of a particular work of influence, such as the *Almagest,* is entirely subsumed by the richness and complexity of its transmission. This movement, in both a literal and therefore an epistemological sense, produced a library of true "originals" in a succession of languages. The *Syntaxis Matematica* in fourth-century Greek was not the same work as existed in sixth-century Syriac, ninth-century

Arabic, or twelfth-century Latin. Indeed, these texts did not share the same title and differed from each other in aspects of style, terminology, organization, factual content, and even theology. Each version served as a primal source in its respective tongue. That no "original" in Ptolemy's own hand exists any longer matters in a purely archival sense. More to the point, it matters only to the degree that such a text could help shed light on its own historical multiplicity, on the earliest changes, adaptations, corruptions, etc. that made the work a continuing textual presence. The meaning of "influence" for such a work, after all, is exactly measured by the number of "originals" gathered under its name, or under related names.

The great translator of early medieval Islam, Ḥunayn ibn Isḥāq, embarked on pilgrimages to search the libraries of Byzantium and Syria for "the treasures of Greek knowledge." Yet the texts he found, often stored for centuries, were even at their composition more than five hundred years removed from the lifetimes of their authors. These works had already passed through many hands, through the uncertain corridors of scribal culture. They were, however, Ḥunayn's "originals," and they were invested with all the sensibility of "riches" that has always been associated with this concept. Such did not prevent Ḥunayn from altering them in certain ways, however. In many cases, he was able to collect different Greek versions of single works and, under the very idea of discovering an inceptive source, took what he considered to be the most authentic pieces from these various copies and stitched them together to form a final rudiment. Thus Ḥunayn produced his own "originals" in an entirely concrete fashion, suggesting (again) that an almost infinite number of such sources could have been made. Nearly four hundred years later, if we can believe existing accounts, Gerard of Cremona made a similar pilgrimage to Spain in order to find and translate for the good of Latin culture "the *Almagest.*" But which *Almagest* was this? It turns out that, like Ḥunayn before him, Gerard used several versions, each with its own history of assembly. Some of these versions had been rendered into Arabic from the Greek; others from Syriac; some possibly from a mixture of the two.

It is crucial to understand that both Ḥunayn and Gerard, as well as many of their kind, were, unlike translators in the modern sense, inducted fully into the canon of *auctoritates,* the pantheon of authorial producers of timeless and fundamental works whose influence remained beyond measure. This was especially true of Ḥunayn, who remained for centuries a model of literary style, rational thought, and scientific authority, as well as a standard for translation technique. Medieval translators, both in Islam and in Europe, were far from being invisible conveyors of text. They, too, were "originals" of a sort. Indeed, this was true in twelfth- and thirteenth-century Europe despite the common demand for literalism in translation, an approach whose basic philosophy (ever since St. Jerome) suggested the translator's activity was tantamount to that of a precious device. The fame achieved by those like Constantinus Africanus or Adelard of Bath indicates that the medieval view of translation was more complex than commonly assumed. The perceived value of such men derived directly from association with the works they bestowed upon Latin culture. Before their day, writing was more often a means to preserve, recycle, and augment knowledge that already existed, sometimes in a confusion of different forms. In the framework of this system, men such as Cassiodorus, Martianus Capella, Duns Scotus, and Bede reached various pinnacles of authorial influence, in large part as writers who had reintroduced, reclarified, and refocused the efforts and products of previous eras. Thus, on the one hand, it can hardly be a surprise that translators such as Adelard or Gerard of Cremona would be venerated at similar levels for having transferred, en masse as it were, the textual archive of a more sophisticated culture.

Yet this was not all. As Roger Chartier has perceptively noted, "the twelfth century marks a critical rupture, for then writing ceased to be strictly a means of conservation and memorialization and came to be composed and copied for reading that was understood as intellectual work" (1995, 16). The unprecedented influx of new textual material revealed to the Europeans that authorship had been a much broader, more diverse and expanding field than they had ever imagined. The great proliferation of writing that took place thereafter—the profusion of new subjects and new writers—was in some sense a direct response to this, a type of necessary imitation. The written word became demystified; book production moved out of the monastery and into cities and towns; the great quantity and variety of written material scholars were now required to consume bore its own demands for commentary and addition. The new type of scholar-writer that eventually emerged—Roger Bacon, Thomas Aquinas, Albertus Magnus, Robert Grosseteste—was many things: a critic, a compiler, an investigator, a theologian or philosopher, a pedagogue, a literary stylist, a pontificator, an illustrator. His reach far exceeded the grasp of the commentator of previous centuries. He produced not merely isolated works or brief handbooks but massive collections that now had a much larger world to try to encompass. No longer bound to individual texts of sacred power, he was free to wander the expansive plains of authorship set out by the textual wealth from the East. It is important to stress that this new type of textual producer—who rests as the foundation of the "author" as we know him today—emerged as a result of translation. Indeed, his first manifestation was in the likes of Adelard of Bath, Constantinus Africanus, Dominicus Gundissalinus, and other translators who themselves populated their translations with explanatory glosses, definitions, and questions, and who wrote their own treatises on many topics. It is really in the immediate wake of the translation episode that the modern conception of the author was born.

Some of the same could be said of early medieval Islamic society, with important differences. Overall, the changes wrought seem equally impressive. Seen in its full significance, translation had two fundamental effects: first, it served as a major force, perhaps *the* major force, urging to completion the movement from oral to written culture; second, it helped establish here too the very idea of the author, as one with the power to choose among many subjects for his efforts of inscription. In Islam, however, there was often a unique sense of modesty attached to this idea: a scholar's work was that of continuation and completion of what had gone before. This was a position somewhere between that of the commentator-preserver and the author-as-original, the two poles of medieval authorship in Europe.

"Incommensurability" and the Foreign

Again and again in the history of science, therefore, one finds this process occurring, the creation of new "originals" from older ones. Where does this leave the long-debated, much elevated theme of incommensurability (whether, or to what degree for example, a particular chunk of fact, theory, or method can be accurately transferred across the alienness of different languages)?[2] In something of a difficult position, to be sure. For if no translation is ever really fixed, if it is always undergoing some degree of reinterpretation, whether formal or informal, from the inside out as it were, then the question of an unalterable "sense" or "meaning" (*sensus*) becomes quite troubled. At the most, it reduces to thin foliage surrounding the quicksand of "cognitive content"—a content presumably solid beyond alteration, rescued from history, separable from any and all forms. The very possibility of translation has been often cited as the primary argument in favor of such content. For if single works can exist in multiple languages, then some epistemological essence must be present that is shared by each of them and effec-

2. An informative discussion of this topic, with respect to traditional notions of translation, is offered by Fuller (1988, 32–40).

tively unites them all. Such an idea, however, can be challenged by the type of perspective argued here, namely that, in any such case, one is dealing not with a single granitic source multiplied any number of times, but instead a collection of changing, evolving sources that are themselves individually capable of being adapted or recreated into still newer versions. Faith in epistemological finality revealed through translation erodes inevitably before the more fluid notion that what really exist in the history of science are a host of "originals," each appropriate to its own linguistic-cultural locus. Moreover, if, as the history of science shows, mutations undergone by a particular work have been only partly the result of intention—if, that is, they have often come as the result of linguistic evolution and manuscript culture itself, the reigning technology of the written word for the great majority of the time that scientific-type knowledge has had an inscribed existence—then "commensurability" must itself be forced to accept a new context for discussion.

In short, "incommensurability," as an issue, needs to be evaluated with respect to these two contexts, manuscript culture and print culture. Possibly it is an issue that should even be discarded in discussions of premodern (manuscript) science; such would constitute an easy way out of the morass. But to reserve the problem for a small chunk of recent history would be false to the universality of mind embedded in the idea itself. One should not feel forced to "save the phenomena" by resorting to tricks of conscience. The belief (theory) that the goal of translation is to transform an alien text into a familiar one is far too confining and, in any case, fares poorly, or at least very unevenly, against the actual record. Many of the more complex Arabic works rendered into Latin by the twelfth-century translators yielded texts that were at first barely usable, even partly illegible, given the foreignness and difficulty of the material they contained, the new and previously unknown terminologies they introduced, the deformations of contemporary Latin they proposed or into which they occasionally collapsed. The central effort of these translators was not in the least to create a friendly, domesticated result, as was common with their Roman counterparts. They were eager, even breathlessly so, to search out and haul back the enormous textual resources of "the Arabs," including the most difficult and complex work available. If their renderings left these resources often in large, partly mined chunks that could only be refined, interpreted, and built upon by others, later on, this was an inevitable consequence of the meeting between the diligence of their enthusiasm and the complexity of the material. The newness and enormity of this knowledge, its qualities of power and distance, made it formidably *resistant* to Latin culture, even in Latin: it required a whole set of new texts, mainly textbooks and epitomes for students, to become friendly. What the translators produced retained many ferric aspects. Moreover, the literalism of medieval translators, such as Gerard of Cremona, must also be wholly reevaluated. This is because, as the texts themselves reveal, the philosophy of fidelity has often resulted in *more* substantive change—more neologisms, corruptions, stretched or deformed syntactic forms, and so forth—than even the most *ad libitum* paraphrase. Literalism has proven itself, time and again, the guarantor of metamorphosis. The reason is simple, and well known: the more a translator seeks loyalty to an "original," the more he or she is forced to create a linguistic hybrid residing somewhere between the two languages at issue.

Similar realities of translation in the history of science suggest that a different type of interpretive framework be called upon than the common dichotomy claiming that versions favor either the originating or the receiving language. Taken as a cultural act with epistemological intent and result, the act of translation can be said to involve two foreign languages—tongues that are otherwise unconnected in any integrated way—whose purposeful contact is embodied in the specific work created by the translator. This work is thus, in the beginning of its life, alien to both languages. Through survival and, often enough, reworking, it eventually loses this quality of foreignness vis-à-vis one of the languages (the receiving one, hopefully). But in its gestation, it is the required expression of the distance between these languages, and its resistance to familiarity is the sign of its originality. This, too, is part of the meaning of "influence." Translation means creating an embrace between languages, the producing of offspring. Viewed from such a vantage point, the issue of "commensurability" loses a good bit of its importance. In a historical sense, it may be of greater help to avoid the solar myth of "cognitive content" and instead accept *in*commensurability as a given. This, at least, leaves one free to admit that the degree of transformation involved in the translation process has always been too large and too significant for considerations of "equivalence" to take precedence over all others.

Eras of Translation: Times of Cultural Change

It is a historical fact, not to be underestimated, that against the background of continual linguistic contact among peoples there have occurred episodes when translation activity has risen to become a major focus of intellectual work and has resulted in immense tides of new textual material entering a particular language. As we have seen, a number of these episodes in a range of different settings have specifically involved scientific (and philosophical) material; indeed, among their central purposes was to create a new base of technical knowledge.

One might therefore ask: what has been the cultural role of such episodes? Certainly each major era of translation—whether from Greek into Syriac, Arabic into Latin, or Dutch into Japanese—had its major evolutionary effects upon the receiving language and thus upon the reservoir of semantic capabilities available to a particular people. But there is much more to be said here. Most of the eras we have examined in this book have themselves been part of—and have acted as pivotal forces behind—periods of large-scale cultural change, particularly involving the creation of principal urban societies—that of eighth- and ninth-century Islam, twelfth- and thirteenth-century Europe, nineteenth-century Japan. Each episode of translation proved itself fundamental to the diversification and institutionalization of textual culture—in the case of Islam, the transition to a primarily written culture and the creation of new forms of writing; in Europe, the building of universities and the rapid spread of authorship; in Japan, the vernacularization of intellectual culture generally.

There are other aspects to consider. Translation has been a powerful means to both encourage and manage, even control in part, the movements of foreign influence. This was obviously the intent of the Tokugawa regime in its support of translations from the Dutch, which then came to form the foundations of modern science in Japan. It was also a working motive for the early ʿAbbasid caliphs, who had a broad vision of the culture they wanted to build and who set up the *Bait al Ḥikma* as a specific means to acquire the necessary knowledge. The case of Arabic-to-Latin translation is somewhat different; no centralized program supporting this effort existed. Yet it seems accurate, on the basis of their own writings, to see the Latin translators as self-conscious gatekeepers, whose individual efforts also included "managing" what they considered to be the most needed, useful, or spectacular foreign works yet unavailable in Europe. Such is what allowed them an almost evangelical attitude of achievement and an aggressive sense of superiority.

Orality, too, has had a variable role in the cultural transformations associated with eras of science translation. In certain cases, such as early Islam, oral traditions embodied in long-standing genres of poetry and astrology, extended into religious domains surrounding memorization of the Qu'rān, acted to oppose the incursion of a dominantly textual and foreign knowledge. Yet, at the same time, the oral dimension was itself central to ensuring both the scale and the qualities of this incursion. It seems significant that for the episodes of translation into Arabic and later into Latin many texts were translated twice on their way into the respective target language. The first stage involved translating the "original" out loud into an intermediary language (Syriac on the one hand; Catalan, Spanish, possibly Italian on the other), which was then rendered into written form in the final target language. The term "intermediary," though often used, is probably a poor choice; the true process is really a matter of double translation (written-to-oral, then oral-to-written) centered on mouth-to-hand transfer. It seems likely that this allowed for more rapid progress in the work; doubtless it encouraged literalism of technique, as choices had to be made on the spot, with little time for reflection. What does all this imply regarding relationships between "source" and "translation"? The complexities are at least as extensive and alluring as any related to literary works of the first rank.

The Great Library of Science

If there have been great thinkers in the history of science, there have also been great texts. Indeed, there have been texts whose long-term influence has been to entirely subsume their authors, to make the once living, biological man or woman a mere exclamation point at the end of a title. The realities of influence are those of transmission; the fate of authorship is inseparable from the fate of specific works, and thus translation. If Aristotle, as an unparalleled embodiment of textual authority, is largely—if not entirely—a construction, *post vivo,* no less is this the case for Ptolemy, Euclid, and Newton, though in a different sense. The presence of Newtonian ideas in Japan, like Ptolemaic astronomy in medieval Baghdād, had nothing to do with original authorship by these respective "greats." Newtonian thought depended for its existence on a creative, neo-Confucian adaptation of a Dutch rendering of a popularization by one of Newton's followers, John Keill. Ptolemaic astronomy entered Islamic society through an even more tortured, convoluted set of passages involving not merely rewritten and translated versions of Ptolemy's own texts, but adaptations of these as well, in at least four different languages. The great library of science is enormously larger than a "five-foot shelf" of canonical works, a dusty collection of museum finalities. Often in its history, scientific knowledge has received its greatest impetus and nourishment for growth from translation. Doubtless there are many instances where this remains true today, though in more local quarters. Translation has proved itself the great multiplier of science over time, and it has always, inevitably, done this by increasing the number, scope, and diversity of its texts and thus its discourses in the greater arena of human language.

As an essential process (or set of processes) in the history of science, translation deserves a much expanded claim among the rightful subjects of sustained inquiry. If nothing else, this book is an effort to lay initial support at the feet of this claim. What remains are innumerable questions: How have the effects of translation been dispersed among the various disciplines? What uses did such thinkers as Copernicus, Galileo, or Kepler make of translated materials? What important similarities, contrasts, or cross-overs might exist between eras of scientific and literary translation? What have been the social structures of translating groups or communities? Are there other cultural-linguistic themes (such as displacement or orality) that might be discovered from comparative studies of different epochs? How did Islamic intellectuals view the translation of their own works into Latin? What role did translation play between 1880 and 1920 in the building of modern particle physics? What effects might the spread of scientific texts have had on the history of the book, on reading, or on specific relationships to the written word? And so forth.

Attempts at answering such questions would result in much more than borrowed sagacities or the reboiling of old bones. The great library of science—realized in all its multitudinous parts—is also a great laboratory. And the purpose of any such setting, hopefully, involves the making of new perceptions. Translation easily becomes a new eye cast upon the delights and difficulties of science scholarship. Through its lens, one discerns that mobilities of knowledge are integral to the substance and crucial to any understanding of the scientific past and present.

REFERENCES

Al-Daffa, A. A., and J. J. Stroyls. 1984. *Studies in the Exact Sciences in Medieval Islam.* Dhahran, Saudi Arabia: University of Petroleum and Minerals.

Al-Andalusi, Saʿid. 1991. *Science in the Medieval World: "Book of the Categories of Nations."* Trans. S. I. Salem and A. Kumar. Austin: University of Texas Press.

Allen, R. 1998. *The Arabic Literary Heritage.* Cambridge: Cambridge University Press.

Amos, F. R. 1920. *Early Theories of Translation.* New York: Columbia University Press.

Aratus. 1955. *Phaenomena.* Trans. G. R. Mair. In *Callimachus, Lycophron, Aratus,* 185–299. Cambridge: Harvard University Press, Loeb Classical Library.

Aristotle. 1939. *De Caelo (On the Heavens).* Trans. W. K. C. Guthrie. Cambridge: Harvard University Press, Loeb Classical Library.

———. 1952. *Meteorologica.* Trans. H. D. P. Lee. Cambridge: Harvard University Press, Loeb Classical Library.

———. 1984. *The Complete Works of Aristotle.* 2 vols. Ed. J. Barnes. Princeton, N.J.: Princeton University Press.

Ashtiany, J., T. M. Johnstone, J. D. Latham, and R. B. Serjeant, eds. 1990. *Abbasid Belles-Lettres.* Cambridge: Cambridge University Press.

Bailey, R. W. 1996. *Nineteenth-century English.* Ann Arbor: University of Michigan Press.

Bakaya, R. M. 1973. The place of English in science translation courses in India. *Language and Literature in Society: Journal of the School of Languages* (winter):115–20.

Barnstone, W. 1993. *The Poetics of Translation.* New Haven, Conn.: Yale University Press.

Bartholomew, J. R. 1993. *The Formation of Science in Japan.* New Haven, Conn.: Yale University Press.

Bartlett, J. R. 1985. *Jews in the Hellenistic World.* New York: Cambridge University Press.

Barton, T. 1995. *Power and Knowledge: Astrology, Physiognomics, and Medicine under the Roman Empire.* Ann Arbor: University of Michigan Press.

Bassnett-McGuire, S. 1991. *Translation Studies.* Rev. ed. London: Routledge.

Bäuml, F. H. 1980. Varieties and consequences of medieval literacy and illiteracy. *Speculum* 55:237–65.

Baumstark, A. 1894. Lucubrationes Syro-Gracae. *Jahrbucher fur classische Philologie Supplement* 21:358–84.

———. 1922. *Geschichte der syrischen Literatur.* Bonn: Marcus and Webers.

Beagon, M. 1995. *Roman Nature: The Thought of Pliny the Elder.* Oxford: Oxford University Press.

Beaujouan, G. 1982. The transformation of the quadrivium. In *Renaissance and Renewal in the Twelfth Century,* ed. R. L. Benson, G. Constable, and C. D. Lanham, 463–86. Cambridge: Harvard University Press.

Beer, J., ed. 1989. *Medieval Translators and Their Craft.* Lansing: Western Michigan University.

———. 1997. *Translation Theory and Practice in the Middle Ages.* Kalamazoo: Western Michigan University.

Beer, J., and K. Lloyd-Jones, eds. 1995. *Translation and the Transmission of Culture Between 1300 and 1600.* Kalamazoo: Western Michigan University.

Benjamin, W. 1977. Die Aufgabe des Übersetzers. In *Illuminationen: Ausgewahlte Schriften,* 50–62. Frankfurt: Suhrkamp.

Benson, R. L., G. Constable, and C. D. Lanham, eds. 1982. *Renaissance and Renewal in the Twelfth Century.* Cambridge: Harvard University Press.

Berggren, J. L. 1996. Islamic acquisition of the foreign sciences: A cultural perspective. In *Tradition, Transmission, Transformation,* ed. F. J. Ragep and S. P. Ragep, 263–83. Leiden: E. J. Brill.

Bergsträsser, G. 1913. *Ḥunain ibn Isḥāk und seine Schule.* Leiden: Brill.

———. 1925. *Ḥunain ibn Isḥāk über die syrischen und arabischen Galen-Übersetzungen.* Abhandlungen für die Kunde des Morgenlandes 17, no. 2.

Bernard, H. 1945. Les adaptations chinoises d'ouvrages européens. Bibliographie chronologique depuis la venue des Portugais à Canton jusqu'à la Mission française de Pékin, 1514–1688. *Monumenta serica* 10:1–57, 309–88.

Bhat, D. K., and A. K. Chatterji. 1979. The Karewa deposits of Kashmir Valley—a reappraisal. *Himalayan Geology Seminar,* Geological Survey of India Miscellaneous Publications 41, Part 1:191–200.

Bischoff, B. 1961. The study of foreign languages in the Middle Ages. *Speculum* 36:209–23.

Blacker, C. 1969. *The Japanese Enlightenment.* Cambridge: Cambridge University Press.

Bloch, H. 1982. The new fascination with ancient Rome. In *Renaissance and Renewal in the Twelfth Century,* ed. R. L. Benson, G. Constable, and C. D. Lanham, 615–36. Cambridge: Harvard University Press.

Bloch, J., and L. Chi. 1995. A comparison of the use of citations in Chinese and English academic discourse. In *Academic Writing in a Second Language,* ed. D. Belcher and G. Braine, 231–74. Norwood, N.J.: Ablex Publishing Corp.

Bloch, R. H., and C. Hesse. 1995. *Future Libraries.* Berkeley: University of California Press.

Bloomfield, L. 1939. *Linguistic Aspects of Science. International Encyclopedia of Unified Science* 1, no. 4. Chicago: University of Chicago Press.

———. 1987. Linguistic aspects of science. In *A Leonard Bloomfield Anthol-*

ogy, 205–19. Chicago: University of Chicago Press. First published in *Philosophy of Science* 2 (1935):499–517.

Böll, F. 1903. Sphaera*: Neue griechische Texte und Untersuchungen zur Geschichte der Sternbilder.* Hildesheim: Georg Olms.

Bonn, A. M., and E. Weber. 1938. *History of Syriac Literature.* Leiden: E. J. Brill.

Borst, A. 1994. Das Buch der Naturgeschichte: Plinius und seine Leser im Zeitalter des Pergaments. In *Abhandlungen der Heidelberger Akademie der Wissenschaften: Philosophisch-historische Klasse.* Abhandlung 2. Heldelberg: Universitätsverlag Carl Winter.

Bosworth, C. E. 1990. Administrative literature. In M. J. L. Young, J. D. Latham, and R. B. Serjeant, *Religion, Learning and Science in the 'Abbasid Period,* 155–67. Cambridge: Cambridge University Press.

Bovie, S. P. 1959. *Satires and Epistles of Horace.* Chicago: University of Chicago Press.

Bowen, J. 1973. *A History of Western Education, Vol. 1: The Ancient World.* New York: St. Martin's.

Bowersock, G. W. 1990. *Hellenism in Late Antiquity.* Ann Arbor: University of Michigan Press.

Braine, G. 1995. Writing in the natural sciences and engineering. In *Academic Writing in a Second Language,* ed. D. Belcher and G. Braine. Norwood, N.J.: Ablex Publishing Corp.

Brock, S. 1975. Some aspects of Greek words in Syriac. In *Synkretismus im syrisch-persischen Kulturgebiet. Abhandlungen der Akademie der Wissenschaften in Göttingen, Philologisch-Historische Klasse, Dritte Folge,* 96, ed. A. Dietrich. Göttingen: Vandenhoeck & Ruprecht. Reprinted in *Syriac Perspectives on Late Antiquity* (London: Variorum, 1984), 80–108.

———. 1977. Greek into Syriac and Syriac into Greek. *Journal of the Syriac Academy* 3:422–39. Reprinted in *Syriac Perspectives on Late Antiquity* (London: Variorum, 1984), 2.1–17.

———. 1982. From antagonism to assimilation: Syriac attitudes to Greek learning. In *East of Byzantium: Syria and Armenia in the Formative Period,* ed. S. Garsoïan, T. Mathews, and R. Thompson. Washington, D.C.: Dumbarton Oaks. Reprinted in *Syriac Perspectives on Late Antiquity* (London: Variorum, 1984), 5.17–34.

———. 1983. A history of Syriac translation technique. *Orientalia Christiana Analecta* 221:1–14.

———. 1984a. Aspects of translation technique in antiquity. In *Syriac Perspectives on Late Antiquity,* vol. 3, 69–87. London: Variorum.

———. 1984b. Syriac attitudes to Greek learning. In *Syriac Perspectives on Late Antiquity,* vol. 7, 20–35. London: Variorum.

———. 1984c. *Syriac Perspectives on Late Antiquity.* London: Variorum.

———. 1992. *Studies in Syriac Christianity: History, Literature, and Theology.* London: Variorum.

———. 1994. Greek and Syriac in Late Antique Syria. In *Literacy and Power in the Ancient World,* ed. A. K. Bowman and G. Woolf, 149–60. Cambridge: Cambridge University Press.

Brockelmann, C. 1943–1949. *Geschichte der Arabischen Literatur.* 2 vols. Leiden: E. J. Brill.

Budge, E. A. W., ed. and trans. 1894. *The Discourses of Philoxenus.* London: Duckworth.

Burke, R. B., ed. and trans. 1928. *Opus Majus of Roger Bacon.* Philadelphia: University of Pennsylvania Press.

Burnett, C. S. F. 1978. Arabic into Latin in twelfth-century Spain: The works of Hermann of Carinthia. *Mittellateinisches Jahrbuch* 13:100–134.

———. 1985. Some comments on the translating of works from Arabic into Latin. *Miscellanea Mediaevalia* 17:161–71.

———. 1997. *The Introduction of Arabic Learning into England.* Toronto: University of Toronto Press.

———, ed. 1987. *Adelard of Bath: An English Scientist and Arabist of the Early Twelfth Century.* Warburg Institute Surveys and Texts 14. London: University of London Press.

Busard, H. L. L. 1977. *The Translation of the* Elements *of Euclid from the Arabic into Latin by Hermann of Carinthia(?), Books VII–XII.* Amsterdam: Mathematisch Centrum.

Busch, P. 1998. Orthogonality and disjointedness in spaces of measures. *Letters in Mathematical Physics* Paper 9804005, 11 pp.

Butzer, P. L., and D. Lohrmann, eds. 1993. *Science in Western and Eastern Civilization in Carolingian Times.* Boston: Birkhauser.

Cameron, A. 1993. *The Mediterranean World in Late Antiquity* A.D. *395–600.* London: Routledge.

Cameron, A., and L. I. Conrad, eds. 1992. *The Byzantine and Early Islamic Near East.* Princeton, N.J.: Darwin Press.

Canfora, L. 1990. *The Vanished Library: A Wonder of the Ancient World.* Berkeley: University of California Press.

Cantor, N. F. 1993. *The Civilization of the Middle Ages.* Revised edition. New York: HarperCollins.

Carmody, F. J. 1956. *Arabic Astronomical and Astrological Sciences in Latin Translation: A Critical Bibliography.* Berkeley: University of California Press.

Carpenter, R. L. 1989. Translation among English, French, German, Russian, and Japanese. *The Social Science Journal* 26, no. 2:199–204.

Carter, M. G. 1990. Arabic grammar. In M. J. L. Young, J. D. Latham, and R. B. Serjeant, *Religion, Learning and Science in the ʿAbbasid Period,* 118–38. Cambridge: Cambridge University Press.

Chakraborty, T., S. Sarkar, A. K. Chaudhuri, and S. D. Gupta. 1996. Depositional environment of Vindhyan and other Purana basins: A reappraisal in the light of recent findings. *Memoir Geological Society of India* 36:101–26.

Chandra Rose, P. 1994. Problems in translating scientific and technical texts with special reference to texts on computer science. In *Art and Science of Translation,* ed. J. V. Sastry, 62–72. Hyderabad: Booklinks Corp.

Chartrier, R., ed. 1987. *Les usages de l'imprimé (XVe–XIXe siècle).* Paris: Fayard.

———. 1995. *Forms and Meanings.* Philadelphia: University of Pennsylvania Press.

Chattopadhyaya, D. 1986. *History of Science and Technology in Ancient India.* Calcutta: Firma KLM.

Chomsky, N. 1957. *Syntactic Structures.* The Hague: Mouton.

Cicero. 1913. *De Officiis.* Trans. W. Miller. Cambridge: Harvard University Press, Loeb Classical Library.

———. 1914. *De Finibus.* Trans. H. Rackham. Cambridge: Harvard University Press, Loeb Classical Library.

———. 1928. *De Re Publica, De Legibus.* Trans. C. W. Keyes. Cambridge: Harvard University Press, Loeb Classical Library.

———. 1933. *De Natura Deorum, Academica.* Trans. H. Rackham. Cambridge: Harvard University Press, Loeb Classical Library.

———. 1942. *De Oratore.* Trans. E. W. Sutton and H. Rackham. Cambridge: Harvard University Press, Loeb Classical Library.

Cipolla, C. M. 1994. *Before the Industrial Revolution: European Society and Economy 1000–1700.* New York: W. W. Norton.

Clagett, M. 1953. The medieval Latin translations from the Arabic of the *Elements* of Euclid, with special emphasis on the versions of Adelart of Bath. *Isis* 44:16–42.

———. 1957. *Greek Science in Antiquity.* London: Abelard-Schuman.

———. 1964–1976. *Archimedes in the Middle Ages.* 5 vols. Madison: University of Wisconsin Press.

———. 1972. Adelard of Bath. In *Dictionary of Scientific Biography,* vol. I:61–64. New York: Scribner's.

Clanchy, M. T. 1979. *From Memory to Written Record: England, 1066–1307.* Cambridge: Harvard University Press.

Classen, P. 1981. Die geistesgeschichtliche Lage Anstösse und Möglichkeiten. In *Die Renaissance der Wissenschaften im 12. Jahrhundert,* ed. P. Weimar, 11–33. Zurich: Artemis.

Cluver, A. D. de V. 1990. The role of the translator in the information society. In *Übersetzungswissenschaft: Ergebnisse und Perspektiven,* ed. R. Arntz and G. Thome, 476–88. Tübingen: Narr.

Colish, M. L. 1997. *Medieval Foundations of the Western Intellectual Tradition, 400–1400.* New Haven, Conn.: Yale University Press.

Copeland. R. 1991. *Rhetoric, Hermeneutics, and Translation in the Middle Ages.* Cambridge: Cambridge University Press.

Corbin, H. 1993. *History of Islamic Philosophy.* Trans. L. Sherrard. London: Kegan Paul.

Coulmas, F. 1989. *The Writing Systems of the World.* Oxford: Blackwell.

Coyaud, M. 1977. *Études sur le lexique Japonais de l'histoire naturelle et de biologie.* Paris: Presses Universitaires de France.

Craig, A. M. 1965. Science and Confucianism in Tokugawa Japan. In *Changing Japanese Attitudes towards Modernization,* ed. M. B. Jansen, 133–60. Rutland, Vt.: Charles E. Tuttle.

———. 1969. Fukuzawa Yukichi: The philosophical foundations of Meiji nationalism. In *Political Development in Modern Japan,* ed. R. E. Ward, 22–39. Princeton, N.J.: Princeton University Press.

Crawford, E., T. Shinn, and S. Sörlin. 1993. *Denationalizing Science.* Dordrecht: D. Reidel.

Crombie, A. C. 1957. *Medieval and Early Modern Science.* Vol. 1. New York: Anchor.

Crystal, D. 1985. How many millions? The statistics of English today. *English Today* 1:1–8.

Curtius, E. 1953. *European Literature and the Latin Middle Ages.* Trans. W. Trask. Princeton, N.J.: Princeton University Press.

Dallal, A. 1995. *An Islamic Response to Greek Astronomy.* Leiden: E. J. Brill.

Dall'Olmo, U. 1982. Latin terminology relating to aurorae, comets, meteors, and novae. *Journal for the History of Astronomy* 11:10–27.

D'Alton, J. F. 1962. *Roman Literary Theory and Criticism.* New York: Russell & Russell.

D'Alverny, M.-T. 1982. Translations and translators. In *Renaissance and Renewal in the Twelfth Century,* ed. R. L. Benson, G. Constable, and C. D. Lanham, 421–61. Cambridge: Harvard University Press.

D'Alverny, M.-T., and C. Burnett, eds. 1994. *La transmission des textes philosophiques et scientifiques au Moyen Age.* London: Variorum.

Daniels, N. 1975. *The Arabs and Mediaeval Europe.* London: Longman.

Davis, W. S., ed. 1912–1913. *Readings in Ancient History: Illustrative Extracts from the Sources.* 2 vols. Boston: Allyn and Bacon.

de Man, P. 1986. *The Resistance to Theory.* Minneapolis: University of Minnesota Press.

Dear, P., ed. 1991. *The Literary Structure of Scientific Argument.* Philadelphia: University of Pennsylvania Press.

Degen, R. 1981. Galen im Syrischen: Eine Übersicht über die syrischen Überlieferung der Werke Galens. In *Galen: Problems and Prospects,* ed. V. Nutton, 131–66. London: Blackwell.

Delhaye, P. 1947. L'organisation scolaire au XXIIe siècle. *Traditio* 5:211–68.

Delia, D. 1992. From romance to rhetoric: The Alexandrian library in classical and Islamic traditions. *American Historical Review* 97:1449–67.

Dembowski, P. F. 1997. Scientific translation and translator's glossing in four medieval French translators. In *Translation Theory and Practice in the Middle Ages,* ed. J. Beer, 113–34. Kalamazoo: Western Michigan University.

Derrida, J. 1985. Des Tour de Babel. In *Difference in Translation,* ed. J. Graham, 196–242. Ithaca, N.Y.: Cornell University Press.

Dicks, D. R. 1970. *Early Greek Astronomy to Aristotle.* Ithaca, N.Y.: Cornell University Press.

Dihle, A. 1994. *A History of Greek Literature.* Trans. C. Krojzl. London: Routledge.

Diringer, D. 1982. *The Book before Printing: Ancient, Medieval, and Oriental.* New York: Dover.

Doke, T. 1973. Yoan Udagawa—a pioneer scientist of early 19th century feudalistic Japan. *Japanese Studies in the History of Science* 12:99–120.

Dorey, T. A., ed. 1965. *Cicero: Studies in Latin Literature and its Influence.* London: Duckworth.

Drijvers, H. 1984. *East of Antioch.* London: Variorum Reprints.

Duhem, P. 1913–1959. *Le système du monde: Histoire des doctrines cosmologiques de Platon à Copernic.* 10 vols. Paris: Hermann.

Dunlop, D. M. 1971. *Arab Civilization to A.D. 1500.* New York: Praeger.

Eastwood, B. 1987. Plinian astronomical diagrams in the early Middle Ages. In *Mathematics and its Applications to Science and Natural Philosophy in the Middle Ages,* ed. E. Grant and J. Murdoch, 141–72. Cambridge: Cambridge University Press.

———. 1993. The astronomies of Pliny, Martianus Capella and Isidore of Seville in the Carolingian world. In *Science in Western and Eastern Civilization in Carolingian Times,* ed. P. L. Butzer and D. Lohrmann, 161–80. Boston: Birkhäuser.

———. 1997. Astronomy in Christian Latin Europe, c. 500–c. 1150. *Journal for the History of Astronomy* 28:235–58.

Edelstein, L., and I. G. Kidd. 1972. *Posidonius: The Fragments.* Cambridge: Cambridge University Press.

Edgren, R., ed. 1945. *Les methaeres d'Aristote; traduction du XIIIe siècle par Mahieu Le Vilain.* Uppsala: Almqvist & Wiksells.

Eisenstein, E. 1979. *The Printing Press as an Agent of Change.* Cambridge: Cambridge University Press.

Ellis, R., ed. 1989. *The Medieval Translator: The Theory and Practice of Translation in the Middle Ages.* Cambridge: D. S. Brewer.

Endress, G. 1982. Die wissenschaftliche Literatur. In *Grundriss der arabischen Philologie,* ed. W. Fischer, 2.400–506, 3.3–152. 3 vols. Wiesbaden: Reichert.

———. 1989. Die Griechisch-Arabischen Übersetzungen und die Sprache der Arabischen Wissenschaften. In *Symposium Graeco-Arabicum II,* ed. G. Endress. Amsterdam: B. R. Grunder.

Endress, G., and D. Gutas, 1992–. *A Greek and Arabic Lexicon. Materials for a Dictionary of the Mediaeval Translations from Greek into Arabic.* Leiden: E. J. Brill.

Fabre, J. 1987. Les séries paléozoique d'Afrique: Une approche. *Journal of African Earth Sciences* 7, no. 1:1–40.

Fakhry, M. 1994. *Philosophy, Dogma and the Impact of Greek Thought in Islam.* London: Variorum.

Febvre, L., and H.-J. Martin. 1958. *L'apparition du Livre.* Paris: Éditions Abin Michel.

Feingold, M. 1996. Decline and fall: Arabic science in seventeenth-century England. In *Tradition, Transmission, Transformation,* ed. F. J Ragep and S. P. Ragep, 441–69. Leiden: E. J. Brill.

Ferruolo, S. C. 1984. The twelfth-century renaissance. In *Renaissances before the Renaissance,* ed. W. Treadgold, 25–47. Stanford, Calif.: Stanford University Press.

Fischbach, H. 1978. Translation, the great pollinator of science. *Babel, Revue Internationale de la Traduction* 38, no. 4:193–202.

Fleisch, H. 1994. Arabic linguistics. In *History of Linguistics, Volume 1: The Eastern Traditions of Linguistics,* ed. G. Lepschy, 164–84. London: Longman.

Freibergs, G. 1989. The knowledge of Greek in Western Europe in the fourteenth century. In J. A. S. Evans and R. W. Unger, *Studies in Medieval and Renaissance History,* 71–85. New York: AMS Press.

Freudenthal, G. 1995. Science in the medieval Jewish culture of southern France. *History of Science* 33:23–58.

Fuhrmann, M. 1973. *Einführung in de antike Dichtungstheorie*. Darmstadt: Wissenschaftliche Buchgesellschaft.

———. 1984. *Die antike Rhetorik: Eine Einfuhrung*. Munich: Artemis.

Fuller, S. 1988. *Social Epistemology*. Bloomington: Indiana University Press.

Furlani, G. 1923. Il trattato di Sergio de Rêšaynâ, sull'universo. *Rivista trimestrale di studi filosofici e religiosi* 4:1–22.

Ganzenmüller, W. 1914. *Das Naturgefühl im Mittelalter*. Leipzig: B. G. Teubner.

Gärtner, H. A., ed. 1988. *Die römische Literatur in Text und Darstellung*. Vol. 5: *Kaiserzeit*. Stuttgart: Akademische Verlagsgesellschaft.

Gasset, J. O. y. 1992. The misery and the splendor of translation. In *Theories of Translation: An Anthology of Essays from Dryden to Derrida*, ed. Rainer Schulte and John Biguenet, 93–112. Chicago: University of Chicago Press.

Gentzler, E. 1993. *Contemporary Translation Theories*. London: Routledge.

Gibb, H. R., et al., eds. 1960–1994. *Encyclopedia of Islam*. Leiden: E. J. Brill.

Gies, F., and J. Gies. 1994. *Cathedral, Forge, and Waterwheel: Technology and Invention in the Middle Ages*. New York: HarperCollins.

Gimpel, Jean. 1983. *The Medieval Machine*. New York: Penguin.

Gluck, C. 1985. *Japan's Modern Myths: Ideology in the Late Meiji Period*. Princeton, N.J.: Princeton University Press.

Goldstein, B. R. 1979. The survival of Arabic astronomy in Hebrew. *Journal for the History of Arabic Science* 3:31–39.

Goodman, L. E. 1990. The translation of Greek materials into Arabic. In M. J. L. Young, J. D. Latham, and R. B. Serjeant, *Religion, Learning and Science in the ʿAbbasid Period*, 477–97. Cambridge: Cambridge University Press.

Goodyear, F. R. D. 1982. Technical writing. In *The Cambridge History of Classical Literature*. Vol. II, part 4: *The Early Principate*, ed. E. J. Kenney, 171–78. Cambridge: Cambridge University Press.

Graf, G. 1944–1953. *Geschichte der christlichen arabischen Literatur*. 5 vols. Vatican City: Biblioteca Apostolica Vaticana.

Grant, E., ed. 1974. *A Sourcebook in Medieval Science*. Cambridge: Harvard University Press.

———. 1984. Science and the medieval university. In *Rebirth, Reform and Resilience: Universities in Transition, 1300–1700*, ed. J. M. Kittelson and P. J. Tansue. Columbus: Ohio State University Press.

———. 1992. *Planets, Stars, and Orbs: The Medieval Cosmos, 1200–1687*. Cambridge: Cambridge University Press.

———. 1996. *The Foundations of Modern Science in the Middle Ages*. Cambridge: Cambridge University Press.

Gray, B. 1994. Saljuq-style painting and a fragmentary copy of al-Sufi's "Fixed Stars." In *The Art of the Saljuqs in Iran and Anatolia*, ed. R. Hillenbrand. Costa Mesa, Calif.: Mazda Publishers.

Green, P. 1990. *Alexander to Actium: The Historical Evolution of the Hellenistic Age*. Berkeley: University of California Press.

Grunebaum, G. E. 1976. *Islam and Medieval Hellenism: Social and Cultural Perspectives*. London: Variorum.

Habein, Y. S. 1984. *The History of the Japanese Written Language*. Tokyo: University of Tokyo Press.

Hadas, M. 1951. *Aristeas to Philocrates; Letter of Aristeas*, ed. and trans. M. Hadas. New York: Harper, for the Dropsie College for Hebrew and Cognate Learning.

Haq, S. N. 1996. The Indian and Persian background. In *History of Islamic Philosophy*, ed. S. H. Nasr and O. Leaman, 1.52–70. London: Routledge.

Hargrove, H. L. 1904. *King Alfred's Old English Version of St. Augustine's Soliloquies, Turned into Modern English*. New York: Henry Holt and Co.

Häring, N. M. 1964. Thierry of Chartres and Dominicus Gundissalinus. *Medieval Studies* 26:271–86.

Harris, M. H. 1995. *History of Libraries in the Western World*. 4th edition. Metchuen, N.J.: Scarecrow Press.

Harris, W. V. 1989. *Ancient Literacy*. Cambridge: Harvard University Press.

Haskins, C. H. 1925. Arabic science in western Europe. *Isis* 7:478–85.

———. 1927. *The Renaissance of the 12th Century*. Cambridge: Harvard University Press.

———. 1929. *Studies in Medieval Science*. Cambridge: Harvard University Press.

———. 1957. *Rise of the Universities*. Ithaca, N.Y.: Cornell University Press.

Heylen, R. 1992. *Translation, Poetics and the Stage*. London: Routledge.

Hiroshige, T. 1973. *Kagaku no Shakai-shi* [A Social History of Japanese Science]. Tokyo: Chuoko-ron.

Hirotal, K. 1988. *Meiji no Kagakusha* [Japanese Chemists in the Meiji Era]. Tokyo: Tokyo Kagaku Dojin.

Hitti, P. K. 1989. *History of the Arabs*. 10th ed. London: Macmillan.

Hodgson, M. G. S. 1974. *The Venture of Islam*. 3 vols. Chicago: University of Chicago Press.

Hollister, C. W., ed. 1969. *The Twelfth Century Renaissance*. New York: Wiley.

Honigmann, E. 1950. The Arabic translation of Aratus. *Isis* 41:30–31.

Horace. 1926. *Satires, Epistles, and Ars poetica*. Trans. H. R. Fairclough. Cambridge: Harvard University Press.

———. 1959. *Satires and Epistles of Horace*. Trans. S. P. Bovie. Chicago: University of Chicago Press.

Hornblower, S., and A. Spawforth, eds. 1996. *The Oxford Classical Dictionary*. 3d ed. Oxford: Oxford University Press.

Horovitz, J. 1927. The origins of "The Arabian Nights." *Islamic Culture* 1:36–57.

Hugonnard Roche, H. 1989. Aux origines de l'exégèse orientale de la logique d'Aristotle: Sergius de Reś'aina. *Journal Asiatique* 277:1–17.

Humphreys, R. S. 1991. *Islamic History: A Framework for Inquiry*. Princeton, N.J.: Princeton University Press.

Hutchins, W. J. 1986. *Machine Translation: Past, Present, Future*. New York: Halsted Press.

Hutchinson, G. O. 1988. *Hellenistic Poetry.* Oxford: Oxford University Press.

Isocrates. 1928. *Discourses.* Trans. G. Norlin. Cambridge: Harvard University Press, Loeb Classical Library.

Jacquart, D., ed. 1994. *La formation du vocabulaire scientifique et intellectuel dans le monde arabe.* Turnhout: Brepols.

———. 1997. *Les voies de la science grecque: études sur la transmission des textes de l'Antiquité au dix-neuvième siècle.* Geneva: Librairie Droz.

Jaeger, W. 1939–1945. *Paideia: The Ideals of Greek Culture.* 3 vols. Oxford: Blackwell.

Jamet, D., and H. Waysbord. 1995. History, philosophy, and ambitions of the Bibliothèque de France. In R. H. Bloch and C. Hesse, *Future Libraries.* Berkeley: University of California Press.

John of Salisbury (d. 1180). 1955. *The Metalogicon, a Twelfth-Century Defense of the Verbal and Logical Arts of the Trivium.* Trans. Daniel D. McGarry. Berkeley: University of California Press.

Johns, A. 1998. *The Nature of the Book: Print and Knowledge in the Making.* Chicago: University of Chicago Press.

Johnson, M. C. 1936. Manuscripts of the Baghdad astronomers, 769–1000 A.D. *The Observatory* 59:215–26.

Jolivet, J. 1988. The Arabic inheritance. In *A History of Twelfth Century Western Philosophy,* ed. P. Dronke, 113–48. Cambridge: Cambridge University Press.

Jones, A. 1990. Ptolemy's first commentator. *Transactions of the American Philosophical Society* 80, pt. 7.

———. 1994. The place of astronomy in Roman Egypt. In *The Sciences in Greco-Roman Society,* ed. T. D. Barnes, 25–51. Edmonton, Alberta: Academic Printing & Publishing.

Joosten, J. 1996. *The Syriac Language of the Peshitta and Old Syriac Versions of Matthew.* Leiden: E. J. Brill.

Kachru, B. B., ed. 1992. *The Other Tongue: English across Cultures.* 2d ed. Urbana: University of Illinois Press.

Kamata, H. 1993. Beppu-wan oyobi henshuchi-iki no shinbun chikakozo to sono sei-in [Deep subsurface geologic structure and genesis of Beppu Bay and adjacent zones]. *Chishitsu-gaku Zasshi* [Journal of the Geological Society of Japan] 99, no. 1:39–46.

Keene, D. 1968. *The Japanese Discovery of Europe, 1720–1830.* Stanford, Calif.: Stanford University Press.

Kelly, D. 1997. *Fidus Interpres:* Aid or impediment to medieval translation and Translatio? In *Translation Theory and Practice in the Middle Ages,* ed. J. Beer. Kalamazoo: Western Michigan University.

Kelly, L. G. 1978. *The True Interpreter: A History of Translation Theory and Practice.* Oxford: Blackwell.

Kennedy, E. S. 1956. *A Survey of Islamic Astronomical Tables.* Philadelphia: American Philosophical Society.

Kennedy, G. 1972. *The Art of Rhetoric in the Roman World.* Princeton, N.J.: Princeton University Press.

Kennedy, H. 1981. *The Early Abbasid Caliphate.* London: Croom Helm Publishers.

Kennedy, H., and J. H. Liebeschuetz. 1987. Antioch and the villages of Northern Syria in the fifth and sixth centuries A.D.: Trends and problems. *Nottingham Medieval Studies* 32:65–90.

Kenner, H. 1989. *Mazes.* San Francisco: North Point.

Kenney, E. J., ed. 1982. *The Cambridge History of Classical Literature.* Vol. II, parts 1–4: *The Early Republic-The Early Principate.* Cambridge: Cambridge University Press.

King, D. 1993. *Astronomy in the Service of Islam.* Aldershot: Variorum.

———. 1996. Islamic astronomy. In *Astronomy before the Telescope,* ed. C. Walker, 143–74. New York: St. Martin's Press.

———. 1997. Astronomy in the Islamic world. In *Encyclop3/4dia of the History of Science, Technology, and Medicine in Non-Western Cultures,* 125–33. Dordrecht: Kluwer.

Kluxen, W. 1981. Der Begriff der Wissenschaft. In *Die Renaissance der Wissenschaften im 12. Jahrhundert,* ed. P. Weimar, 273–93. Zurich: Artemis.

Knight, D. 1972. *Natural Science Books in English, 1600–1900.* London: Portman Books.

Kobori, A. 1964. Un aspect de l'histoire de la diffusion des sciences européennes au Japon. *Japanese Studies in the History of Science* 3:1–5.

Kren, C. 1983. Astronomy. In *The Seven Liberal Arts in the Middle Ages,* ed. David L. Wagner, 218–47. Bloomington: Indiana University Press.

Kritzeck, J. 1964. *Peter the Venerable and Islam.* Princeton, N.J.: Princeton University Press.

Kunitzsch, P. 1959. *Arabische Sternnamen in Europa.* Wiesbaden: Otto Harrasowitz.

———. 1974. *Der Almagest: die* Syntaxis Mathematica *des Claudius Ptolemaus in Arabisch-Lateinischer Überlieferung.* Wiesbaden: Otto Harrasowitz.

———. 1983. How we got our "Arabic" star names. *Sky and Telescope* 65:20–22.

———. 1986a. *Der Sternkatalog des* Almagest. *Die arabisch-mittelalterliche Tradition von Claudius Ptolemäus.* Wiesbaden: Otto Harrassowitz.

———. 1986b. The star catalogue commonly appended to the Alfonsine Tables. *Journal for the History of Astronomy* 17:89–98.

———. 1986c. John of London and his unknown Arabic source. *Journal for the History of Astronomy* 17:51–57.

———. 1987. Peter Apian and "Azophi" Arabic constellations in Renaissance astronomy. *Journal for the History of Astronomy* 18:117–24.

———. 1989. *The Arabs and the Stars: Texts and Traditions on the Fixed Stars and Their Influence in Medieval Europe.* Northampton: Variorum.

———. 1993. Arabische Astronomie im 8. Bis 10. Jahrhundert. In *Science in Western and Eastern Civilization in Carolingian Times,* ed. P. L. Butzer and D. Lohrmann, 205–20. Basel: Birkhauser.

Laffranque, M. 1964. *Poseidonios d'Apamée: Essai de mise au point.* Paris: Corti.

Lapidus, I. M. 1988. *A History of Islamic Societies.* Cambridge: Cambridge University Press.

Large, A. 1985. *The Artificial Language Movement.* Oxford and New York: Basil Blackwell.

Le Boeuffle, A. 1977. *Les noms latins d'astres et de constellations.* Paris: Société d'Édition "Les Belles Lettres."

———. 1987. *Astronomie, astrologie, lexique latin.* Paris: Picard.

Lecomte, G. 1965. *Ibn Qutayba (mort en 276/889); l'homme, son œuvre, ses idées.* Damascus: Institut français de Damas.

Lefevere, A. 1992a. *Translation, History, Culture: A Sourcebook.* London: Routledge.

———. 1992b. *Translation, Rewriting, and the Manipulation of Literary Fame.* London: Routledge.

Lefevere, A., and S. Bassnett-McGuire, eds. 1990. *Translation, History, and Culture.* London: Pinter.

Le Goff, J., ed. 1997. *The Medieval World.* Trans. L. G. Cochrane. London: Parkgate.

Lemay, R. 1962. *Abu Ma'shar and Latin Aristotelianism in the Twelfth Century.* Beirut: American University of Beirut.

———. 1976. The teaching of astronomy in medieval universities, principally at Paris in the 14th century. *Manuscripta* 20:197–217.

———. 1978. Gerard of Cremona. In *Dictionary of Scientific Biography,* 15.173–92. New York: Scribner's.

Levy, T. 1997. The establishment of the mathematical bookshelf of the medieval Hebrew scholar: Translations and translators. *Science in Context* 10, no. 3:431–51.

Lindberg, D. C. 1978. The transmission of Greek and Arabic learning to the West. In *Science in the Middle Ages,* ed. D. C. Lindberg. Chicago: University of Chicago Press.

———. 1992. *The Beginnings of Western Science: The European Scientific Tradition in Philosophical, Religious, and Institutional Context, 600 B.C. to A.D. 1450.* Chicago: University of Chicago Press.

Lloyd, G. E. R. 1970. *Early Greek Science: Thales to Aristotle.* New York: W. W. Norton.

———. 1987. *The Revolutions of Wisdom: Studies in the Claims and Practice of Ancient Greek Science.* Berkeley: University of California Press.

———. 1991. *Methods and Problems in Greek Science.* Cambridge: Cambridge University Press.

Long, P. O., ed. 1985. Science and technology in medieval society. *Annals of the New York Academy of Sciences* 441.

Lorimer, W. L. 1924. *The Text Tradition of Pseudo-Aristotle "De Mundo."* St. Andrews University Publications 18. London: Oxford University Press.

Loveday, L. J. 1996. *Language Contact in Japan: A Sociolinguistic History.* Oxford: Oxford University Press.

Low, M. F. 1989. The butterfly and the frigate: Social studies of science in Japan. *Social Studies of Science* 19:313–42.

MacLean, J. 1974. The introduction of books and scientific instruments into Japan, 1712–1854. *Japanese Studies in the History of Science* 13:9–68.

Macrobius. 1952. *Commentary on the Dream of Scipio.* Trans. W. H. Stahl. New York: Columbia University Press.

Maës, H. 1970. *Hiraga Gennai et son Temps.* Paris: École Française d'Extrême Orient.

Mahadevan Kochi, T. M. 1995. Deep continental structure of north-western and central Indian peninsular shield—a review. *Memoir Geological Society of India* 31:1–35.

Maier, A. 1949. *Die Vorläufer Galileis im 14. Jahrhundert. Studien zur Naturphilosophie der Spätscholastik.* Rome: Edizioni di Storia e Letteratura.

Mair, G. R., ed. and trans. 1955. *Callimachus, Lycophron, Aratus.* Cambridge: Harvard University Press, Loeb Classical Library.

Makdisi, G. 1981. *The Rise of Colleges: Institutions of Learning in Islam and the West.* Edinburgh: Edinburgh University Press.

Mamiya, F. 1952. *Toshkan Dai Jiten* [Complete Dictionary of Library Terms]. Tokyo: Japan Library Bureau.

Manilius. 1977. *Astronomica.* Trans. G. P. Goold. Cambridge: Harvard University Press, Loeb Classical Library.

Marrou, H. 1956. *A History of Education in Antiquity.* Trans. G. Lamb. New York: Sheed and Ward.

Martin, H.-J. 1988. *L'histoire et pouvoirs de l'écrit.* Paris: Librarie Académique.

Martin, J. 1982. Classicism and style in Latin literature. In *Renaissance and Renewal in the Twelfth Century,* ed. R. L. Benson, G. Constable, and C. D. Lanham, 537–66. Cambridge: Harvard University Press.

Mattock, J. N. 1989. The early translations from the Greek into Arabic: An experiment in comparative assessment. In *Symposium Graeco-Arabicum II,* ed. G. Endress. Amsterdam: B. R. Grunder.

Mayhoff, K. 1933. *Naturalis Historia von Pliny.* Books I–VI. Munich: Teubner.

McCluskey, S. C. 1998. *Astronomies and Cultures in Early Medieval Europe.* Cambridge: Cambridge University Press.

McKnight, G. H. 1968. *The Evolution of the English Language: From Chaucer to the 20th Century.* New York: Dover.

McMurtrie, D. C. 1943. *Book: The Story of Printing and Bookmaking.* New York: Dorset Press.

McVaugh, M. 1973. Constantine the African. In *Dictionary of Scientific Biography,* 3.393–95. New York: Scribner's.

Menut, A. D., and A. J. Denomy, eds. 1968. *Nicolas Oresme: le livre du ciel et du monde.* Madison: University of Wisconsin Press.

Meyerhof, M. 1930. Von Alexandrien nach Baghdad: Ein Beitrag zur Geschichte des philosophischen und medizinischen Unterrichts bei den Arabern. *Sitzungsberichte der preussischen Akademie der Wissenschaften, Berlin,* Philos.- Histor. Klasse, 389–429.

Michel, J. 1982. Linguistic and political barriers in the international transfer of information in science and technology. *Journal of Information Science* 5:131–35.

Millás Vallicrosa, J. M. 1963. Translations of Oriental scientific works. In *The Evolution of Science,* ed. G. S. Métraux and F. Crouzet, 128–67. New York: Mentor.

Miller, R. A. 1967. *The Japanese Language.* Chicago: University of Chicago Press.

Minio-Paluello, L. 1961. *Aristoteles Latinus, codices; supplementa altera.* Bruges: De Brouwer.

Montgomery, S. L. 1996. *The Scientific Voice*. New York: Guilford.
Moraux, P. 1951. *Les listes anciennes des ouvrages d'Aristote*. Louvain: Éditions universitaires de Louvain.
———. 1973. *Der Aristotelismus bei den Griechen: von Andronikos bis Alexander von Aphrodisias*. Berlin and New York: de Gruyter.
Moraux, P., and J. Wiesner, eds. 1983. *Zweifelhaftes im Corpus Aristotelicum*. Akten des 9. Symposium Aristotelicum. Berlin: de Gruyter.
Morelon, R. 1996. General survey of Arabic astronomy. In *Encyclopedia of the History of Arabic Science*, ed. R. Rashed, 1–19. London: Routledge.
Moussa, G. M. 1980. *Questions on Medicine for Scholars by Hunayn ibn Ishaq*. Critical edition. Cairo: al-Ahram Center for Scientific Translations.
Murdoch, John. 1984. *Album of Science: Antiquity and the Middle Ages*. New York: Scribner's.
———. 1968. The medieval Euclid: Salient aspects of the translations of the *Elements* by Adelard of Bath and Campanus of Novara. *Revue de Synthèse*, series 3, nos. 49–52:69–74.
Nadīm, al-. 1970. *The Fihrist of al-Nadim: A Tenth-Century Survey of Muslim Culture*. Trans. B. Dodge. 2 vols. New York: Columbia University Press.
Nagazumi, A. 1983. The diffusion of the idea of social Darwinism in east and southeast Asia. *Historia Scientiarum* 24:1–17.
Nakayama, S. 1964. Edo jidai ni okeru jusha no kagakukan [The scientific views of Confucian scholars during the Edo Period]. *Kagakusi Kenkyu* 72:157–68.
———. 1969. *A History of Japanese Astronomy: Chinese Background and Western Impact*. Cambridge: Harvard University Press.
———. 1977. *Characteristics of Scientific Development in Japan*. New Delhi: Centre for the Study of Science, Technology, and Development.
———. 1987. Japanese scientific thought. In *Dictionary of Scientific Biography*, Supplement 1, 15.728–58. New York: Scribner's.
———. 1992. Kindai seiyo kagaku yogo no chu-hi taishaku taishohyo [Comparison of Chinese and Japanese translations of modern Western scientific terms]. *Kagakusi Kenkyu*, series 2, vol. 31, no. 181:1–8.
Nakosteen, M. 1964. *History of Islamic Origins of Western Education, A.D. 800–1350*. Boulder: University of Colorado Press.
Nasr, S. H. 1976. *Islamic Science: An Illustrated History*. Kent: World of Islam Festival Publishing and Westerham Press.
———. 1987. *Science and Civilization in Islam*. 2d edition. Cambridge: Islamic Texts Society.
Nasr, S. H., and O. Leaman, eds. 1996. *History of Islamic Philosophy*. 2 vols. London: Routledge.
Nau, F. 1910. La cosmographie au VIIe siècle chez les Syriens. *Revue d'Orient Chrétien* 15:249.
———. 1929–1932. Le traité sur les constellations écrit en 661 par Sévère Sebokt. *Revue d'Orient Chrétien* 27:327–410, 28:85–100.
Netton, I. R. 1992. *Al-Farabi and His School*. London: Routledge.
Neugebauer, O. 1969. *The Exact Sciences in Antiquity*. 2d ed. New York: Dover.

———. 1975. *A History of Ancient Mathematical Astronomy.* 3 vols. Berlin: Springer Verlag.

Newman, W. R. 1991. *The* Summa Perfectionis *of Pseudo-Geber.* Leiden: E. J. Brill.

Newmeyer, F. 1980. *Linguistic Theory in America.* New York: Academic Press.

Nida, E. 1964. *Towards a Science of Translating.* Leiden: E. J. Brill.

Nöldeke, T. 1973. *Geschichte der Perser und Araber zur Zeit der Sasaniden.* Leiden: E. J. Brill.

North, J. 1992. The quadrivium. In *A History of the University in Europe I: Universities in the Middle Ages,* ed. H. de Ridder-Symoens, 337–59. Cambridge: Cambridge University Press.

———. 1995. *Norton History of Astronomy and Cosmology.* New York: W. W. Norton.

O'Donnell, J. J. 1998. *Avatars of the Word: From Papyrus to Cyberspace.* Cambridge: Harvard University Press.

Oettinger, A. G. 1959. Automatic (transference, translation, remittance, shunting). In *On Translation,* ed. R. Brower, 240–67. Cambridge: Harvard University Press.

Ohmori, M. 1964a. A study on the Rekishō Shinso, part 1. *Japanese Studies in the History of Science* 2:18–26.

———. 1964b. A study on the Rekishō Shinso, part 2. *Japanese Studies in the History of Science* 3:81–88.

O'Leary, D. L. 1949. *How Greek Science Passed to the Arabs.* London: Routledge and Kegan Paul.

Ong, W. J. 1977. *Interfaces of the Word.* Ithaca, N.Y.: Cornell University Press.

Opelt, I. 1959. Zur Übersetzungstechnik des Gerhard von Cremona. *Glotta* 38, no. 1/2:135–60.

Pedersen, O. 1978. Astronomy. In *Science in the Middle Ages,* ed. D. C. Lindberg. Chicago: University of Chicago Press.

———. 1993. *Early Physics and Astronomy.* Rev. ed. Cambridge: Cambridge University Press.

Peters, F. E. 1968. *Aristotle and the Arabs.* New York: New York University Press.

———. 1973. *Allah's Commonwealth: A History of Islam in the Near East, 600–1100 A.D.* New York: Simon & Schuster.

———. 1996. The Greek and Syriac background. In *History of Islamic Philosophy,* ed. S. H. Nasr and O. Leaman. 2 vols. London: Routledge.

Pfeiffer, R. 1968. *History of Classical Scholarship from the Beginnings to the End of the Hellenistic Age.* Oxford: Clarendon Press.

Pingree, D. 1968. *The Thousands of Abu-Mashar.* London: Warburg Institute.

———. 1970. The fragments of the works of al-Fazari. *Journal of Near Eastern Studies* 29:103–23.

———. 1971. On the Greek origin of the Indian planetary model employing a double epicycle. *Journal for the History of Astronomy* 2:80–85.

———. 1973. The Greek influence on early Islamic mathematical astronomy. *Journal for the American Oriental Society* 93, no. 1:32–43.

———. 1976. The recovery of early Greek astronomy from India. *Journal for the History of Astronomy* 7:109–23.

———. 1978. History of mathematical astronomy in India. In *Dictionary of Scientific Biography,* 15.533–633. New York: Scribner's.

———. 1981. *Jyotihsastra: Astral and Mathematical Literature.* Vol. IV of *A History of Indian Literature.* Wiesbaden: Otto Harrassowitz.

———. 1989. Classical and Byzantine astrology in Sassanian Persia. *Dumbarton Oaks Papers* 43:227–39.

———. 1990. Astrology. In M. J. L. Young, J. D. Latham, and R. B. Serjeant, *Religion, Learning and Science in the 'Abbasid Period,* 290–300. Cambridge: Cambridge University Press.

Pinto, O. 1959. Libraries of the Arabs during the time of the Abbasides. *Pakistan Library Review* 2:44–72.

Pirenne, H. 1952. *Medieval Cities.* Trans. F. D. Halsey. Princeton, N.J.: Princeton University Press.

Plato. 1957. *Plato's Cosmology. The Timaeus.* Trans. F. M. Cornford. New York: Liberal Arts.

Pliny the Elder. 1942a. *Pliny: Natural History.* Vol. 1, books I–II. Trans. H. Rackam. Cambridge: Harvard University Press, Loeb Classical Library.

———. 1942b. *Natural History.* Vol. 2, books III–VII. Trans. H. Rackam. Cambridge: Harvard University Press, Loeb Classical Library.

Plunket, E. 1903. *Calendars and Constellations of the Ancient World.* London: John Murray.

Poulle, E. 1988. The Alfonsine Tables and Alfonso X of Castile. *Journal for the History of Astronomy* 19:97–113.

Prakash, S. 1968. *A Critical Study of Brahmagupta and His Works.* New Delhi: Indian Institute of Astronomical and Sanskrit Research.

Ptolemy (Cladius Ptolemius). 1940. *Tetrabiblos.* Trans. F. E. Robbins. Cambridge: Harvard University Press, Loeb Classical Library.

———. 1967. The Arabic version of Ptolemy's *Planetary Hypotheses.* Trans. B. R. Goldstein. *Transactions of the American Philosophical Society* 57, pt. 4.

———. 1984. *The Almagest.* Trans. G. J. Toomer. New York: Springer-Verlag.

Quine, W. V. O. 1959. Meaning and translation. In *On Translation,* ed. R. Brower, 148–72. Cambridge: Harvard University Press.

Quintilian. 1920. *Institutio Oratoria.* Books I–III. Trans. H. E. Butler. Cambridge: Harvard University Press, Loeb Classical Library.

———. 1922. *Institutio Oratoria.* Books X–XII. Trans. H. E. Butler. Cambridge: Harvard University Press, Loeb Classical Library.

Ragep, F. J., and S. P. Ragep, eds. 1996. *Tradition, Transmission, Transformation.* Leiden: E. J. Brill.

Rashdall, H. 1936. *The Universities of Europe in the Middle Ages.* 3 vols. Oxford: Clarendon Press.

Rashed, R. 1989. Problems of the transmission of Greek scientific thought into Arabic: Examples from mathematics and optics. *History of Science* 27:199–209.

———, ed. 1996. *Encyclopedia of the History of Arabic Science.* 3 vols. London: Routledge.

Reynolds, L. D., and N. G. Wilson. 1991. *Scribes & Scholars: A Guide to the*

Transmission of Greek and Latin Literature. 3d ed. Oxford: Clarendon Press.

Ridder-Symoens, H., ed. 1992a. *A History of the University in Europe*. Cambridge: Cambridge University Press.

———. 1992b. Mobility. In *A History of the University in Europe,* ed. H. Ridder-Symoens. Cambridge: Cambridge University Press.

Rosenthal, F. 1970. *Knowledge Triumphant: The Concept of Knowledge in Medieval Islam*. Leiden: E. J. Brill.

———. 1975. *The Classical Heritage in Islam*. London: Routledge and Kegan Paul.

Ross, J. B., and M. M. McLaughlin. 1949. *The Portable Medieval Reader*. New York: Penguin.

Rouse, R. H., and M. A. Rouse. 1982. *Statim invenire:* Schools, preachers, and new attitudes to the page. In *Renaissance and Renewal in the Twelfth Century,* ed. R. L. Benson, G. Constable, and C. D. Lanham, 201–27. Cambridge: Harvard University Press.

Ruegg, W. 1992. The rise of humanism. In *A History of the University in Europe*. Volume I: *Universities in the Middle Ages,* ed. H. de Ridder-Symoens, 442–68. Cambridge: Cambridge University Press.

Ryssel, V. 1880–1881. *Über den textkritischen Werth der syrischen Übersetzungen griechischer Klassiker* I (1880), pp. 4–48 and II (1881), pp. 1–29. Leipzig: B. G. Teubner.

Sabra, A. I. 1987. The appropriation and subsequent naturalization of Greek science in medieval Islam: A preliminary statement. *History of Science* 25:223–43.

Sachau, E., ed. 1870. *Inedita Syriaca*. Vienna: Thyxs.

Sacrobosco (John of). 1949. *The "Sphere" of Sacrobosco and Its Commentators*. Trans. L. Thorndike. Chicago: University of Chicago Press.

Saenger, P. 1982. Silent reading: Its impact on late medieval script and society. *Viator: Medieval and Renaissance Studies* 13:367–414.

Saigusa, H. 1962. Die Entwicklung der Theorien vom "Ki" (Ch'i), also Grundproblem der Natur-"Philosophie" im alten Japan. *Japanese Studies in the History of Science* 4, 51–56.

Saigusa, S., and I. Shimizu, eds. 1956. *Nihon testsugaku shiso zensho* [Source book in Japanese philosophy and early modern science]. Tokyo: Heibonsha.

Sakaguchi, M. 1964. Studies on the *Seimi Kaiso,* part 2: The original of translation. *Kagakushi Kenkyu* 72:145–51.

———. 1968. On the chemical nomenclature in the *Seimi Kaiso*. *Kagakushi Kenkyu* 85:10–21.

Sakigawa, J., ed. 1975. *Ei-wa kagaku yogo jiten* [English-Japanese dictionary of scientific terms]. Tokyo: Kodansha.

Sale, W. 1966. The popularity of Aratus. *Classical Journal* 61:160–64.

Saliba, G. 1982. The development of astronomy in medieval Islamic society. *Arab Studies Quarterly* 4:211–25.

———. 1990. Al-Biruni and the sciences of his time. In M. J. L. Young, J. D. Latham, and R. B. Serjeant, *Religion, Learning and Science in the 'Abbasid Period,* 405–23. Cambridge: Cambridge University Press.

———. 1994. *A History of Arabic Astronomy.* New York: New York University Press.

Salmon, D. 1939. The medieval translations of Alfarabi's works. *New Scholasticism* 13:245–61.

Salmon, V. 1979. *The Study of Language in 17th-century England.* Amsterdam: John Benjamins.

Samso, J. 1994. *Islamic Astronomy and Medieval Spain.* London: Variorum.

Saxl, F. 1932. The zodiac of Qusayr Amra. In *Early Muslim Architecture,* ed. K. C. Creswell, 1.289–95. London: Oxford University Press.

Schall, A. 1960. *Studien über griechische Fremdwörter im Syrischen.* Darmstadt: W. Mies.

Schulte, R., and J. Biguenet, eds. 1992. *Theories of Translation: An Anthology of Essays from Dryden to Derrida.* Chicago: University of Chicago Press.

Schwartz, W. 1944. The meaning of *fidus interpres* in medieval translation. *Journal of Theological Studies* 45:73–78.

Seneca. 1971. *Naturales Quaestiones,* book 1. Trans. T. H. Corcoran. Cambridge: Harvard University Press, Loeb Classical Library.

Seshadri, T. R. 1968. *Three Lectures on Chemistry.* Mysore: Prasaranga Manasagangotri.

Sezgin, F. 1967–. *Geschichte des arabischen Schriftums.* 9 vols. Leiden: E. J. Brill.

Shalaby, A. 1954. *History of Muslim Education.* Beirut: Dar al-Kashshaf.

Shayegan, Y. 1996. The transmission of Greek philosophy to the Islamic world. In *History of Islamic Philosophy,* ed. S. H. Nasr and O. Leaman, 89–104. 2 vols. London: Routledge.

Sherwood, P. 1952. Sergius of Reshaina and the Syriac versions of the Pseudo-Denis. *Sacris Erudiri* 4:171–84.

Shore, L. A. 1989. A case study in medieval non-literary translation: Scientific texts from Latin to French. In *Medieval Translators and Their Craft,* ed. J. Beer, 297–327. Kalamazoo: Western Michigan University.

———. 1995. The continuum of translation as seen in three middle French treatises on comets. In *Translation and the Transmission of Culture Between 1300 and 1600,* ed. J. Beer and K. Lloyd-Jones, 1–54. Kalamazoo: Western Michigan University.

Shute, R. 1888. *On the History of the Process by which the Aristotelian Writings Arrived at Their Present Form.* Oxford: Clarendon Press.

Sinha-Roy, S., G. Malhotra, and D. B. Guha. 1995. A transect across Rejasthan Precambrian terrain in relation to geology, tectonics and crustal evolution of south-central Rajasthan. *Memoir Geological Society of India* 31:63–89.

Sivin, N. 1995. *Science in Ancient China: Researches and Reflections.* Aldershot, U.K.: Variorum.

Snell-Hornby, M. 1988. *Translation Studies: An Integrated Approach.* Amsterdam: John Benjamins.

Sourdel, D. 1954. Bayt al-hikma. In *Encyclopedia of Islam,* 1.1141. Leiden: E. J. Brill.

Southern, R. W. 1962. *Western Views of Islam in the Middle Ages.* New York: Vintage.

Speer, A. 1995. *Die entdeckte Natur: Untersuchungen zu Begründungsversuchen einer "scientia naturalis" im 12. Jahrhundert.* Leiden: E. J. Brill.

Stahl, W. H. 1962. *Roman Science: Origins, Development, and Influence on the Later Middle Ages.* Madison: University of Wisconsin Press.

Stahl, W. H., R. Johnson, and E. L. Burge. 1972, 1977. *Martianus and the Seven Liberal Arts.* New York: Columbia University Press.

Steiner, G. 1975. *After Babel.* London: Oxford University Press.

Steiner, T. R. 1975. *English Translation Theory: 1650–1800* (Amsterdam: Van Gorcum).

Steinschneider, M. 1889–1893. *Die arabischen Uebersetzungen aus dem Griechischen.* Leipzig: O. Harrassowitz.

———. 1956. *Die europaischen Ubersetzungen aus dem Arabischen bis Mitte des 17. Jahrhunderts.* Graz: Akademische Druck-u. Verlagsanstalt.

———. 1966. *Al-Farabi (Alpharabius); des arabischen Philosophen Leben und Schriften, mit besonderer Rucksicht auf die Geschichte der griechischen Wissenschaft unter den Arabern.* Amsterdam, Philo Press.

Stiefel, T. 1985. *The Intellectual Revolution in Twelfth-Century Europe.* London: Croom Helm.

Stock, B. 1978. Science, technology, and economic progress in the early Middle Ages. In *Science in the Middle Ages,* ed. D. C. Lindberg, 1–51. Chicago: University of Chicago Press.

———. 1983. *The Implications of Literacy: Written Language and Models of Interpretation in the Eleventh and Twelfth Centuries.* Princeton, N.J.: Princeton University Press.

Strevens, P. 1992. English as an international language: Directions in the 1990s. In *The Other Tongue: English Across Cultures,* 2d ed., 27–47. Urbana: University of Illinois Press.

Subbarayappa, B. V., and K. V. Sarma. 1985. *Indian Astronomy: A Source Book.* Bombay: Nehru Centre.

Sudhoff, K. 1914. Die kurze "Vita" und das Verzeichnis der Arbeiten Gerhards von Cremona. *Archiv für Geschichte der Medizin* 8:73–84.

Sufi, Abd al-Rahman ibn Umar (A.D. 903–986). 1953. *Kitab Suwar al-kawakib al-Thamaniyah a-al-Arbain.* India: Dairat al-Maarif al-Uthmaniyah.

Sugawara, K. 1984. Mizaki Shosuke no Kagaku-sha to shite no Katsudo [The Chemical Works of Mizaki Shosuke]. *Kagakushi Kenkyu,* series 2, 23:20–27.

Sugawara, K., and K. Itakura. 1989. Bakufu, Meiji Shoki ni okeru Nihongo no Gensomei (I) [Names of the Elements in Japanese during the late Edo and Early Meiji Periods]. *Kagakushi Kenkyu,* series 2, 28:193–202.

———. 1990a. Bakufu, Meiji shoki ni okeru nihongo no gensomei (II) [Names of the elements in Japanese during the late Edo and early Meiji periods]. *Kagakushi Kenkyu,* series 2, vol. 29, no. 175:193–202.

———. 1990b. Tokyo Kagaku ka ni okeru Gensomei no Toitsu Katei [The Process of Standardizing Japanese Element Names by the Tokyo Chemical Society]. *Kagakushi Kenkyu,* series 2, 29:136–49.

Sugawara, K., N. Kunimitsu, and K. Itakura. 1986. Atom no Yakugo no Keiseikako [The Process of Translating the Term "Atom" into Japanese]. *Kagakushi Kenkyu,* series 2, 25:41–45.

Swales, J. 1985. English as the international language of research. *RELC Journal* 16:1–7.

Swerdlow, N. S. 1996. Astronomy in the Renaissance. In *Astronomy before the Telescope,* ed. C. Walker, 187–230. New York: St. Martin's.

Taher, M. 1992. Mosque libraries: A bibliographic essay. *Libraries and Culture* 27:43–48.

Tanaka, M. 1964. Hundert Jahre der Chemie in Japan, Studien über den Prozess der Verpflanzung und Selbständigung der Naturwissenschaften also wesentlicher Teil des Werdegangs modernen Japans (Mitteilung I). *Japanese Studies in the History of Science* 3:89–107.

———. 1965. Hundert Jahre der Chemie in Japan (Mitteilung II): Die Art und Weise der Selbständigung chemischer Forschungen während der Periode 1901–1930. *Japanese Studies in the History of Science* 4:162–76.

———. 1967. Einige Probleme der Vorgeschichte der Chemie in Japan. Einführung und Aufnahme der modernen Materienbegriffe. *Japanese Studies in the History of Science* 6:96–114.

———. 1976. Rezeption chemischer Grundbegriffe bein dem ersten Chemiker Japans, Udagawa Yoan (1798–1846), in seinem Werk, *Seimi Kaiso.* Beiträge zur Geschichte der Chemie in Japan. *Japanese Studies in the History of Science* 15:97–109.

Taylor, G., and T. G. Chen. 1989. Linguistic, cultural, and subcultural issues in contrastive discourse analysis: Anglo-American and Chinese scientific texts. *Applied Linguistics* 12:319–36.

Theorica planitarum (anonymous). 1974. Trans. O. Pedersen. In *A Source Book in Medieval Science,* ed. E. Grant, 451–65. Cambridge: Harvard University Press.

Thompson, J. W. 1929. The introduction of Arabic science into Lorrain in the 10th century. *Isis* 12, no. 38:184–93.

———. 1939. *The Medieval Library.* Chicago: University of Chicago Press.

Thorndike, Lynn. 1923. *A History of Magic and Experimental Science.* 2 vols. New York: Columbia University Press.

Tibawi, A. L. 1954. Muslim education in the golden age of the Caliphates. *Islamic Culture* 28:418–38.

Tithon, A. 1993. L'astronomie à Byzance à l'époque iconoclaste (VIIIe–IXe siècles). In *Science in Western and Eastern Civilization in Carolingian Times,* ed. P. L. Butzer and D. Lohrmann, 181–203. Basel and Boston: Birkhauser.

Tolan, J. 1993. *Petrus Alfonsi and His Medieval Readers.* Gainesville: University Press of Florida.

Toomer, G. J. 1996. *Eastern Wisdom and Learning: The Study of Arabic in Seventeenth-Century England.* Oxford: Clarendon Press.

La traduction littéraire, scientifique, et technique. 1991. Paris: La Tilv.

Tritton, A. S. 1953. Muslim education in the Middle Ages (circa 600–800 A.H.). *Muslim World* 43:82–94.

Tsunoda, R., W. T. de Bary, and D. Keene. 1958. *Sources of Japanese Tradition.* Vol. 2. New York: Columbia University Press.

Turki, M. M., J. Delteil, R. Truillet, and C. Yaich. 1988. Les inversions tectoniques de la Tunisie centro-septentrionale. *Bulletin Géologique de la Société de la France,* series 8, vol. 4, no. 3:399–406.

Unger, J. M. 1996. *Literacy and Script Reform in Occupation Japan: Reading between the Lines.* Oxford: Oxford University Press.

Van Bekkum, W., J. Houben, I. Sluiter, and K. Versteegh, eds. 1997. *The Emergence of Semantics in Four Linguistic Traditions: Hebrew, Sanskrit, Greek, Arabic.* London: John Benjamins.

Van Houten, F. B., and R. B. Hargraves. 1985. Palaeozoic drift of Gondwana: Palaeomagnetic and stratigraphic constraints. *Geological Journal* 22:341–59.

Van Leuven-Zwart, K., and T. Naaijkens, eds. 1991. *Translation Studies: The State of the Art.* Amsterdam: Rodopi.

Vaux, R. de. 1933. La première entrée d'Averro's chez les latins. *Revue des Sciences Philosophiques et Théologiques* 22:193–245.

Venuti, L. 1992. *Rethinking Translation: Discourse, Subjectivity, Ideology.* London: Routledge.

———. 1995. *The Translator's Invisibility.* London: Routledge.

Versteegh, C. H. 1977. *Greek Elements in Arabic Linguistic Thinking.* Leiden: Brill.

Waddell, H. 1934. *The Wandering Scholars.* 7th edition. London: Constable.

Watanabe, M. 1990. *The Japanese and Western Science.* Trans. O. T. Benfey. Philadelphia: University of Pennsylvania Press.

Weimar, P., ed. 1981. *Die Renaissance der Wissenschaften im 12. Jahrhundert.* Zurich: Artemis.

Weissbord, D., ed. 1989. *Translating Poetry: The Double Labyrinth.* Iowa City: University of Iowa Press.

Welborn, M. C. 1931. Lotharingia as a center of Arabic and scientific influence in the eleventh century. *Isis* 16:188–99.

White, Lynn Jr. 1948. Natural science and naturalistic art in the Middle Ages. *American Historical Review* 52, no. 3:421–35.

———. 1962. *Medieval Technology and Social Change.* London: Oxford University Press.

Wickens, G. M., ed. 1952. *Avicenna: Scientist and Philosopher.* London: Luzac and Co.

Witty, F. J. 1974. Reference books of antiquity. *Journal of Library History* 9:101–19.

Wright, W. 1966. *A Short History of Syriac Literature.* Amsterdam: Philo Press.

Yoshikawa, H., and J. Kauffman. 1994. *Science Has No National Borders.* Cambridge, Mass.: The MIT Press.

Young, M. J. L., J. D. Latham, and R. B. Serjeant. 1990. *Religion, Learning and Science in the ʿAbbasid Period.* Cambridge: Cambridge University Press.

INDEX

'Abbasid caliphs, 80, 95–96, 102, 106; court society under, 103–8, 115
abstracts, scientific, translation of, 266–67
Abū Hanīfa al-Dīnawarī, *Kitāb al-Anwā*, 98
Abū Yaḥya ibn al-Biṭrīq, 110, 158
adab, 128
adaptation process, 149, 279
Adelard of Bath, 141, 144, 146–48, 163–64, 166
Adrastus of Aphrodisias, 53
Aelfric, 152
afterlife of texts, 284–86
alchemy, 109, 115
Alexandria, 86; library of, 5–6
Alexandrian curriculum, 72
Alfonsine Tables, 180–81
Alfonso X, King of Spain, 180
Alfred, 152–53
Almagest (Ptolemy), 20–21, 35–36, 98, 116–17, 175, 286–87; Arabic translations, 110–12; Gerard of Cremona's translation, 155, 158–59, 161–62, 169, 179
Andronicus of Rhodes, 9
Anglo-Saxon, 152–53
Anleitung zur Qualitativen Chemischen Analyse (Fresenius), 245
Apian, Peter, 181
Arabic: as language of learning, 106–7; linguistic changes in, 129–32; scientific terminology in, 115, 131; study of, in medieval Europe, 171–73
Arabic-Greek synthesis, 159–61, 169. *See also* Greek science; Islamic science
Arabicisms, in Latin translations, 148–49, 161–65, 169–70
Aramaic, 65
Aratus of Soli, 36, 102; *Phaenomena*, 27–29, 36, 38–45, 74
Ardashīr I, 79–80
Aristotelian corpus, 7–10; in medieval universities, 170–71
Aristotle, 5–10, 25, 98–99, 173, 177; *De caelo*, 158–61, 173–74; *Meteorologica*, 173
Ars Poetica (Horace), 32
Āryabhaṭa I, 84–87
Āryabhaṭasiddhānta (Āryabhaṭa I), 84
Āryabhaṭīya (Āryabhaṭa I), 84–85
Ashab al-ḥadīth, 94
as-Ṣafadī, 122
astrolabe, Theon of Alexandria's treatise on, 73–74
astrologers, Roman, 30
astrology, 21, 30, 39, 75, 86, 110, 115
astronomy: Arabic, 124–26; Chinese, 207; "folk," 117; Greek, 18–19, 60, 71–77, 118; Indian, 83–85; Islamic, 77, 79–85, 87–88, 117–18; Japanese, 203, 206; mathematical, 22, 35–37, 83–85, 98; mathematical-observational, 117–18; in medieval Europe, 21, 167–70; "powers" of, 86; Roman, 21–22, 30, 36–37, 39, 47–48, 51–53; Western, 183–85. *See also* Ptolemy
astronomy, translation of term, in Japanese, 229–30
Athanasios of Balad, 69
Atlas de la Navigation et du Commerce . . . (Renard), 227
Atlas van Zeevaert en Koophandel . . ., 227
auctoritates system, in medieval Europe, 151–52, 174–79, 287–88

author, concept of, 177–78, 288–89
"authoritative editions," 282–83
Averroes (Ibn Rushd), 135
Avienus, 36

Bacon, Roger, 171–73, 177; *De linguarum cognitio,* 171–72; *Opus Majus,* 173
Baghdad, 96–97, 104
Baien, Miura, 208–9
Bait al Ḥikma, 129, 142, 292
Banū-Mūsā brothers, 115–16
Bardaisân, *Book of the Laws of the Countries,* 65
Bartholomew, J. R., 219
Benjamin, Walter, "Task of the Translator," 283–86
Berggren, J. L., 91
Bernard of Chartres, 141, 145
Bibliothèque de France, 10–11
big science, in Japan, 224–25
biology, 232–35
Blacker, C., 204–5, 218
Bloch, R. H., 11
Bloomfield, Leonard, 274
Boethius, Anicius Manlius Severinus, 150; *Consolation of Philosophy,* 152–53
book, future of, 11–12. *See also* manuscript culture; print culture
Book of the Laws of the Countries (Bardaisân), 65
books, European, in Japan, 202–5, 207–10, 212, 215
book trade: in Hellenistic age, 22–23; in Islam, 106
borrowing, linguistic, in Syriac, 70–71
botany, 203, 238
Bowersock, G. W., 65
Brahmagupta, 87; *Khandakhadyaka,* 85
Braine, G., 268
Brock, Sebastian, 67–69
Bruni, Leonardo, 175
Buddhism, in Japan, 191–92, 211, 214
Buijs, Johannes, *Natuurkundig Schoolboek,* 242
Burgundio of Pisa, 142
Burnett, C. S. F., 157, 164
Byzantium, 60–61, 79

calligraphy, Japanese, 194. *See also* writing systems, Japanese
Cameron, A., 63, 79
Carolingian renaissance, 55–56, 152
Cassiodorus, 55n
Catasterisms (Eratosthenes), 38
Chalcidius, *Commentary* on Plato's *Timaeus,* 53–54
Chandra Rose, P., 276
Charles V, King of France, 174
Chartier, Roger, 288
chemical nomenclature: Japanese, 1–2, 236, 239–42; schemes of, 239–49
chemistry, in Japan, 235–49
chemistry, translation of term, in Japanese, 246
Chinese influence on Japanese: intellectual, 203, 212–13, 237; limits of, 213–15, 246–47; linguistic, 191–96, 228, 232, 241, 243–44. *See also* nationalism, Japanese; *rangaku* (Dutch studies)
Chomsky, Noam, *Syntactic Structures,* 276
Chosroes I Anūshirwān, 79
Christianity, 65–66, 142, 166
Christian texts, Syriac translations of, 68–71
Chu-Hsi, 205–6
Cicero, 31–32, 34, 36, 38; *De Natura Deorum,* 43; *De Officiis,* 34–35; *De Oratore,* 33; *Dream of Scipio,* 36, 40n; translation of Aratus, 38–45
citation patterns, in Chinese, 267–69
civil servant, in Islam, 100–101
classical literature, preservation and transfer of, 8n
Coelum stellatum christianum (Schiller), 55n
coinage. *See* neologism
commentaries, 147, 151, 177–78, 280
Commentary on Cicero's *Dream of Scipio* (Macrobius), 53–55
Commentary on Plato's *Timaeus* (Chalcidius), 53–54
compilation/standardization, as linguistic process, 18
compilers, and handbook tradition, 25–26. *See also* Pliny the Elder
computers, and future of book, 11–13
Consolation of Philosophy (Boethius), 152–53
Constantine the African, 144, 148
constellations, 27–28, 42–43, 74–75, 121, 126–27, 132–33, 170
Copeland, R., 29, 35

"corpus vetustius" and "corpus recentius," 170–71
Craig, A. M., 206
Crombie, A. C., 171
culture, knowledge and, 2–5

D'Alverny, M.-T., 143, 157
Daniel, Norman, 177
Daniel of Morely, 145–46, 156, 164; *Philosophia,* 164–65
Daniels, N., 166
David bar Paulos, 76
De anima (Dominicus Gundissalinus), 145
De Architectura (Vitruvius), 41n
De caelo (Aristotle), 158n; Gerard of Cremona's translation, 158–61; Oresme's French translation, 173–74
De cursu stellarum (Gregory of Tours), 55n, 57
De linguarum cognitio (R. Bacon), 171–72
Demetrius Phalereus, 5–7
De Natura Deorum (Cicero), 43
De Officiis (Cicero), 34–35
De Oratore (Cicero), 33
De Rerum Natura (Isidore of Seville), 56
De rerum natura (Lucretius), 32–33
Dihle, A., 22–23
"discovery," translation as, 156–57
displacement process, 184, 280–81
Dōbutsu Shinka-ron (Ishikawa Chiyomatsu), 233–34
Dominicus Gundissalinus, 141, 148–49; *De anima,* 145
Dream of Scipio (Cicero), 36, 40n
Dunlop, D. M., 100, 104
Dutch influence on Japanese, 197, 213, 215–16, 227. See also *rangaku* (Dutch studies)

Eastwood, Bruce, 56–57
Edessa, 63, 66–67
education: in Hellenistic age, 23–25; in Islam, 94, 101, 104–5, 130; in Japan, 219, 246; in medieval Europe, 140, 145–46, 151, 158, 167
Egypt, Roman, 30
electricity, introduced to Japan, 209–11
Elementa (al-Farghānī), 178
elements, chemical. *See* chemical nomenclature
Elements (Euclid), 148
Endress, G., 109
English, 255–67; as language of Japanese science, 221, 224–25
Enlightenment, Japanese, 218–20, 234, 246
Ensei Iho Meibutsu Ko (Udagawa Genshin), 236
Ephrem, 63, 65–66
Epistulae (Pliny the Younger), 34
Epitome of Chemistry (Henry), 239
Epitome of the Almagest, 180–81
equivalence, notion of, 285–86
Eratosthenes, *Catasterisms,* 38
Euclid, *Elements,* 148
Eudoxus, 24, 27
Europe, medieval, as textual culture, 139–41, 143–44, 175–79. *See also* manuscript culture; renaissance, twelfth-century
"Europeanness," of Greek thought, 60–65
evolution, Darwinian, introduced to Japan, 232–35
evolution, translation of term, in Japanese, 233–34

Fabre, J., 264
al-Fārābī, 105, 165, 175; *Kitab al-Sa'ada,* 128; *Kitab Ihsa al-Ulūm,* 128, 157
al-Farghānī, *Elementa,* 178
al-Fāzārī, 82, 110
Fenellosa, Ernest, 234
fidus interpres, translator as, 149–52, 160
Fihrist (al-Nadīm), 87, 106, 110, 115–17, 123–24, 127
foreign languages, study of, in medieval Europe, 171–73
foreignness, and incommensurability, 289–91
"foreign sciences," in Islam, 94–103, 116–18, 123–24, 126, 129, 135–37. *See also* Islamic science
foreign teachers, in Japan, 217–20, 232–34, 247
France, as target for Japanese study of Western science, 221
Frederick II, 143
French, 173–75, 221, 263–67
Fresenius, C. R., *Anleitung zur Qualitativen Chemischen Analyse,* 245
Fukuzawa Yukichi, 197, 218
fuqaha, 94

Galenic corpus, 62, 71–72, 98
Galib, 157
geology, 259–67
Gerard of Cremona, 141, 146–49, 155–62, 179, 287
Gerbert of Aurillac, 138
German, 221, 240–42, 245, 247–48
Germanicus Caesar, 36
Germany, as target for Japanese study of Western science, 220–21, 224, 246, 248
al-Ghazzālī, 135
Gluck, Carol, 218–19
Grant, Edward, 178n
Gratama, Frederik, 245
"great books," 293–94
Great Britain, as target for Japanese study of Western science, 221, 246, 248
Greco-Arabic, use of term, 92–93
Greek, 5–6; astronomical terminology in, 41n, 54–55; as language of Christian texts, 68; in Near East, 64–66; as "original" version, 174–79; Romans and, 32–33, 37
Greek culture: Romans and, 29–31; Syrians and, 61–71, 76–77
Greekisms, 70–71, 148–49, 174
Greeks, Arabic misconceptions of, 127–28
Greek science: Arabic appropriation of, 120, 124–26; in Hellenistic age, 21–25, 28; Romans and, 29–31, 42; and "science in Greek," 62; versions of, 182–85. *See also* Arabic-Greek synthesis
Gregory of Tours, *De cursu stellarum,* 55n, 57
Grosseteste, Robert, 172; *Summa philosophiae,* 175

al-Ḥajjāj ibn Matar, 110, 119, 159, 161
ḥakīm, 129
handbook tradition: in Hellenistic age, 23, 25–26; in medieval Europe, 167–70; in Roman world, 31, 49–50. *See also* Martianus Capella; Pliny the Elder
Handy Tables (Ptolemy), 117
Hargraves, R. B., 264
Harrān, 127
Hārūn al-Rashīd, 104
Haskins, C. H., 138n
al-Haytham, 165
Hebrew, 5–6, 171–73
heliocentrism, introduced to Japan, 227–29
Hellenization, concept of, 62–63, 89–93
Henry, William, *Epitome of Chemistry,* 239
Henry of Langenstein, 180
Hermann of Carinthia, 142, 165n
Hesse, C., 11
Hilary, St., 151
Hipparchus, 28–29
Hiraga Gennai, 209–11
hiragana, 191, 193–94, 243–44
Hirata Atsutane, 214
Hitti, P. K., 102n
Holland, and Japan, 197, 202–5, 213–15, 221. See also *rangaku* (Dutch studies)
Horace, *Ars Poetica,* 32
Hugh of Santalla, 141, 148
humanism, 64
Ḥunayn ibn Isḥāq, 98, 110, 112, 116, 121–22, 159, 161, 287
Huxley, Thomas, *Lectures on the Origin of Species,* 233
Hyginus, 43

Ibn al-Muqaffa', 103
Ibn al-Salah, 110
Ibn-Khaldūn: *Muqaddīmah,* 103
Ibn Qutayba, 99–100; *Kitab Adab al-Katib,* 100–101
Ibn Rushd (Averroes), 135
Ibn Sīnā, 175
Ichikawa Morisaburo, 242, 245
illustration, scientific, 56, 125, 133, 201–2, 228
'ilm, 128–29
imitation, Roman concept of, 31–34, 45–46
incommensurability, and foreignness, 289–91
India, scientific writing in, 258–62
influence, issue of, 3, 86, 292
Inleidinge tot de Waare Natuuren Sterrekunde (Lulofs), 229
instability, of texts, 281–86
Institutio Oratoria (Quintilian), 33–34
intermediaries, translators' use of, 4, 156–57, 292–93
internationalism, of Japanese science, 249
Internet, and scientific discourse, 262–63

Introductiones ad veram Physicam et veram Astronomiam (Keill), 229
Isagoge (Porphyry), 99, 150
Ishikawa Chiyomatsu, *Dōbutsu Shinkaron,* 233–34
Isidore of Seville, 55n; *De Rerum Natura,* 56
Islam, 89–93, 107–8, 142, 166; and "foreign sciences," 94–103, 116–18, 123–24, 126, 129, 135–37; image of, 138, 144, 146, 155–56, 166–67
Islamic conquests, 67, 70n, 77–78
Islamic culture, under ʿAbbasids, 103–8
Islamic science: de-Arabicizing of, 169–71, 175–79, 280; Eastern influences, 77–81, 92–93; Western interest in, 138, 144, 146, 155–56. *See also* Arabic-Greek synthesis
Islamic sciences, 64, 94, 102
Itakura, K., 241
Italy, 141, 182–83
Izawa Shuji, 233

Jacob of Edessa, 69
al-Jāḥiz, 98, 100, 102n, 131
James of Nisibis, 69
Japan: Imperial Rescript on Education (1890), 235; Ministry of Education, 248–49; National Research Council, 223–24; Office for Translation of Foreign Books, 215; in postwar period, 224–25
Japanese, 190–202, 221, 239; scientific terminology in, 205–6, 208–9, 211–13, 221–23, 228–32, 236, 239–49
Japanese science, 202–11, 216–26, 249, 272. *See also* Western science
Jerome, St., 149–50
Jesuits, 204, 206–8
Jews, as translators, 6, 155
Jinken Shinsetsu (Kato Hiroyuki), 235
Johannes of Sacrobosco, *Tractatus de Sphaera,* 167–70, 178
John of Gmunden, 180
John of Salisbury, 141
John of Seville, 141, 146–47, 149
John Scotus Eriugena, 150
Jones, Alexander, 30
Jundishapur, 66, 79–81, 105
Jurjīs ibn Bakhtīshūʿ, 105
Justinian, Emperor, 63, 79

Kachru, B. J., 259
Kaitai Shinsō, 215
kalām, 128
kanji, 191–92, 194, 199, 228
katakana, 191, 193–94, 228
kātib, in Islam, 100–101
Kato Hiroyuki, *Jinken Shinsetsu,* 235
Kawamoto Komin, *Kikai Kanran Kōgi,* 242
Kawano Tadashi, *Seimi Benran,* 244
Keene, D., 215
Keill, John, *Introductiones ad veram Physicam et veram Astronomiam,* 229, 293
Kelly, Louis, 274–75
Kenner, H., 17
al-Khaldūn, *Muqaddīmah,* 127
Khandakhadyaka (Brahmagupta), 85
Kikai Kanran (Rinso Aochi), 236, 241–42
Kikai Kanran Kōgi (Kawamoto Komin), 242
al-Kindī, 105, 125, 131, 165
King, David, 93
Kitāb Adab al-Kātib (Ibn Qutayba), 100–101
Kitāb al-Anwā (Abū Hanīfa al-Dīnawarī), 98
Kitāb al-Saʿāda (al-Fārābī), 128
Kitāb at-Tanbīh (al-Masʿūdī), 99
Kitab Iḥṣā al-Ulūm (al-Fārābī), 128, 157
Kiyohara Michio, 243
Kiyomizu Usaburo, *Mono-wari no Hashigo,* 243–44
knowledge: classification of, in Islam, 128–29; encouragement toward, in Qu'rān, 107–8, mobility of, 2–5, 103–8, 145–46
Kunitzsch, P., 19, 118, 132, 159, 161
al-Kwarīzmī, 164–65

languages. *See names of languages*
Lapidus, I. M., 95–96
Latin, 32–33, 38–40; astronomical terminology in, 38–42, 46–47, 52–55, 58, 149
Lavoisier, Antoine-Laurent, 240
Le Boeuffle, A., 38–41
Lectures on the Origin of Species (Huxley), 233
Lehrreise, in ʿAbbasid society, 104–5
Lemay, Richard, 156–57

liberal arts, 37, 50–51, 151, 157
Liber contra sectam sive haeresim Saracenorum (Peter the Venerable), 144
library: of Alexandria, 5–6; dream of "great library," 5–6, 10–13; Islamic, 112, 144, 166; personal, of Aristotle, 7–10; private, in Roman world, 37n; of science, 293–94
Lindberg, D. C., 148–49
literacy: in Hellenistic age, 22–23; in Islam, 95, 100–101; in Japan, 196, 198; in medieval Europe, 140
literalism, 184, 281, 291; in Latin translations, 148–49, 151–52, 158, 160–61, 291; in Syriac translations, 69–71, 73
literary style: Arabic, 102–3, 131; in Japanese scientific discourse, 199–201
literature, classical, preservation and transfer of, 8n
literature, Persian, influence on Arabic, 102–3
Lloyd, G. E. R., 19n
Lovati, Lovato, 182
Lucretius, *De rerum natura,* 32–33
Lulofs, Johan, *Inleidinge tot de Waare Natuuren Sterrekunde,* 229

Macrobius, *Commentary* on Cicero's *Dream of Scipio,* 53–55
madrasas, 94, 101
Mahadevan, T. M., 261
Mahieu le Vilain, 173
Manchurian incident, 224
al-Manṣūr, 95–96, 105, 110
manuals. *See* handbook tradition
manuscript culture, 279, 282, 290
manuscript production, in medieval Europe, 140–41
maʿqulāt, 129
maʿrifa, 128–29
Marriage of Mercury and Philology (Martianus Capella), 49–53
Martianus Capella, 57; *Marriage of Mercury and Philology,* 49–53
al-Masʿūdī, 104; *Kitāb at-Tanbīh,* 99
mathematics, as literature, 254–55
medicine, 72, 105, 115, 236–38, 246
Mesue, 87
Meteorologica (Aristotle), 173
Michael Scot, 142–43
"Mirrors for Princes" literature, 103
"Misery and the Splendor of Translation" (Ortega y Gasset), 273–74
Mitterrand, François, 10
Mizaki Shosuke, 243, 245–46
mobility, of knowledge, 2–5, 103–8, 145–46
Monophysites, 66–67
Mono-wari no Hashigo (Kiyomizu Usaburo), 243–44
Morse, Edward, 232–34
Motoki Ryoei, *Oranda Chikyū Zusetsu,* 227–29
Mueller, Johannes (Regiomontanus), 180–81, 185
Muʿjam al-Buldān (Yaqūt al-Hamawī), 96–97
Muqaddīmah (Ibn-Khaldūn), 103, 127
mu-takallimūn, 94
Muʿtazilites, 94

al-Nadīm, Muhammad ibn Isḥāq, *Fihrist,* 87, 106, 110, 115–17, 123–24, 127
Nagasaki, college of interpreters in, 204
Nakayama, S., 204, 213, 220–21, 228, 249
Naṣīr al-Dīn al-Ṭūsī, 111
Nasr, S. H., 128
nationalism, Japanese, 214–15, 219, 223, 234–35, 248
National Learning movement, in Japan, 214
nativization, 184, 280–81; of Arabic-Greek synthesis in Renaissance, 180–81; of Greek authors in Islam, 124; of Islamic science in medieval Europe, 167–70; of Western science in Japan, 217–20
Naturalis Historia (Pliny the Elder), 36–38, 45–49, 55–58, 282
natural philosophy, in Japan, 203–9, 213–14
nature, Pliny's concept of, 46
Natuurkundig Schoolboek (Buijs), 242
Near East: linguistic complexity of, 63–64; translation activity in late antiquity, 85–88. *See also* Islam
Neleus, 7
neo-Confucianism, 203, 205–6, 208–9, 213
neologism, 18, 33, 189, 228–32
Neoplatonism, 53–55, 99

Nestorians, 9, 60–62, 66–67, 79–81; as translators, 77, 110, 112–15
Neugebauer, O., 20, 87n
Nine Books of the Disciplines (Varro), 31
Nisibis, 66–67
nomenclature, scientific. *See* scientific terminology

O'Donnell, J. J., 281–82
Oettinger, A. G., 276–77
On the Universe (Pseudo-Aristotle), 72
Opelt, I., 158n, 159
Opus Majus (R. Bacon), 173
oral traditions, 292–93
Oranda Chikyū Zusetsu (Motoki Ryoei), 227–29
Oresme, Nicolas, 173–74
"original," concept of, 284–90
Ortega y Gasset, José, "The Misery and the Splendor of Translation," 273–74
orthography: Arabic, 130–31; Japanese, 194
Osrhoene, 62
oyatoi. See foreign teachers, in Japan

paper making, introduction of, 106, 140
paraphrase, 281
patrons, and vernacular translations, 173–75
Paul of Tella, 69
Pedersen, Olaf, 21, 86, 169, 180
periodic table of the elements, in Japanese, 1–2. *See also* chemical nomenclature
Persian, influence on Arabic, 119–20
Peshawar, 86
Peters, F. E., 76–77
Peter the Venerable, 142, 165n; *Liber contra sectam sive haeresim Saracenorum,* 144
Petrus Alfonsi, 144–45
Peurbach, Georg, 180–81
Phaenomena (Aratus of Soli), 27–29, 36, 38–45, 74
pharmacology, 236–38
Philosophia (Daniel of Morely), 164–65
Philoxenos of Mabbug, 66–67
physics, 223, 229–32
physics, translation of term, in Japanese, 211–12
Pingree, D., 80–81
plagiarism, 267–69; and handbook tradition, 25–26
Plato of Tivoli, 146
Pliny the Elder, 46, 57; *Naturalis Historia,* 36–38, 45–49, 55–58, 282
Pliny the Younger, *Epistulae,* 34
popularization, in Roman astronomy, 39. *See also* Aratus of Soli; handbook tradition
Porphyry, 9; *Isagoge,* 99, 150
Posidonius, 25, 31
power, political, astronomy and, 86
print culture, 281, 283, 290
printing, introduction of, 178n, 185
Pseudo-Aristotle, *On the Universe,* 72
Ptolemaic corpus, 20–21
Ptolemy, 19–21, 98, 120; *Almagest,* 20–21, 35–36, 64, 80, 87, 98, 110–12, 116–17, 155, 158–59, 161–62, 169, 175, 179, 286–87; *Handy Tables,* 117
Ptolemy Soter I of Egypt, as patron, 5–6

Quintilian, *Institutio Oratoria,* 33–34
Qu'rān, 107–8, 129, 142

Rackham, Howard, 282
rangaku (Dutch studies), 204, 212–13, 216–17, 227–29, 237
rationalism, European, 145–46
Raymond, Archbishop of Toledo, 143
Razan, Hayashi, 205
al-Rāzī, 165
Regiomontanus (Johannes Mueller), 180–81, 185
Rekishō Shinsho (Shizuki Tadao), 228–32
Renaissance, 180, 182
renaissance: Carolingian, 55–56, 152; twelfth-century, 138–41, 288–89
Renard, Louis, *Atlas de la Navigation et du Commerce . . . ,* 227
rhetoric, study of, 23–25, 33, 35
Ricci, Matteo, 206–7
Riccioli, Giambattista, 181n
Rinso Aochi, *Kikai Kanran,* 236, 241–42
risāla, 131
Robert of Ketton, 142, 165n
Robert of Lincoln, 174
Romans, and Greek science, 29–31, 42
Roman science, 279–80
Rosenthal, Franz, 90, 116

Ryssel, V., 73
ryū gakusei, 217–20

Sabra, A. I., 92, 94
"sage," concept of, 280–81
Sa'id al-Andalusi, 124–27; *Kitab Tabaqat al-Umam,* 108
Samarqand, 106
Sanskrit, 84
Sarma, K. V., 84
Saturae Menippeae (Varro), 37n
Schiller, Julius, *Coelum stellatum christianum,* 55n
science. *See* Greek science; Islamic science; Japanese science; Roman science; Western science
science, translation of term in Japanese, 211–12
scientific advance, translation and, 278–80
scientific community, international, 273
scientific discourse, nature of, 273–75
scientific knowledge, localism of, 277–78
scientific societies, Japanese, 221–22
scientific terminology: astronomical, 17–18, 38–42, 46–47, 52–55, 58, 74–75, 84, 161–62, 179–81; chemical, 1–2, 236, 239–49
Seimi Benran (Kawano Tadashi), 244
Seimi Kaisō (Udagawa Yoan), 236–41
Seimi Kyoku Hikkei (Ueno Hikoma), 244
Seisetsu Botanika-Kyo (Udagawa Yoan), 238
selection: in Gerard of Cremona's corpus, 156–58; in Roman translations, 30–31; in Syriac/Arabic translations, 64, 98–100, 115–18
Sergius, son of Elias, 110
Sergius of Reshaina, 62, 71–73, 158n, 159
Seshadri, T. R., 258
Severus Sebokht, 69–70, 73–74, 117; "Treatise on the Constellations," 74–75
Shāpūr I, 79–80
shinka, 233–35
Shintoism, 214
Shizuki Tadao, *Rekishō Shinsho,* 228–32
Shore, L. A., 173
Shute, Richard, 8n, 79
Sicily, as center for translation, 141, 143, 182–83
siddhānta tradition, 83–85
Siebold, Philip Franz von, 217n
Siebold Affair, 217
Sindhind, 117
Sivin, Nathan, 207
Southern, Richard, 166
Spain, 139, 141, 182–83
Spanish, 157
Spencer, Herbert, 234–35
Stahl, W. H., 50–51
standardization, of Japanese, 197–98, 223
star names, Arabic, 179–81
Stock, B., 151
Strabo, 7–8
Strevens, P., 256
students, Japanese, sent abroad, 217–20
Subbarayappa, B. V., 84
al-Ṣufi, 125, 181; *Kitāb Ṣuwar al-Kawākib,* 132–33
Sufi mysticism, 94, 99
Sugawara, K., 241
Summa philosophiae (Grosseteste), 175
summarizing, 280
Swerdlow, N. S., 180–81
Syntactic Structures (Chomsky), 276
Syntaxis Mathematica (Ptolemy). *See Almagest* (Ptolemy)
Syriac, 61–68, 74–75, 112, 119–20
Syriac Church, 66, 69, 76

tables, astronomical, 80, 164, 179–81. See also *zij*
Tabulae Anatomicae, 215
Tachibana Sensaburo, 234
Takahashi Kageyasu, 217n
Tanaka Minoru, 237, 249
"target" nations, for Japanese study of Western science, 220–25
Ta'rīkh (al-Ya'qūbī), 99
"Task of the Translator" (Benjamin), 283–86
text: afterlife of, 284–86; definitive, 48–49; instability of, 281–86; primacy of, 185, 278
textual analysis, in Hellenistic age, 23–25
textual community, concept of, 19–20
textual criticism, 282–83
textual culture: Arabic, 115; in medieval Europe, 139–41, 143–44, 175–79
Thābit ibn Qurra, 110, 120, 161, 164–65
Theon of Alexandria, 73–74, 117
Theon of Smyrna, 53

Theophilus, 72
Theophrastus, 7, 25, 27
Theoricae novae planetarum, 180
Theorica Planetarum, 167–70, 178
Thomas Aquinas, 171
Thorndike, Lynn, 177
Thousand and One Nights, 103
Tokyo Chemical Society, 222, 247
Toledan Tables, 180–81
Toledo, 141, 143
Tōyō Kanji, 198, 249
Tractatus de Sphaera (Johannes of Sacrobosco), 167–70, 178
transfer of learning, 2–5
translation, machine (MT), 276–77
translation, definition of term, 3–5, 130
translation movement: Arabic, 93–94, 98–100, 107–9, 112, 115–18, 129–35; and cultural change, 291–93; in Near East in late antiquity, 85–88; in twelfth-century Europe, 141–47, 172, 290–91
translatio nova, 171
translation process, 18, 278–80, 283–86, 291; into Arabic, 80, 82, 102–3; from Arabic to Latin, 156–57; into European vernaculars, 173–75; from Greek to Latin, 33–36, 39–42; into Japanese, 204, 215–16, 219, 223, 227–35, 238–41, 272
translation program, of Gerard of Cremona, 156–57
translation readings, in Japanese, 191, 193
translations: *vs.* "originals," 284–85; second-hand, 112; of translations, 227–32, 239, 245; as true originals, 14
translation techniques: in Arabic, 118–29; in Latin, 39–42, 147–49, 152, 158–62; in Syriac, 68–71
translation theory: in medieval Europe, 149–54; Roman, 31–36
translators, 4; Arabic, 108–18, 121–23; as builders, 153; as individual workers, 142–43, 152; Jews as, 6, 155; Nestorians as, 77, 110, 112–15; as travelers, 104–5, 144–45; as ultimate librarians, 13; use of intermediaries by, 4, 156–57, 292–93; views on technical translation, 276–77. *See also names of individuals*
transmission, issue of, 60–65
travel, translators and, 104–5, 144–45
"Treatise on the Constellations" (Severus Sebokht), 74–75
Turki, M. M., 266

Udagawa Genshin, *Ensei Iho Meibutsu Ko,* 236
Udagawa Yoan: *Seimi Kaisō,* 236–41; *Seisetsu Botanika-Kyo,* 238
Ueno Hikoma, *Seimi Kyoku Hikkei,* 244
Umayyads, 95, 105–6
United States, as target for Japanese study of Western science, 221, 225, 248
universalism, of scientific discourse, 253–54
universalist tradition, 271–78
universities, of medieval Europe, 141, 158, 167–71, 175

Van Houten, F. B., 264
Varro: *Nine Books of the Disciplines,* 31, 37–38; *Saturae Menippeae,* 37n
Venuti, Lawrence, 284n
Vitruvius, *De Architectura,* 41n
vocabulary. *See* scientific terminology

Watanabe, M., 234n
Western science: in China, 206–7; in Japan, 212–49
William of Conches, 146, 148
William of Moerbeke, 142, 147, 171–74
word, dominance of, 12
World War I, 248
World Wide Web, 262
Wright, W., 70n
writing systems: Chinese, 191; Japanese, 190–99, 220, 222–23

Yāḥyā ibn Māsawayh (Mesue), 87
Yamagawa Kenjiro, 223
al-Yaʿqūbī, *Taʾrīkh,* 99
Yaqūt al-Hamāwī, *Muʿjam al-Buldān,* 96–97

Zeno, Emperor, 79
zij, 80–84, 99, 110
zij al-Sindhind, 81–82
zij-Shahriyaran, 117
Zik-i Shahriydran, 87